I0199501

SALISH ARCHIPELAGO

Environment and Society in the Islands Within and Adjacent to the Salish Sea

SALISH ARCHIPELAGO

Environment and Society in the Islands Within and Adjacent to the Salish Sea

EDITED BY MOSHE RAPAPORT

ANU PRESS

ASIA-PACIFIC ENVIRONMENT MONOGRAPH 17

Dedicated to the courageous students worldwide who support
the rights of occupied and displaced people

'Injustice is not naturally remedied by history.
It is challenged and fought by people who mobilise
to dismantle systems of oppression.'
Gerald Nesmith Jr

Published by ANU Press
The Australian National University
Canberra ACT 2600, Australia
Email: anupress@anu.edu.au

Available to download for free at press.anu.edu.au

ISBN (print): 9781760466374
ISBN (online): 9781760466381

WorldCat (print): 1424831751
WorldCat (online): 1424832787

DOI: 10.22459/SA.2024

This title is published under a Creative Commons Attribution-NonCommercial-NoDerivatives 4.0 International (CC BY-NC-ND 4.0) licence.

The full licence terms are available at
creativecommons.org/licenses/by-nc-nd/4.0/legalcode

Cover design and layout by ANU Press. Cover photograph: 'Dodd Narrows' by Moshe Rapaport.

This book is published under the aegis of the Asia-Pacific Environment Monographs editorial board of ANU Press.

This edition © 2024 ANU Press

Contents

Abbreviations

AET	actual evapotranspiration
BC	British Columbia
CFIA	Canadian Food Inspection Agency
CFU	colony forming units
CO_2	carbon dioxide
CPR	common pool resources
DFO	Department of Fisheries and Oceans
EF	ecological footprint
ENSO	El Nino-Southern Oscillation
ESA	*Endangered Species Act* (US)
FWC	freshwater catalogue
GHWC	Gabriola Health and Wellness Collaborative
GINPR	Gulf Islands National Park Reserve
GMA	*Growth Management Act* (WA)
HBC	Hudson's Bay Company
HTG	Hul'qumi'num Treaty
ICE	Indigenous Circle of Experts
ICOM	International Council of Museums
IR	Indigenous Reserve Land
IUCN	International Union for Conservation of Nature
JIRC	Joint Indian Reserve Commission
MAP	mean annual precipitation
MOCAP	Medical On-Call Availability Program
NAGPRA	*Native American Graves Protection and Repatriation Act*

OCP	Official Community Plan
OFM	Office of Financial Management
SSMSP	Salish Sea Marine Survival Project
TRC	Truth and Reconciliation Commission
UNDRIP	United Nations Declaration of the Rights of Indigenous People
WRIA	Water Resource Inventory Areas
WSA	*Water Sustainability Act* (BC)
WSC	Water Survey of Canada

Contributors

Diana M. Allen is Professor of Hydrogeology in the Department of Earth Sciences at Simon Fraser University (SFU). Her research focuses on the processes that take place as natural groundwater systems respond to stressors like climate change, developing strategies to assess risks to water security, and ultimately, informing decision makers and policies. She was a member of the Groundwater Advisory Board for the Province when British Columbia (BC) had no legislation about groundwater. That process culminated in the creation of the Groundwater Protection Regulation under the *Water Sustainability Act*. She is a Past President of the Canadian National Chapter of the International Association of Hydrogeologists.

Bill Angelbeck is an archaeologist and anthropologist who focuses on the cultures of Salishan peoples of the Northwest Coast and Interior. Since 2000, he has worked throughout the Northwest on academic and applied projects, concerning archaeology, ethnography and ethnohistory. His interests include archaeological theory, sociopolitical organisation, religion, ideation and heritage. He has published his research in *Current Anthropology*, *World Archaeology*, and *Ethnohistory*, among others. Currently, he is a faculty member of the Department of Anthropology and Sociology at Douglas College in New Westminster, BC.

Chris Arnett is an independent archaeologist, ethnohistorian and heritage consultant who lives on Salt Spring Island, BC. His parents, grandparents and great-grandparents instilled in him a keen interest in North American history and in the people and culture of his ancestral homelands of New Zealand, Cornwall and Norway. He is a registered member of Ngai Tahu, the fourth-largest Māori tribe in New Zealand, holds a PhD (2016) in anthropology from the University of British Columbia and has authored and co-authored books and articles on Indigenous rock art and history. He has lectured at Malaspina University College, the University of British

Columbia and the University of Victoria. Chris continues self-directed and institutional ethnographic, archaeological and historical research and writing on Indigenous BC, New Zealand and Cornwall.

Russel L. Barsh is executive director of Kwiáht (Center for the Historical Ecology of the Salish Sea), a conservation biology laboratory in the San Juan Islands, where he leads studies of marine fish and food webs, pollinator networks, bat ecology and peat core-based reconstructions of paleo climates and vegetation. He has taught at the University of Washington and University of Lethbridge, and served in advisory roles in the United Nations, the Government of Canada, and First Nations relating to Indigenous peoples and the environment, including as Science Adviser to the Indigenous Peoples Commission at the United Nations Earth Summit.

Nelly Bouevitch works on climate resilience for the government of Yukon. Throughout the past five years, she has been working on projects that weave together Indigenous knowledge and science for taking action on climate change. She led the Yukon Climate Risk and Resilience Assessment, which brought together knowledge holders, territorial and Indigenous governments for informing how to build resilience to climate change in Yukon. She has also worked on co-developing climate change policies and programs for the federal government, in partnership with the Métis Nation and the Assembly of First Nations. She completed her Masters research at SFU, focusing on how Indigenous and non-Indigenous co-managers can work together in parks and protected areas. Having immigrated to Canada as a young girl, she has always been interested in learning about the stories that shape our culture and our heritage. She lives in Whitehorse, and spends as much time as she can in the surrounding mountains.

Oliver Brandes is an economist and lawyer by training and the co-director of the POLIS Project on Ecological Governance, based at the University of Victoria, and also serves as the Associate Director at the Centre for Global Studies. He is an Adjunct Professor at the Faculty of Law and School of Public Administration and is a fellow of the Environmental Law Centre. Oliver is a founding member and current Chair of the national Forum for Leadership on Water (FLOW), an adviser to numerous national, provincial and local water organisations, including the BC Ministry of Environment and Climate Change.

Randy Christensen a lawyer with a focus on water law and policy. He has been a lawyer with Ecojustice Canada (formerly the Sierra Legal Defence Fund) for nearly 15 years and served as managing lawyer for its Vancouver office for several years. Randy has been a lead counsel in numerous cases in Canadian courts. He has also made many appearances before administrative and international tribunals. Randy served as a member of the Canadian delegation to the United Nations Commission on Sustainable Development from 2007 to 2010. He began working as a research associate with the POLIS Water Sustainability Project in 2012.

John Clague is Emeritus Professor at SFU and an expert in glacial geology, natural hazards and climate change. Clague worked as a research scientist with the Geological Survey of Canada from 1975 until 1998, and in the Department of Earth Sciences at SFU from 1998 until 2016. Clague is a Fellow of the Royal Society of Canada, former President of the Geological Association of Canada, and Past-President of the International Union for Quaternary Research and the Association of Professional Engineers and Geoscientists of British Columbia. He received an honorary PhD from the University of Waterloo in 2017.

Dyan Dunsmoor-Farley has a PhD from the University of Victoria with undergraduate degrees in music (University of British Columbia) and interdisciplinary studies including sociology, political science, history and human geography (Athabasca University and University of Victoria). Her interdisciplinary work examines small community self-governance strategies in response to the ruptures triggered by global capital. Prior to starting her doctoral studies Dyan held senior positions in the BC government and had a successful consulting practice. Dyan has lived on the shores of the Salish Sea and sailed extensively through its archipelagos for over 40 years.

Steven Earle has lived on the edge of the Salish Sea for over 30 years. He holds a BSc in earth science from the University of British Columbia and a PhD in geochemistry from Imperial College, London. He worked in the mineral exploration industry for 15 years but has been teaching earth science at the post-secondary level since 1990, both at Vancouver Island University and Thompson Rivers University. Much of his understanding of the geology of the coast is based on research projects and field trips with students and colleagues on Vancouver Island and on many of the smaller islands in the Salish Sea.

Richard Hebda is Curator Emeritus at the Royal BC Museum and adjunct faculty at the University of Victoria (UVIC). He curated the museum's climate change permanent exhibit and travelling fossil displays. Has was the first faculty coordinator of the Restoration of Natural Systems Program at UVIC in which he teaches. He studies vegetation and climate history of BC, climate change impacts, alpine ecosystems and heritage potatoes. Richard Hebda is (co-)author of 135+ scientific papers 250+ popular articles on native plants and climate change; and (co-)author/editor of several books and reports (grasses, ethnobotany).

Soudeh Jamshidian is a conservation expert working on Indigenous Protected and Conserved Areas (IPCAs) in Canada. Soudeh is the director of education and international relations at the IISAAK OLAM Foundation, a main partner for the Conservation through Reconciliation Partnership. She is the coordinator for the Pacific IPCA knowledge hub, hosted at Tla-o-qui-aht Tribal Parks in the west coast of Vancouver Island, where she is with the Nuu-chah-nulth elders and activists to support Indigenous-led conservation. Soudeh is the coordinator of IPCA Knowledge Basket and an adjunct professor at Vancouver Island University teaching the IPCA planning certificate. Her PhD research at SFU focused on Indigenous conservation in management of protected areas in India, Iran and Afghanistan.

Sophia Johannessen is a geochemical oceanographer. She has worked for Fisheries and Oceans Canada, at the Institute of Ocean Sciences in Sidney, BC, since 2001. Her research interests range from light and underwater weather at the top of the ocean to burial and reworking of sediments at the bottom. She has worked in the Mid-Atlantic Bight, the Bering Sea and the Arctic Ocean. Currently she works in the coastal waters of BC. For recent research projects and a list of publications, see sophiajohannessen.net.

Richard Kool is a professor in the School of Environment and Sustainability at Royal Roads University where, in 2003, he founded the MA program in Environmental Education and Communication. He has worked as a secondary schoolteacher, as educator and head of public programs at the Royal BC Museum, as head of Interpretation for BC Parks and in the Biodiversity section of the BC Ministry of Environment. His research has ranged from the ecology of protozoa to Nuu-chah-nulth whaling; effectiveness of museum exhibits; religion, environment and education; and hope and despair in environmental education.

Marie Mauzé is an emeritus CNRS[1] researcher in anthropology and a member of the Laboratoire d'anthropologie sociale (Paris, France). She has conducted fieldwork with the Kwakwa̲ka̲'wakw since 1980. In addition to numerous articles published in French and English, she is the author of *Les Fils de Wakai: Une histoire des Lekwiltoq* (ERC,1992) and the editor of *Present is Past: Some Uses of Tradition in Native Societies* (University Press of America, 1997). She co-published, with Marine Degli, *Arts premiers* (Gallimard 'Découvertes', 2000). She is also co-editor of *Coming to Shore: Northwest Coast Ethnology, Traditions, and Visions* (University of Nebraska Press, 2004).

Alan D. McMillan is an adjunct professor in archaeology at SFU and emeritus faculty in anthropology at Douglas College. He holds a PhD in archaeology from Simon Fraser. His publications include *Since the Time of the Transformers: The Ancient Heritage of the Nuu-chah-nulth, Ditidaht, and Makah* (UBC Press, 1999) and *First Peoples in Canada* (with Eldon Yellowhorn; Douglas & McIntyre, 2004), as well as several monographs and numerous articles and reviews. His archaeological and ethnographic research has focused on the Nuu-chah-nulth peoples of western Vancouver Island.

R.D. (Dan) Moore is Emeritus Professor in the Department of Geography at the University of British Columbia, Vancouver, Canada. He has a BSc (Hons) in physical geography (climatology) from the University of British Columbia and a PhD in physical geography (hydrology) from Canterbury University in Christchurch, New Zealand. Dan's research addresses a range of topics related to hydrology, climatology and glaciology, including the effects of forest management and disturbance on downstream hydrology, water quality and fish habitat

Joan Morris (Sellemah) lived on Chatham Island (Tl'ches) until 1957 when the well went dry. She is a survivor of BC residential schools and Indian hospitals and advocate for those who have been impacted by these experiences. In collaboration with a field school led by UVIC faculty member Darcy Mathews, she has helped students see the Songhees Islands (Chatham and Discovery Islands, near Oak Bay, Victoria) through her eyes—superimposing past and present. Sellemah has helped students learn about the wider diversity that once flourished at Tl'ches, including the traditional gardens and orchards that have disappeared in intervening decades.

1 *Centre national de la recherche scientifique*, the French National Centre for Scientific Research.

Madrona Murphy, born and raised on a farm on Lopez Island and trained in genetics and botany at Reed College, was a technician at the University of Washington's Center for Cell Dynamics before joining Kwiáht as laboratory manager, field botanist and director of its research garden and 'neglected foods' program. She has also led development of Kwiáht's Soundscapes program. She has also contributed to the revival of traditional indigenous food plants such as camas in tribal and First Nations food programs, and the preservation and use of legacy orchards and tree-fruit varieties in the San Juan Islands.

Isobel A. Pearsall is the Director of the Pacific Salmon Foundation's Marine Science Program. Formerly, she was the Project Coordinator (Canada) for the Salish Sea Marine Survival Project, a transboundary $20M program set to address declines in Chinook, coho and steelhead in the Salish Sea. Over 60 organisations, representing diverse philosophies and encompassing most of the region's fisheries and marine research and management complex, were involved in this massive transboundary effort. The Pacific Salmon Foundation and Long Live the Kings worked together to coordinate efforts in the Strait of Georgia and Puget Sound, respectively.

Moshe Rapaport is a geographer specialising in environments and populations of Oceania and the Pacific Northwest. He is editor of *The Pacific Islands: Environment and Society* (Bess Press, 1999; University of Hawai'i Press, 2013), and has published articles in *Island Studies Journal*, *Journal of Pacific History* and other periodicals.

Michael W. Schmidt is the Center Director, Western Fisheries Research Center, United States Geological Survey, Seattle, Washington. Previously, Michael worked as deputy director for the salmon recovery non-profit, Long Live the Kings, where he established and led collaborative research.

Brian Thom is an associate professor in the Anthropology department at UVIC, where in 2010 he founded the Ethnographic Mapping Lab (ethnographicmapping.uvic.ca). From 1994 to 1997, and from 2000 to 2010, he worked as researcher, senior adviser, and negotiator for several Coast Salish First Nations (Canada) engaged in treaty, land claims and self-government negotiations. His research is focused on issues of Indigenous territoriality, knowledge and governance; revealing contemporary practices of Indigenous law; and clarifying the ontological imperatives behind Indigenous political strategies.

Nancy J. Turner, who shares the name Sellemah with Joan Morris and her grandmother, is Distinguished Professor Emeritus, School of Environmental Studies, UVIC. Turner has worked with Indigenous elders in western Canada for over 50 years, helping to document their rich knowledge of plants and environments. She has authored or co-authored/co-edited over 20 books and over 135 book chapters and papers. Her book *Ancient Pathways, Ancestral Knowledge* (McGill-Queen's University Press, 2014) received the Canada Prize for Social Sciences. Other awards include the Order of British Columbia (1999), the Order of Canada (2009), election to the Royal Society of Canada, honorary degrees from Vancouver Island University, University of British Columbia, University of Northern British Columbia and SFU, and the Society of Ethnobiology's Distinguished Ethnobiologist of the Year (2019).

Lissa K. Wadewitz is Associate Professor and the Beekman Chair of Northwest and Pacific History at the University of Oregon. Her first book, *The Nature of Borders: Salmon and Boundaries in the Salish Sea* (University of Washington Press, 2012) explores the social and ecological effects of imposing cultural and political borders on a critical West Coast salmon fishery. Dr Wadewitz has also published several peer-reviewed articles and book chapters with academic journals and presses. Her current research is focused on the intersections of race, sexuality, labour and the environment in the nineteenth-century US Pacific whaling fleet.

John R. Welch is a professor jointly appointed in the Department of Archaeology and the School of Resource and Environmental Management at SFU in BC. Welch also serves as an expert witness and advocate for Indigenous peoples and directs Archaeology Southwest's Landscape and Site Protection Program. Welch's projects, which have included field schools with the Tla'min Nation and Nelly Bouevitch's investigation of shellfishery co-management in Gulf Islands National Park, focus on integral links between Indigenous Peoples' sovereignty—rights and responsibilities derived from authority over people and territory—and stewardship—sustainable and broadly beneficial uses of sociocultural and biophysical inheritances.

1

Introduction

Moshe Rapaport with contributions by Nancy J. Turner, Richard Hebda and John R. Welch

The Salish Sea

The Salish Sea is an overarching, internationally recognized name for three connected inland marine waters on the Northwest Coast of North America: the Strait of Georgia, Strait of Juan de Fuca, and Puget Sound. The name – to honor the numerous Indigenous Peoples living in the region speaking related languages and dialects of the Salish Language Family – was first proposed by Bert Webber, geographer and professor at Huxley College of the Environment, Western Washington University in Bellingham.

A long-time researcher with strong environmental interests, Bert was convinced that these separately named bodies of water would be better understood and better protected if they were recognized as a single, integral ecosystem, which he defined as an 'estuary ecosystem', distinct from the other waters of the Northwest. He was deeply concerned about the potential consequences to these enclosed marine waters of accidental spills from oil tankers and other marine traffic, as well as the need for unified efforts to protect, conserve and restore the deteriorating wildlife populations and their habitats, including the numerous islands distributed throughout the Sea.

In March 1989, Bert filed an application with the Washington State Board on Geographic Names for approval of the name 'Salish Sea', not to replace the names for the component parts, but to include them within the broader name. It took several attempts to gain acceptance of 'Salish Sea' as a legitimate geographical name. He was supported in his efforts by the Salish Peoples of the area themselves. Finally, in 2009 and 2010 respectively, the geographic boards in the United States and Canada formally approved its use and started putting the name 'Salish Sea' on their maps. For additional information on this topic see Norman (2012) and Endter (2015).

Islands as Archipelago

Current maps of the Salish Sea view the region as a watershed, excluding most of Vancouver Island. Recent books on the Salish Sea region (Beamish and McFarlane 2014; Benedict and Gaidos 2015; Stewart 2017) largely ignore the islands. An island-centric perspective (Stratford et al. 2011) would consider all islands within the region, including Vancouver Island, as an archipelago. This perspective does not neglect the sea which connects these embedded islands and societies, as shown throughout this volume.

Regional precedents do exist for an island-centric perspective. Charles Forward (1979) produced an edited collection titled *Vancouver Island: Land of Contrasts* (which also discusses the Gulf Islands). While at the time the population of Vancouver Island was half what it is today, the prospect of forest decline, conflicts over water use, the attraction of the island for recreation, extensive subdivision and the potential dangers of environmental damage were already becoming evident.

Located at the edge rather than within the Salish Sea, Vancouver Island is integrally connected with islands to its east in key respects: (1) It is the mountainous spine of Vancouver Island that creates the rain shadow in its south-east coast and outlying islands (MacKinnon 2003). (2) Gulf Island populations are reliant on Vancouver Island for education, work and governance. Ferry and air connections to a nearby large island provide a lifeline to small island populations here, as in other parts of the world (Vannini 2011).

Figure 1.1 The Salish Archipelago
Source: Google Maps.

Environmental and Social Dynamics

The chapters of this book address a theme common to islands today—the evolving relationships between nature and society: the natural and human challenges confronting ecosystems and communities, the environmental and social consequences following disruptions, and the resilience of ecosystems and societies to cope and adapt to change (Angelbeck and McLay 2011; Rorabaugh 2023; Mora-Soto et al 2024).

Past events have shaped and continue to shape the Islands today. Sustained for millennia by Indigenous Peoples, only remnants now exist of the tall forests and rich meadows that once covered this region. Recently logged the land is becoming increasingly urban and suburban. The bountiful salmon of surrounding waters are on the decline with disturbances on lands.

Historic colonial settlement and disease altered Indigenous lifeways leading to decline in language use, food gathering practices and social traditions in Indigenous communities. Despite these impacts these ecologically adapted cultures are now beginning to recover. The extent to which historical disruptions can be undone is contentious.

Should invasive, naturalised species be accepted as part of emerging novel ecosystems (Glen et al. 2013; Gomes 2013)? What are the best pathways for reconciliation with Indigenous communities (Claxton 2015; Thom 2019; Townsend 2022)? Can Indigenous understandings of the island world be incorporated into future natural and human communities (Lepofsky et al. 2020; Turner 2020)?

Organization and Scope

Authors have been encouraged to define their own locational and thematic focus depending on the needs of the chapter and individual research specialisation. Most authors have decided to limit the scope of their chapter to one or more of the small island groups within the Salish Sea rain shadow zone, commonly the Gulf and San Juan Islands, as in the chapters on biogeography and water (Chapters 5 and 17).

Some have focused entirely on one or two islands. For example, a chapter on dislocation focuses on the Lyackson community on the east coast of Vancouver Island and their original home on Valdes in the Gulf Islands, the chapter on sustainability on Vancouver Island, and the local governance chapter on Gabriola. Some topics cannot be easily confined to any one island or island group, which is why the chapters on climate, oceanography, geology, water, pre-contact history and post-contact history cover the entire Salish Sea region.

Fidalgo, Whidbey, Camano and Bainbridge are connected via bridges to the mainland, and the ecological and sociocultural comparisons with islands in the San Juans and further north linked only by sea and ferry would have been instructive. It is unfortunate that there are fewer references to such islands. Similarly, it would have been desirable to have more attention devoted to bio-climatically distinctive Quadra and Discovery Islands to the North, or the Lord Howe Islands to the east.

Many of the chapters confront 'messy' real world problems that lack any simple solutions. Does conservation always favour native over introduced species? Barsh and Murphy question this assumption. Would the Lyackson community on Vancouver Island return to their homeland on Valdes? Not any time soon, suggests Thom. Should the 'leanness of the state' on small islands be solved by administrative support from Vancouver Island? No, says Dunsmoor-Farley.

Chapter summaries are presented below, abstracted from the respective chapters. I have grouped these in a way that seemed most logical, beginning with the natural environment, following by history, society, and environmental management.

The Natural Environment

Chapter 2 by Richard Hebda shows that weather and climate of the Salish Sea Islands have always been changing and, in comparison to the past, today's conditions are exceptionally amicable. Projections from climate models suggest that unprecedented rapid warming will transform the region into typical Mediterranean conditions characteristic of latitudes further to the south. Forest cover will decline, and new ecosystems will arise as summer drought intensifies.

Chapter 3 by Sophia Johannessen outlines the ways global changes manifest themselves in coastal waters of Salish Sea Islands. Sea level rise and storms can interact with coastal development, placing low-lying estuaries, intertidal zones and mudflats at risk. Decreases and earlier peaking in zooplankton biomass can cascade to higher trophic levels such as fish and birds. Acting to control these stressors will support resilience of biota in the face of inevitable global changes.

Steve Earle and John Clague (Chapter 4) distinguish between geological entities of the region that migrated from elsewhere, and others formed in situ. Mountain building associated with continental collisions and volcanism resulted in sedimentation into the basin that is the precursor to the Salish Sea. The Salish Sea itself is a product of ongoing tectonic activity, and the unique geomorphological features of its islands result from the wide diversity of the rocks of the region.

Russel Barsh and Madrona Murphy (Chapter 5) describe the way plants and animals have established in the San Juan and Gulf Islands over the past 12,000 years. The earliest pioneers were lichens and mosses on a few ice-free mountain peaks. When the ice began retreating, the San Juan group was flooded with melt water. Swimming and rafting drove colonisation once saltwater replaced fresh water in the Salish Sea basin. Humans returned to the islands roughly 8,000 years ago, driving further changes.

History

There are two origin stories for the Salish Archipelago—one Indigenous and the other non-Indigenous. Non-Indigenous stories of place represent ways of looking at the world imported from elsewhere. These ways of knowing complement the origin stories that predate European notions of history and science. In Chapter 6, Chris Arnett suggests a framework based on the work of Nuu-chah-nulth scholar Richard Atleo for understanding Indigenous ways of knowing and 1930s-era Coast Salish origin stories.

As in oral histories, Bill Angelbeck's archaeological overview (Chapter 7) reveals immense changes over long periods of time in both environment and culture. Yet, strong continuities of culture persist throughout these changes. Peoples still seasonally access resources throughout their territories, as done since the earliest inhabitants. Spiritual practices evident in Clovis period, 13,000 years ago, continue to appear in burials 4,000 years ago and later, and are even integrated into contemporary rituals.

The southward migration of the Lig̱wiłda'xw into the Salish Sea has a long history. We do not know all the causes of this expansion movement, but we do understand the complexity of their relations with their Coast Salish neighbours, which were marked by open hostilities and occasional alliances, according to circumstances. In Chapter 8, Marie Mauzé shows how what started as warfare for control of fisheries and obtaining economic wealth resulted in the exchange of intangible wealth between the Lig̱wiłda'x̱w and the K'ómoks.

The closely related Nuu-chah-nulth, Ditidaht and Makah are people of the west coast, with an economy and world-view focused on the open ocean. Networks of kin ties and trade, however, linked the people of the west coast via land and sea with those across the island and beyond. In Chapter 9 Alan McMillan discusses distinct features of Nuu-chah-nulth culture prior to European arrival, along with the networks of contacts and kin ties that linked them to their neighbours to the east.

As Lissa Wadewitz shows (Chapter 10), Salish Sea Islands and the nearby mainland have a long and complicated history, but the relationship between human beings and the natural world has always been at its core. Recent recognition of Coast Salish legal rights to resources and land is finally being factored into ecological management plans. Although the dispossession of Native peoples will forever mar this area's history, there may be hope if all hands are willing to work together for restorative purposes.

Society

Chapter 11, co-authored by Nancy Turner and Songhees elder Joan Morris, shows what life was like for the 'Old Ones' (*s'áliyluxw*) a generation ago on Tl'ches, a group of small islands offshore from the city of Victoria, and the transitions that ensued. Since the early 2000s, Joan has been working with her Nation and a team of archaeologists and ethnologists from the University of Victoria to document oral history and undertake ethnoecological restoration.

What sorts of demographic changes have occurred? What impact have these events led to? Chapter 12 by Moshe Rapaport affirms the deep-time rootedness of the Coast Salish, based on archaeological findings and oral traditions. It then explores the displacement following contact, successive waves of post- contact settlement, census data for the general and Indigenous populations, health and well-being, the human footprint and reducing our footprint.

Chapter 13 by Moshe Rapaport reviews the struggle for sovereignty and territorial rights lost during Euro-American settlement and following periods. Following this, selected examples of self-determination and cultural revival are presented: the concept of 'tsawalk', language revitalization, canoe journeys, reef net fishing, and Indigenous stewardship.

Few Lyackson First Nation members currently live on any of the reserve lands on Valdes Island, in spite of these having been ancestral village sites for millennia. Chapter 14 by Brian Thom, based on interviews with Lyackson First Nation members, traces their experiences of migration from their ancestral village sites and reveals policies and responses which dramatically shifted the economies and livelihoods of Indigenous peoples in the Gulf Islands and beyond.

It is not unusual for people to perceive governance as formal processes through which political representatives act on behalf of the citizenry. Digging deeper into other spheres of influence sheds light on the influences, decisions and outcomes outside the narrow ambit of the formal 'political'. Chapter 15 by Gabriola resident Dyan Dunsmoor-Farley examines how place has shaped the political economy of one Salish Sea Island as a site of autonomous local action.

Environmental Management

The Salish Sea was once abundant with wild Pacific salmon which supported fresh and saltwater ecosystems, thriving fisheries and Indigenous cultures. Beginning in the late 1970s, marine survival rates for chinook, coho and steelhead sharply declined. Chapter 16 by Isobel Pearsall and Michael Schmidt relates these declines to changes in the food web, estuarine habitat loss, pollution and disease, and climate change in the Salish Sea and North Pacific Ocean, and outlines recommended steps for salmon recovery.

Chapter 17 by Dan Moore, Diana Allen, Oliver Brandes and Randy Christensen provides an overview of water sources, supply and governance. Demand peaks in late summer, when part-time residents and tourists augment the islands' populations, and residents can face problems with groundwater supply due to excessive drawdown and well interference. Management arrangements are increasingly being augmented by community-based initiatives focused on monitoring, watershed restoration and other activities.

Humans have modified habitats and facilitated the transfer of species between mainland habitats and the islands for millennia. Chapter 18 by Barsh and Murphy shows how the relatively small geographic area of the islands, the number of pre-contact settlements and the islands' role as an early point of post-contact extractive development have affected natural ecosystems. Conflicting paradigms of Coast Salish and Euro-American conservation remain a barrier to effective, collaborative action in the San Juan and Gulf Islands.

Richard Kool (Chapter 19) was stunned to find that regional and municipal strategies on Vancouver Island failed to consider the dependence of growth on continual provisioning of an island population whose ecological footprint is an order of magnitude greater than the agricultural land available, whose energy-producing capacity falls far short of present demand and that has limited capacity for solid waste disposal.

The distinctive convergence of deep Indigenous histories with relentless assertions of cultural, economic and legal interests has set the stage for initiatives aimed at sharing rights and responsibilities for portions of the Salish Sea. Chapter 20 by Nelly Bouevitch, Soudeh Jamshidian and John Welch examines Indigenous-led conservation and co-management experiments as both important dynamics in recent decades and as windows into possible futures of collaborative governance of lands and waters co-owned by Indigenous and non-Indigenous peoples.

Note on Orthography

Many of the South Coast Salish languages come from a predominantly oral tradition, so attempts to transcribe the language into written texts have evolved and changed as record-keeping practices have changed. The text endeavours to be as consistent as possible, but remains faithful to original texts where necessary: as a result, different spellings for the same underlying words sometimes emerge.

The main Indigenous writing systems used are SENĆOŦEN, with its characteristic all-capitals text, and Hul'q'umi'num'. SENĆOŦEN also makes use of a 'practical orthography' with minimal diacritics, SENCOTEN. This is especially useful for online publications, such as the Tsawout First Nation website.[1]

1 See: tsawout.ca/about-tsawout/.

The following guide to SENĆOŦEN and Hul'q'umi'num' orthography (Tables 1.1 and 1.2), and accompanying overview map of Coast Salish and neighbouring languages and communities (Figure 1.2) are reproduced with permission from Barnett Richling's *The WSANEC and their Neighbours* (2016), based on Montler (1986) and Harvey (2008), as listed below.

Table 1.1 Elliott's SENĆOŦEN alphabet (with phonetic equivalents)

vowels						
A / Á (e)	Ⱥ	E (ə)	I (i)	Í (ə̕)	O (á)	U (u)

consonants					
B (p')	C (k)	Ć (č)	Ȼ (k^{w})	D (t')	H (h)
K (q')	Ꝁ (q'^{w})	K̲ (q)	Ḱ (q^{w})	L (l)	Ł (ɬ)
M (m)	N (n)	N̲ (ŋ)	P (p)	Q (k'^{w})	S (s)
Ś (š)	T (t)	Ⱦ (t'^{θ})	T̲ (ƛ')	Ŧ (θ)	W (w)
W̲ (x^{w})	X (x)	X̲ (x)	Y (y)	, (ʔ) (glottalised)	

Source: Richling (2016: 184), based on Montler (1986).

Table 1.2 Island Hul'q'umi'num'

vowels					
a	aa	e	ee	i	ii
o	oo	ou	u		

consonants					
ch	h	hw	k	kw	l
lh	m	n	p	p′	q
q′	qw	qw′	s	sh	t
t′	th	tl′	ts	ts′	tth
tth′	w	x	xw	y	′ (glottalised)

Source: Richling (2016: 184), based on Harvey (2008).

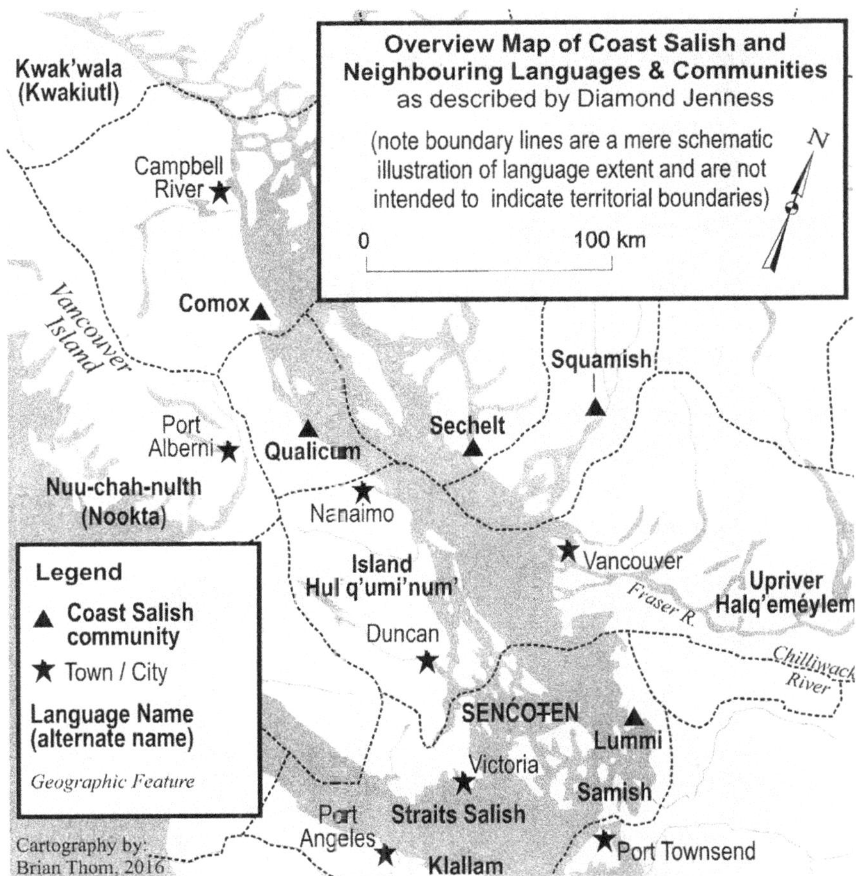

Figure 1.2 Overview map of Coast Salish and neighbouring languages and communities

Source: Courtesy Barnett Richling.

Acknowledgements

Special thanks to Nancy Turner and Richard Hebda for sharing advice and encouragement throughout the proposal and review process; to Chris Arnett for commenting on many of the chapters in this book; to Russel Barsh for sharing advice despite a taxing schedule at Kwiáht Institute; to Richard Hebda for a thoughtful and timely Epilogue; to Tim Bayliss-Smith for advice on publishers; to Barnett Richling, Brian Thom and other copyright holders for permissions; to Colin Filer and the editorial staff at Asia-Pacific Environment Monographs; to Beth Battrick for meticulous copy editing and the note on orthography; and to the contributors, colleagues, friends and anonymous reviewers for their helpful suggestions. Not least, thanks to Kathy. Without her encouragement and sage advice this publication would not have come to fruition.

References

Angelbeck, B. and E. McLay, 2011. The battle at Maple Bay: the dynamics of Coast Salish political organization through oral histories. *Ethnohistory* 58(3): 359–92. doi.org/10.1215/00141801-1263821

Beamish, R. and G. McFarlane, 2014. *The Sea Among Us*. Madeira Park: Harbour Publishing.

Benedict, A.D. and J.K. Gaydos, 2015. *The Salish Sea: Jewel of the Pacific Northwest*. Seattle: Sasquatch Books.

Claxton, N., 2015. To Fish as Formerly: A Resurgent Journey Back to the Saanich Reef Net Fishery. Victoria: University of Victoria (PhD thesis).

Endter, A.L., 2015/6. 'The Naming of the Salish Sea: A Legal, Historical, and Cultural Exploration.' *NW Law,* December/January.

Forward, C., 1979. *Vancouver Island: Land of Contrasts*. Vancouver: Department of Geography, University of Victoria.

Glen, A.S., R. Atkinson, K.J. Campbell, E. Hagen, N.D. Holmes, B.S. Keitt, J.P. Parkes, A. Saunders, J. Sawyer and H. Torres, 2013. 'Eradicating Multiple Invasive Species on Inhabited Islands: The Next Big Step in Island Restoration?' *Biological Invasions* 15: 2589–603. doi.org/10.1007/s10530-013-0495-y

Gomes, T.C., 2013. 'Novel Ecosystems in the Restoration of Cultural Landscapes of Tl'chés, West Chatham Island, British Columbia, Canada.' *Ecological Processes* 2(15): 1–13. doi.org/10.1186/2192-1709-2-15

Harvey, C., 2008. 'Halkomelem.' Viewed 11 July 2023 at: languagegeek.com/salishan/halkomelem.html

Lepofsky, D., C.G. Armstrong, D. Mathews and S. Greening, 2020. 'Understanding the Past for the Future: Archaeology, Plants, and First Nations' Land Use and Rights.' In N.J. Turner (ed.), *Plants, People, and Places: The Roles of Ethnobotany and Ethnoecology in Indigenous Peoples' Land Rights in Canada and Beyond*. Montreal: McGill-Queen's University Press. doi.org/10.1515/9780228003175-011

MacKinnon, A., 2003. 'West Coast, Temperate, Old Growth Forests.' *The Forestry Chronicle* 79(3): 475–84. doi.org/10.5558/tfc79475-3

Montler, T., 1986. 'An Outline of the Morphology and Phonology of Saanich, North Straits Salish'. *Occasional Papers in Linguistics 4*. Missoula: University of Montana Linguistics Laboratory. Viewed 11 July 2023 at: cas.unt.edu/~montler/Saanich/Outline

Mora-Soto, A., S. Schroeder, L. Gendall, et al., 2024. Kelp dynamics and environmental drivers in the southern Salish Sea, British Columbia, Canada. *Frontiers in Marine Science* 11. doi.org/10.3389/fmars.2024.1323448

Norman, E.S., 2012. Cultural politics and transboundary resource governance in the Salish sea. *Water Alternatives* 5(1): 138–60.

Richling, B. (ed.), 2016. *The WSANEC and their Neighbours: Diamond Jenness on the Coast Salish of Vancouver Island, 1935*. Oakville: Rock's Mills Press.

Rorabaugh, A., 2023. Assessing population dynamics in the Central Salish Sea, Pacific Northwest Coast of North America. *Plos one, 18*(8), p.e0285021. doi.org/10.1371/journal.pone.0285021

Stewart, H.M., 2017. *Views of the Salish Sea*. Madeira Park: Harbour Publishing.

Stratford, E., G. Baldacchino, E. McMahon, C. Farbotko and A. Harwood, 2011. 'Envisioning the Archipelago.' *Islands Studies Journal* 6(2): 113–30. doi.org/10.24043/isj.253

Thom, B., 2019. 'Leveraging International Power: Private Property and the Human Rights of Indigenous Peoples in Canada.' In I. Bellier and J. Hays (eds), *Indigenous People and the Law: Scales of Governance and Indigenous Peoples' Rights*. London: Routledge. doi.org/10.4324/9781315671888-8

Townsend, J., 2022. Indigenous and Decolonial Futurities: Indigenous Protected and Conserved Areas as Potential Pathways of Reconciliation. University of Guelph (PhD thesis).

Turner, N.J., 2020. 'From "Taking" to "Tending": Learning About Indigenous Land and Resource Management on the Pacific Northwest Coast of North America.' *ICES Journal of Marine Science* 77(7-8): 2472–82. doi.org/10.1093/icesjms/fsaa095

Vannini, P., 2011. 'The Techne of Making a Ferry: A Non-representational Approach to Passengers' Gathering Taskscapes.' *Journal of Transport Geography* 19: 1031–6. doi.org/10.1016/j.jtrangeo.2010.10.007

Part One: Environment

2

Weather and Climate and the Islands of the Salish Sea: Everything in Moderation, with a Few Extremes for Interest

Richard Hebda

Few island archipelagos on earth can say they owe their existence to climate … the Salish Sea Islands can! Thousands of years before the warm ocean waters and gentle breezes touched the islands, they rose as uplands on a vast plain of silt, sand and gravel stretching from the continental mainland to Vancouver Island and southward toward Puget Sound. Progressive global cooling fostered the advance of thick glacial ice that scoured out the intervening soft sediments between the emergent bedrock masses (Clague and James 2002). The ice left scarcely 15,000 years ago, the ocean flooded the Salish Sea and the islands as we know them formed.

Today the islands of the Salish Sea experience an amicable or friendly climate by most measures, especially for those who live in the northern half of the northern hemisphere. The Salish Sea is located after all north of 47 degrees north latitude. The summers are not especially hot, the winters not especially cold. The rain is usually plentiful but falls mostly in the winter and the summers are relatively dry, affording ideal conditions to be outside. Snow is something one mostly sees on nearby mainland and Vancouver Island mountains.

The climatic spice is provided by unusual and extreme events, short-lived, powerful and often transformative. Drenching downpours, ripping winds, extended browning droughts and rare massive snowfalls. This variation takes place on all time scales from those typically known as weather, what you see from day-to-day, to the longer intervals of many decades used to describe climate. The expression often heard in the Canadian Gulf Islands and adjacent Vancouver Island is 'If you don't like the weather now just wait a few minutes'. This usually means the rains will stop and there will be sunshine soon.

On a month-to-month and year-to-year scale, extreme events can be expected, and many residents look forward to them, as long as they are not too inconvenient. On longer time scales of thousands and millions of years, wide climatic variation has been the normal pattern; today's relatively brief (geologically speaking) mild and gentle respite being a highly unusual feature in the past several hundred thousand years. The progressing climatic change of the next few decades will certainly bring more variation and conditions not experienced for millions of years.

I explore the climate of the islands from several perspectives; from the conditions as they occur today, then as they have been in the past and finally as we may expect them in future decades. I include an island resident's annual weather account to personalise the climate experience. I do not write from the perspective of a meteorologist, which I am not, but that of a palaeoclimatologist interested in long-term trends and changes, and their relationship to the landscape. This perspective is especially important as major weather and climate changes are imminent (Mote and Salathé 2010; Pinna Sustainability 2020). Before I launch into an account of today's conditions, it is essential to appreciate the important role climate and weather have on the island landscape and island lives.

Climate Rules

As individuals and communities, we appreciate and adapt to the day-to-day and yearly features of weather and climate. Our buildings and infrastructure protect us from it, or moderate it for our comfort. Roads and bridges are constructed to 100- and 1,000-year extremes. Agriculture depends on and sometimes suffers from it. Seasonal rains provide the life-giving water we drink. Weather-driven fires sear our landscapes and cause misery and death. Folks who venture onto the sea live and sometimes die by it (Lange 2003).

Climate and weather shape the living and physical environment. Glaciers driven by climatic variation sculpt the land surface. Climate acting through ambient temperatures, rain and snowfall drives geological processes and the formation of glacial and slope deposits, and sand and gravel of flood plains. Further geological processes transform raw ground into the base upon which we build our homes and communities and upon which ecosystems develop. Beaches, old lake beds, terraces, gentle slopes, craggy rocks, river bottoms and the shape of valleys are all the result directly or indirectly of the action of weather and climate.

These processes and physical features collaborate with the life forms resulting in the plant and animal communities of the Salish Sea Islands. In British Columbia (BC) this inextricable relationship was recognised many decades ago and led to the ecological classification system of Biogeoclimatic Zones (Meidinger and Pojar 1991; Moore et al. 2010) by which landscapes are managed. Plant communities in the region can be recognised on a much finer scale according to their moisture regime and accumulation of soil organic matter, all driven by climate (Green and Klinka 1994). And these individual plant communities interact with the physical substrate and the climate to generate soils.

It is no exaggeration that we and our surroundings exist and thrive with the permission of climate. This reality shapes the islands of the Salish Sea as much as anywhere else on the planet. And this fundamental relationship has special importance when considering the climate of the near future. Unprecedented climatic changes in terms of rate, amplitude and direction are certain to occur before the end of this century (Mote and Salathé 2010; Pinna Sustainability 2020).

Weather and Climate Today

Large-scale Factors and Patterns

The region's climate is classified as modified Mediterranean, technically the Csb climatic class in the Köppen-Geiger system (Peel et al. 2007). In simple terms, the islands experience dry warm summers and mild damp winters. Mediterranean climates are relatively temperate, with average coldest month temperatures above 0° Celsius and summer months in general having less the 40 mm of rain per month. The Salish Sea Islands are located somewhat north of the Mediterranean zone which reaches about 45 degrees latitude.

The summers are generally cooler than typical with the warmest summer months, July and August in the Salish Sea region having average temperatures much less than 22° Celsius. Broadly such climates occur on the western side of continents under strong moderating influence from ocean waters (e.g. southern Chile, Argentina, New Zealand in Peel et al. 2007).

Island climate is shaped by regional factors familiar to the residents. Nearby ocean waters moderate temperatures and the occurrence of large mountain ranges to the east divert cold Continental Arctic air masses. Foremost, the climate and weather are driven by broad global atmospheric patterns and especially phenomena in the north Pacific Ocean.

The region occurs in the zone where solar energy absorbed by the atmosphere at the equator and adjacent subtropical zone is transferred to northern and polar regions. The way these mechanisms work and how they link to weather and climate in BC (and the Salish Sea Islands) are explained in Pigott and Hume (2009: Chapter 1) and for coastal BC by Lange (2003). Briefly, two strong vertically circulating cells, the Hadley (tropical/subtropical latitudes) and Polar (polar latitudes) cells, move air and energy in the atmosphere. They interact through a mid-latitude transition region with weak vertical cycling but well-developed horizontal air movements at high and low elevations in the atmospheric column. High- and low-pressure systems develop, travel in this transitional zone and interact with each other along fronts. Cold or warm fronts occur as cold or warm air pushes into the adjacent air mass. Low pressure often forms along fronts and is associated with clouds and rain. The movement of the air masses and associated storm systems is steered by the jet stream. The storm track shifts with the seasons, its position sometimes being north of the region (summer), sometimes south of or over the islands (winter). This broad pattern of the shifting storm track creates the region's weather directly and is responsible for its annual variation.

Another way of looking at regional climate is through the pattern of air masses generated as atmospheric features of the north Pacific and North American continent (Pigott and Hume 2009). Air masses are large bodies of atmospheric air of similar temperature and relative humidity several hundreds to thousands of kilometres across that develop over specific features of the earth. Air masses then jostle over the earth's surface according to the dominant wind direction and the behaviour of adjacent air masses. Because of the considerable size of these atmospheric air bodies, people over large regions experience the same general weather.

The Maritime Polar air mass of the north Pacific Ocean predominates in summer. Originating over the mid-latitudes of the ocean, a series of low- and high-pressure troughs and ridges are brought to the region by prevailing westerlies. The jet stream is usually northward of the region in the summer and there is little rain. The Pacific tropical air mass occasionally penetrates northward and brings warm, even hot, generally dry air and a 'blocking high' pressure ridge. Persistence of the blocking high results in drought and water supply issues on the islands in the summer months. Maritime Polar air predominates in spring and summer during which the jet stream passes near or over the islands. Rain and moderate temperatures are typical.

During the 'winter', which for our region includes November, two air masses shape weather and climate: the Maritime Arctic air mass and the Continental Arctic air mass. Northwest flow brings air originating over the cold lands of Alaska and Yukon, which is moderated by passing over north Pacific waters. These air masses are cooler than average and showery.

The coldest winter temperatures occur when the Salish Sea is exposed to Continental Arctic air from the northern heart of North America that occasionally breaks out westward through the Coast Mountains. The skies are often clear, and cold east and northeast winds descend out of the mainland coastal inlets buffeting several Salish Sea Islands directly. When Continental Arctic air encounters moist air in the Salish Sea heavy snowfalls can occur.

One fall-winter, air mass interaction brings a special phenomenon recently recognised and described as an 'atmospheric river' (National Ocean and Atmospheric Administration 2015). Southwesterly winds bring warm and moist Pacific tropical air. A narrow front develops extending from mid or even western Pacific waters and generates river-like streams of clouds and heavy rain to the Salish Sea Islands. Regionally called the 'Pineapple Express' (for its origins near Hawai'i), the atmospheric river phenomenon can cause heavy flooding in watersheds of large mainland rivers and big Island streams.

Despite occurring thousands of kilometres away, the El Niño–Southern Oscillation (ENSO) impacts island weather and results in unusual climatic conditions. The inverse condition, La Niña, also affects the region. The strongest impacts of these short-lived phenomena are experienced from February to June following the beginning of an ENSO event. Generally, temperatures are higher than normal in El Niño years and lower than normal in La Niña years (Fleming et al. 2007). An El Niño winter may represent what the Salish Sea Islanders might expect in the future as climate change advances, the mild winter of 2015–16 being an excellent example. Under El Niño conditions enhanced southwest flow brings more Pacific tropical air masses into the region (ibid.). Warm El Niño conditions generally are wetter than normal and La Niña conditions drier than normal in winter and spring.

The Pacific Decadal Oscillation is another climate-related pattern expressed as an index derived from sea surface temperatures (Mass 2008; Whitfield et al. 2010). This phenomenon varies somewhat systematically on a decadal scale with positive ('warm') and negative ('cold') index values lasting 2–3 decades and a cycle of 50–60 years. The mechanism is not yet fully understood. During a positive index period the sea surface of the central and western Pacific is relatively cool and the eastern Pacific, adjacent to western North America, is relatively warm. The opposite pattern occurs during a negative index period. Changes from positive to negative phases tend to be abrupt. A marked change from a negative to positive index periods took place in the mid-1970s. The index has varied widely from positive to negative values in the 2000s, showing strong positive and negative intervals. A positive index generally brings warmer than normal winter temperatures to the Salish Sea Islands (Whitfield et al. 2010).

Climate of the Islands from North to South

Two geographic features need to be kept in mind when looking at the climate and weather across the region. Northern Salish Sea Islands are surrounded by, or near, large areas of ocean water compared to southern ones because the Salish Sea is broad in the north. Northern islands are also closer to the climate of the continental interior, being near the mouths of long fjords such as Howe Sound and Jervis and Bute Inlets. In the south, the Fraser River Valley provides a conduit for cold inland air into the Salish Sea (Mass 2008). The region experiences a climatic regime that is broadly common to the coast of BC and Washington State (see Lange 2003: Chapter 4). For example, almost all coastal locations in BC exhibit a precipitation dip in mid-November on average followed by a sharp decline in mid-December (Lange 2003: 108). Annual maximum monthly temperature profiles are similar too.

Table 2.1 Climate data for selected stations of the Salish Sea

Climate station	Elevation (m)	Mean annual temperature	Mean July temperature (daily range)	Mean December temperature (daily range)	Mean annual precipitation (mm)	Mean annual snowfall (cm)
Campbell River	109	9.0	17.3 (11.5–23.0)	2.1 (-0.8–2.9)	1,489	84
Comox	26	10.0	18.0 (13.3–22.8)	3.5 (0.9–5.9)	1,097	55
Merry Island	7	11.0	18.2 (14.9–21.4)	5.2 (3.5–6.8)	1,006	23
Mayne Island	28	10.2	16.8 (11.0–22.5)	4.4 (1.8–6.9)	842	29
Friday Harbour	33	9.8	15.6 (9.4–21.7)	4.0 (1.8–7.1)	634	unavailable
Whidbey Island	14	9.9	15.2 (11.2–19.2)	4.6 (1.7–7.5)	515	13
Tacoma Narrows	89	11.1	18.1 (12.2–24.0)	4.8 (2.3–7.2)	1,093	unavailable
Olympia	57	10.3	17.7 (10.4–24.9)	3.6 (0.3–6.7)	1,270	27

Notes: Temperature in degrees Celsius, 1981–2010 normals. See citations in text for sources. See Figure 2.1 for station locations.

Source: Author's tabulation.

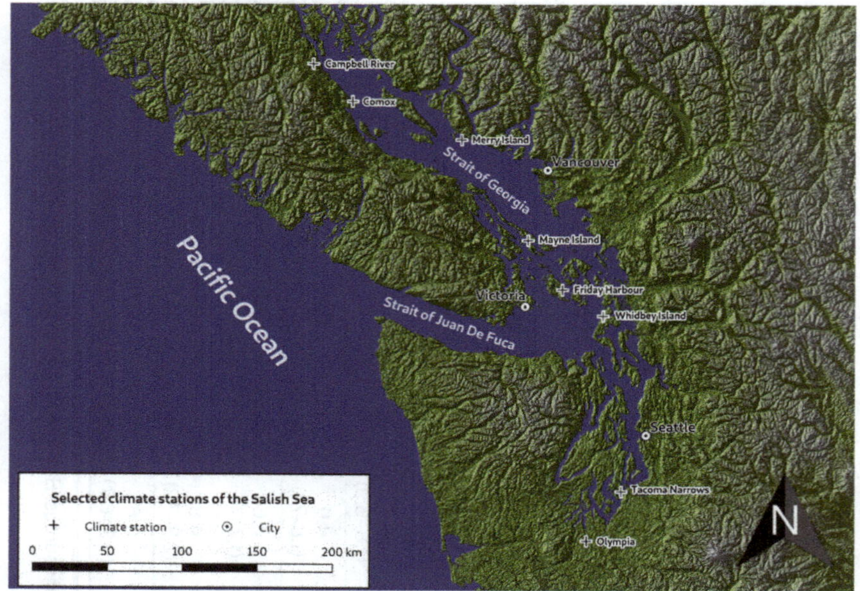

Figure 2.1 Representative climate stations of the Salish Sea Islands
Source: Google Earth.

Climatic data from observing stations, mostly located in major population centres on the continental mainland or on Vancouver Island, provide insight into conditions specific to the islands (Table 2.1 and Figure 2.1) because there are few island stations especially in the northern Salish Sea. The following data come largely from the Environment Canada website for Canadian Climate Normals website (Government of Canada 2021) for the years 1981–2010 and the US National Oceanic and Atmospheric Administration (n.d.) climate data website with normals also for 1981–2010.

The climate at Campbell River, at the north end of the Salish Sea, adjacent to Quadra Island, has a mean annual temperature of 9.0° C. Mean July temperatures are 17.3° C with daily high and low means ranging from 10.3° to 23.1° C. The coldest month is December, averaging 2.1° C and its daily range is from –2.6° C to a high of 4.2° C. Most of the mean annual precipitation of 1,489 mm falls as rain in November and December. Summers are dry, having about 40 mm of rain per month. The annual snowfall averages about 85 cm a year. This station is at 109 m above sea level and represents the slightly higher and damper portions of the Salish Sea Islands, as one might encounter on Quadra, Texada and Lasqueti islands.

Only 40 km to the southeast, conditions at Comox, on Vancouver Island near the shoreline, represent the more inhabited parts of the northern islands. Mean annual temperature is 10° C with mean July temperature of 18.0° C and a range of 13.3° to 22.8° C. The December average is 3.5° C with a range of –0.9° to 5.9° C and a general absence of frost, a condition typical of the islands. The mean annual precipitation is 1,097 mm, much of it falling in November and December. Summers are very dry (about 30 mm in July). The mean annual snowfall of 55 cm is notably less than at Campbell River.

Across the Salish Sea tucked up against the mainland coast, a lighthouse station on Merry Island demonstrates a strong oceanic effect. Mean annual temperature is 11.0° C. Summer conditions are similar to those at Comox with July and August mean temperatures at about 18.2° C, but the daily range is notably less from a mean low of 14.9° C to a mean high of 21.4° C. The January conditions are mild having a mean of 5.2° C and a daily range of 3.5–6.8° C. The area has a mean annual precipitation of 1,006 mm, slightly less than at Comox. Assuming similar cloud cover, the mild conditions are the direct result of the tiny island being surrounded by moderating warm ocean waters year-round.

Mayne Island, 100 km southeast of Comox and one of Canada's Southern Gulf Islands, represents conditions in the middle of Salish Sea (Figure 2.2). Mean annual temperature is 10.2° C with mean July temperatures at 16.8° C. The winters are slightly warmer than in the north, with a mean December temperature of 4.4° C. Mayne Island is drier than the northern Salish Sea Islands with mean annual precipitation of 842 mm, 22 cm of this being contributed by snow. The lower mean annual precipitation compared to northern islands is the result of less rainfall in the winter months. The dry summer months of July, August and September have from 21 to 28 mm of monthly rain on average and in some years no rainfall.

Crossing the international boundary, Friday Harbor, one of the larger towns of the Salish Sea Islands, is located 45 km southeast of Mayne Island, near the point where the Strait of Juan de Fuca and the Strait of Georgia and Puget Sound join. The mean annual temperature is 9.8° C, similar to that throughout the region. Summers are warm, with July and August means of 15.5° C. December is the coldest month, with mean monthly temperature of 4.4° C, being slightly colder than January, as is typical for the islands. Friday Harbour is one of the driest places among the islands with mean annual precipitation of only 634 mm and like Mayne Island much less winter rainfall than the northern Islands. November and January are the wettest months. Summers are very dry at 21 mm July monthly rainfall.

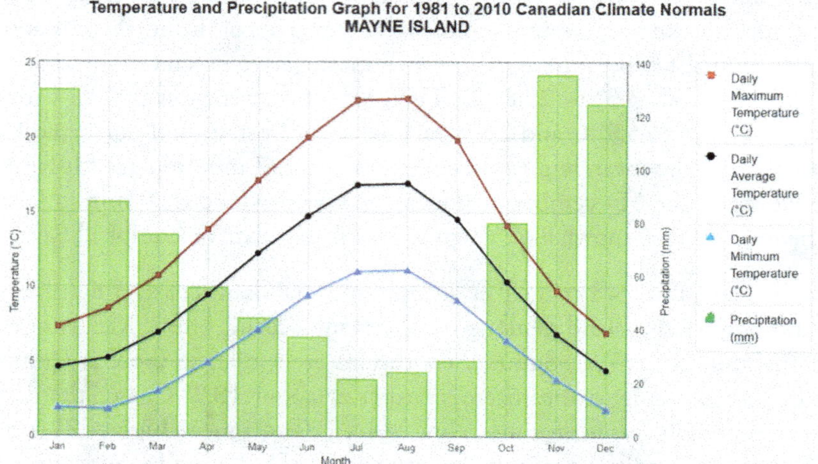

Figure 2.2 Temperature precipitation diagram for Mayne Island

Source: Derived from Government of Canada (2021) data.

Whidbey Island sits up against the eastern shore of the Salish Sea, 40 km south-southeast of Friday Harbor, and in the rain shadow of the 2,000-m tall Olympic Mountains. It is directly eastward of the long fetch of open water of Juan de Fuca Strait. Mean annual temperature is 9.9° C and December is the coldest month, with a mean of 4.6° C. July and August have a mean daily temperature of 15.2° C. The island is dry, with mean annual precipitation of only 515 mm, and November and January are the wettest months. November gets only 86 mm of precipitation, less than half of that for northern islands. July is dry (19 mm) and drought usually persists until the end of September.

Tacoma Narrows airport climate station is located near a narrow channel of the Salish Sea. It provides a good example of weather and climate for the southern Salish Sea, where islands such as Anderson Island are separated from the mainland by narrow stretches of water. The mean annual temperature of 11.1° C is notably warmer than islands to the north. December mean monthly temperatures are 4.8° C, with July monthly mean of 18.1° C the warmest of all the stations. The mean annual precipitation of 1,093 mm is much greater than at Whidbey Island, illustrating a general trend of increasing winter precipitation southward among the US Salish Sea Islands.

The climate station at Olympia, Washington State represents the southernmost Salish Sea Islands. The mean annual temperature of 10.3° C is notably cooler than at Tacoma Narrows and similar to that of most of the other Salish Sea stations. Mean monthly December temperature is 3.6° C and in July 17.7° C. Olympia is wetter still than Tacoma Narrows, with a mean annual precipitation of 1,270 mm and heavy rains in November to January.

Extreme Weather

Mass (2008) makes the case that despite the generally mild climate of the Pacific Northwest, the region experiences extremes, the most notable being strong winds and rare but deep snow falls.

Islanders are familiar with wind and most islands have their very windy spot often on a headland pointing into one of the channels. The windswept trees unambiguously point away from the prevailing wind direction. Prevailing winds blow from the west, but incredibly strong and destructive winds come from the south and southeast or from the north or northeast out of valleys connecting to the interior.

Lange (2003: 111) describes the typical pattern of wind in the Salish Sea as a weather front advances from the northwest. Southeast winds develop and intensify, generally being strongest in the north. Other Salish Sea wind patterns occur under a range of conditions, some even having local names such as the 'Qualicum' wind; they are of short duration but highly predictable and characteristic of portions of the Salish Sea.

Mass (2008) relates several cases of powerful wind storms and the associated meteorological conditions. Destructive winds are associated with October to April storms that track northeast, usually aiming for the northwest corner of the Olympic Peninsula and south Vancouver Island. The 'Columbus Day Storm' of October 1962 provides a destructive example (ibid.). In the US Pacific Northwest scores of people died, tens of thousands of homes were damaged and millions of board feet of timber lost. The storm began as a tropical typhoon near the Philippines then transitioned into a powerful mid-latitude storm before hitting the coast of Washington. The strongest winds occurred on the outer coast with sustained velocities of 241 kilometres per hour (kph). In the Salish Sea wind gusts reached 129 kph on Whidbey Island and 182 kph in Bellingham. In December 2006 a major windstorm hit the Salish Sea, cutting off electrical service to more than a million homes and knocking down huge amounts of timber. Winds reached 137 kph at the eastern end of the Strait of Juan de Fuca and 119 kph over the Hood Canal (ibid.).

Strong winds in the region also blow from the north when a high-pressure Continental Arctic air mass slides into southern BC. Under these conditions cold air flows into the Salish Sea through the northern inlets and especially down the Fraser Valley. Winds can gust to 97–129 kph and on 28 December 1990, 129–145 kph winds occurred widely, reaching more than 161 kph on Fidalgo and Lummi islands (Mass 2008). During these cold windy outbreaks, trees topple and moored boats wash onto shore.

Snow is uncommon or light in the Salish Sea especially for the southern Islands. However extreme snowfall events occur and can cause much damage. Everyone living in the region at the time has stories of the December 1996 big snow. Snow and strong winds brought cold and blizzard-like conditions on 26 December, closing roads in the region. Then two days later a massive snow event dropped even more snow with up to 61 cm on the ground. Roofs collapsed, power went out and roads were closed for days. Hundreds of boats were overloaded by wet snow and sank (Mass 2008).

Climate over the Long Term

The Salish Sea Islands climate has not always been as it is today, varying from subtropical to Arctic and alpine. Subtropical climates prevailed millions of years ago in the Age of the Dinosaurs (Mesozoic Era). Ancient fossil leaves studied from the shales and coals on the east side of Vancouver Island, 4 kilometres from Mudge and Gabriola Islands, contain palms and breadfruit (Pearson and Hebda 2006). Even after the great extinction event at 65.5 million years ago, palms luxuriated in Washington State near Bellingham (Mustoe and Gannaway 1995).

The greatest swings in climate occurred during the Ice Age, or Pleistocene Epoch, between 2.6 million and about 12,000 years ago. For much of this interval conditions were cooler and drier than today. Mountain icefields formed and glaciers flowed seaward toward today's Salish Sea. Cool subalpine climates occurred 70,000 years ago, then warming followed until about 40,000 years ago when progressive cooling and drying began. Cold whistling winds descended from growing ice fields in BC's Coast Mountains and the climate resembled that of a northern tundra. Beautiful grassy meadows developed on the lands of the Salish Sea (Hebda et al. 2016). These meadows supported Ice Age creatures such as the woolly mammoth, whose remains have been recovered on Gabriola and Whidbey Islands.

As the cold persisted, ice extended to Olympia, Washington State, covering the islands of the Salish Sea (see Earle and Clague, this volume) and a few of the islands may have stood above ice rather than sea water. With glacier retreat, mixed coniferous forest featuring lodgepole pine (*Pinus contorta*) occupied the region (Leopold et al. 2016).

The story of the last 11,000–10,000 years is of more than historical interest because it provides insight into what the Salish Sea Islands may experience in the warming decades ahead. Sudden warming of several degrees Celsius snapped the islands into a new climatic regime, even warmer than today. Several studies of pollen, spores, charcoal and other remains in sediment cores reveal how the climate and vegetation varied (Pellatt et al. 2001, Saanich Inlet; Leopold et al. 2016, Orcas Island; Lucas and Lacourse 2013, Pender Island). This warming brought on widespread open Douglas-fir (*Pseudotsuga menziesii*) forests, mixed coniferous woodlands and even grasslands (Pellatt et al. 2001; Lucas and Lacourse 2013; Leopold et al. 2016). Bracken fern (*Pteridium aquilinum*) and shrubs featured in the woodland openings and grasses and forbs in the grasslands as summer temperatures rose to levels higher than today. Alder was particularly widespread, and fire occurred frequently (Brown and Hebda 2002).

Forests began to close after 7,000 years ago in response to cooling and increasing moisture. Douglas-fir, western hemlock (*Tsuga heterophylla*) and bigleaf maple (*Acer macrophyllum*) variously became more abundant. Garry oak (*Quercus garryana*) occurred widely and abundantly for a thousand years or so. Fires were less widespread under this cooler moister climate than before. Beginning about 5,000 years ago, closed forests dominated by Douglas-fir and including western red cedar (*Thuja plicata*) and western hemlock developed in the region as climate cooled and moistened. Initially fires in the region declined, however about 2,000 years ago fire activity increased, possibly a result of human intervention (Brown and Hebda 2002).

The public may have the impression that climate change will be gradual. However, Zhang and Hebda's (2005) study of fossil tree rings immediately adjacent to the Salish Sea region demonstrates that rapid and strong climatic changes have occurred as recently as 4,000 years ago. A marked, high amplitude growth change reveals that a sudden and persistent shift in climatic regime occurred in a decade or less about 3,850 years ago.

The last 12,000 years of landscape and vegetation history inform us about potential impacts of projected warming. Warmer climates such as those of 11,000–9,000 years ago will result in major alteration of the composition and structure of ecosystems. Climate-linked processes such as fire regime will also change, and increased fire activity must be expected especially with so many people on the islands. Finally, climate change and the resulting responses in the landscape may occur rapidly as a result of a regime shift rather than gradually. Now let's turn to projections for the region's climate, one that will be very different than what Islanders experience now.

Back to the Future

Atmospheric greenhouse gas concentrations are rising, and the climatic consequences are becoming evident (Hansen et al. 2017). The Salish Sea Islands now experience the early effects of these changes such as record temperatures of the 2021 'Heat Dome' (Philip et al. 2022) and will be subject to greater effects in the decades to come. In 2012 Washington State prepared a comprehensive strategic document about climate change, which outlines climatic trends and projects future conditions (Adelsman and Ekrem 2012). The Pacific Climate Impacts Consortium (2020) has developed a 'projection tool' called Plan2Adapt, which helps users view regional future conditions, and Pinna Sustainability (2020) prepared a report specific to the Islands Trust area (see Moore et al., this volume).

Future climate projections vary according to factors such as the structure of the model being used and the amount of forcing by greenhouse gases. Future climate forcing will depend on social and economic development and successes in reducing greenhouse gas emissions. Future greenhouse gas concentrations also depend on other earth system factors such as the release of greenhouse gasses from permafrost. Global climate models work at coarse geographic scales and in block-like units. The Salish Sea region is small relative to these modelling units. Most regional projections of future conditions use groups or ensembles of models usually choosing a midpoint for future conditions and including a range from high to low outcomes.

The Salish Sea Islands future climate is summarised from two sources, one Canadian and one American, both based on information available to about 2012. The Washington State projections use a comparison to 1970–1999 normals whereas the Plan2Adapt comparison is to 1981–2010 normals (see Moore et al., this volume). Washington State values use the state as the projection area whereas those in in Plan2Adapt are for the Georgia Depression, which specifically includes the northern and central Salish Sea Islands. Regardless, the projections are similar for the time horizons of about 2040–2050 and the 2080s. Notably, the current rate of CO_2 emissions is tracking at the upper limit of that used in many previous global climate models. Hansen et al. (2017) point out that the global climate system has yet to catch up with the extra CO_2 now in the atmosphere; accordingly projections maybe conservative. I use the Plan2Adapt values because they are more regionally specific and use the 1981–2010 climate normal comparison, the same interval used earlier to describe modern-day climate. Where the Washington State values are notably different, they are included in parentheses.

In general, the Salish Sea Islands face a warmer future with wetter winters and drier summers than now. For the 2040s and 50s the mean annual temperature is expected to increase about 1.5° C (1.8° C) with a range of 0.9–2.3° C (2.8° C). For the 2080s, mean annual temperature is projected to be 2.5° C (3.0° C) warmer with a range of 1.3° to 3.7° C (5.4° C).

Projecting future precipitation is less certain because precipitation varies on a finer geographic scale than temperature. Changes in winter precipitation are most important because this is the rainy season in the islands. Precipitation is likely to increase about 10 per cent in the region by the 2080s within a range of 0–42 per cent. Summer rainfall is small in the drought-prone region so even a decline of about 10 per cent (range 0–33 per cent) will mean little in terms of moisture supply. However, increased summer temperatures and decreased cloud cover will result in increased evaporation and deepen the summer drought and lengthen the drought interval.

Snow will become a progressively rare feature of the region's climate. Exceptional snowfalls may occur in the northern Salish Sea Islands associated with outbreaks of Continental Arctic air, but these outbreaks will be extremely rare as winter temperatures warm. Increasing winter temperatures and likely increased cloud cover mean that frost may occur rarely. Plan2Adapt values for the Georgia Depression project an increase of 18 (11–27) frost-free days by the 2050s. Victoria immediately adjacent to the central Salish Sea only gets 12 days with frost now. Even northern reaches of the Salish Sea such as Campbell River can expect only 26 (16–37) frost days by the 2080s.

Projections only provide averages and ranges. What they do not reveal are the extremes, a noted feature of the regional climate. We may lose our extreme snowfall events, but in addition to strong winds and likelier heavy winter downpours, we may have to face much higher summer temperatures and prolonged summer droughts. Salish Sea Islanders already experience climate change impacts as western red cedar trees die and summer drought intensifies. Residents are already responding with rain water collection and water storage systems (see Moore et al., this volume).

Human Perspectives

Weather and climate are best appreciated through the eyes of people with many years of exposure to typical and unusual conditions. Long-time observers pick out trends and identify specific examples that illustrate them. They integrate their observations and impressions into their way of life. Two descriptions of the human–weather experience are presented in the following table, one by the Indigenous Salish People and one by a long-time rural island resident. The similarities are striking (Table 2.2).

For thousands of years the Indigenous Salish people of the Salish Sea Islands and adjacent lands have experienced the region's climate and environment and integrated these experiences into their culture. Climate was either directly or indirectly the basis of the Salish Indigenous names and descriptions of 13 lunar months of the year ('moons'), as summarised from Earl Claxton's work in Turner and Hebda (2012: 25–28). The settler calendar presented below comes from a 40-year resident on Salt Spring Island rooted in observations at a large rural agricultural property.

Typical climate and weather seem to be changing in the past few years. Ice occurs much less regularly than before and the rains that used to come in June and nourish a second cut of hay are now rare. People's fields dry early, and wells dry out in the summer as they never did before. Warm summer evenings when one could sit outside have increased from a handful to sometimes a week or more in a row. People are now acutely concerned about fires and worry about water supply.

Table 2.2 Salish and settler weather calendar

Moon	Salish weather	Settler weather	Month
Moon of the Child (Birth Moon)	The Moon of the Child or Birth Moon (January) sees warming and even a few sunny days, though cold, wet windy weather still prevails.	January begins the year in the middle of what islanders typically call winter, the months from November to February. This is the cool and moist time of the year. January and February are mostly damp, cloudy or with broken cloud, but interrupted by beautiful sunny days, sometimes with spotty fog. Strong frontal winds, especially southeasters still occur. January is often the snow month and the snow can stay on the ground for a week. However, the lengthening days are welcomed.	January–February
Frog Moon	During the Frog Moon (February), the awakening frogs chorus as the days warm and the rains lessen.		
Moon of Opening Hands (Blossoming Out)	During the Moon of Opening Hands or Blossoming Out (March–April) there is more sun and enough warmth to dry the harvested food.	March ushers in the transition to spring and flowers begin to bloom. The days are warmer and drier than they were earlier and weather alternates from cloudy with rain, to bright and sunny. The nights however are cool and killing frosts (–2° C lows) still occur.	March
Bullhead Moon	Big winds come during the Bullhead Moon (April) when large fish appear and are caught near shore. The good weather is sometimes interrupted by sudden thunderstorms.	April to June are the spring months, the 'transition to summer' as it was phrased by the observer. Days and nights warm and storms still pass through, however the soil remains cool. This is the time of peak spring bloom. Fog occurs rarely. This transition varies, sometimes wetter and cooler, and sometimes drier and warmer.	April–May
Moon of the Camas Harvest	During the fine weather of the Moon of the Camas Harvest (May) and Sockeye Moon (May–June), people travelled widely among the islands, gathering and drying many kinds of food.	June is particularly variable and the rain, especially in the past few years, rarely penetrates the soil.	May-June

Moon	Salish weather	Settler weather	Month
Humpback Salmon Return to Earth Moon	People worried about forest fires in the hot, dry Humpback Salmon Return to Earth Moon (June–July).	Summer includes July and August, stretching more and more into September. Summers are dry and warm with bright blue skies and long warm evenings when it is possible to sit outside. Temperatures may reach the 30° C range. Rain falls rarely, and many weeks pass without it. People worry about forest fires.	July–August
Coho Salmon Return to Earth Moon	The first rains begin falling in the Coho Salmon Return to Earth Moon (August) and people hunted deer as the temperatures cooled.		
Dog Salmon Return to Earth Moon	The rains of Dog Salmon Return to Earth Moon (September) bring bountiful fish to rushing swollen rivers.	September begins the transition to winter which takes place largely in October. Storms return and there are more cloudy days than in the summer. Nights cool and winds return with the storms. October can be a nice month, usually having storms alternate with sunshine and warmth. Frost is possible later in the month.	September–October
Moon That Turns the Leaves White	In the Moon That Turns the Leaves White (October) the nights begin to get cold and frost comes to the uplands.		
Moon of the Shaker Leaves	People keep close to shore and their winter villages in the Moon of the Shaker Leaves as winter starts. Snowfalls allow hunters to track game.		
Moon of Putting Your Paddle Away in the Bush	Indoor times and story-telling returns with the Moon of Putting Your Paddle Away in the Bush (November–December).	November starts the winter and cloudy days predominate. Sometimes the rain seems to fall steadily, and cold air outbreaks bring snow. The storms lash island homes with sheets of rain and winds blow hard enough to make a solid house creak. Power lines go down as trees blow over. Killing frost usually occurs in this month. December continues this pattern, though sunny days may occur toward then end of the month.	November–December
Elder or Winter Moon	The Elder or Winter Moon (December) has short stormy and rainy days. This is an indoor time, when elders taught children through stories.		

Source: Author's tabulation.

Conclusions

The weather and climate of the Salish Sea Islands have always been changing and, in comparison to the past, today's conditions are exceptionally amicable. Extreme wind and snow bring 'interest' and sometimes damage to the islands, generally though conditions are mild and vary within a relatively narrow range. Representing a modified Mediterranean climate, winters are cool and rainy rarely experiencing heavy snowfall or even frost. Summers are dry and droughty, but temperatures rarely reach 30° C. Over millions of years, the region has experienced subtropical conditions and much more recently Arctic climates. Landscapes and ecosystems have responded accordingly, adjusting rapidly. Projections from climate models reveal that unprecedented rapid warming will turn the region into typical Mediterranean conditions characteristic of latitudes further to the south. Forest cover will decline, and new ecosystems arise as summer drought intensifies.

Island residents can see and feel the climate changing. They are already adapting to a new climatic future as they install rain water tanks and plant vineyards and olive groves in anticipation of challenges and new opportunities.

Acknowledgements

Thanks to Bill Hamilton for assistance and Moshe Rapaport for advice during the preparation of this chapter. Special thanks to Ted Baker of Salt Spring Island for his Islander's account of a typical weather year.

References

Adelsman, H. and J. Ekrem, 2012. *Preparing for a Changing Climate: Washington State's Integrated Climate Response Strategy*. Olympia: Department of Ecology (Publication No. 12-01-004).

Brown, K.J. and R.J. Hebda, 2002. 'Ancient Fires on Southern Vancouver Island, British Columbia, Canada: A Change in Causal Mechanisms at About 2,000 Ybp.' *Environmental Archaeology* 7: 1–12. doi.org/10.1179/env.2002.7.1.1

Clague, J.J. and T.S. James, 2002. 'History and Isostatic Effects of the Last Ice Sheet in British Columbia.' *Quaternary Science Reviews* 21: 71–87. doi.org/10.1016/S0277-3791(01)00070-1

Fleming, S. W., P.H. Whitfield, R.D. Moore and E.J. Quilty, 2007. 'Regime-Dependent Streamflow Sensitivities to Pacific Climate Modes Cross the Georgia–Puget Transboundary Ecoregion.' *Hydrological Processes* 21: 3264–87. doi.org/10.1002/hyp.6544

Government of Canada, 2021. 'Canadian Climate Normals.' Viewed 31 May 2023 at: climate.weather.gc.ca/climate_normals/

Green, R.N. and K. Klinka, 1994. *A Field Guide for Site Identification and Interpretation for the Vancouver Forest Region.* Victoria: Research Branch, Ministry of Forests, Province of British Columbia (Land Management Handbook 28).

Hansen, J. et al., 2017. 'Young People's Burden: Requirement of Negative CO2 Emissions.' *Earth System Dynamics* 8: 577–616. doi.org/10.5194/esd-8-577-2017

Hebda, R.J., O.B. Lian and S.R. Hicock, 2016. 'Olympia Interstadial: Vegetation, Landscape History, and Paleoclimatic Implications of a Mid-Wisconsinan (MIS3) Nonglacial Sequence from Southwest British Columbia, Canada.' *Canadian Journal of Earth Sciences* 53(3): 304–20. doi.org/10.1139/cjes-2015-0122

Lange, O.S., 2003. *Living with Weather Along the British Columbia Coast: The Veil of Chaos.* Vancouver: Environment Canada.

Leopold, E.B., P.W. Dunwiddie, C. Whitlock, R. Nickmann and W.A. Watts, 2016. 'Postglacial Vegetation History of Orcas Island, Northwestern Washington.' *Quaternary Research* 85: 380–90. doi.org/10.1016/j.yqres.2016.02.004

Lucas, J.D. and T. Lacourse, 2013. 'Holocene Vegetation History and Fire Regimes of *Pseudotsuga menziesii* Forests in the Gulf Islands National Park Reserve, Southwestern British Columbia, Canada.' *Quaternary Research* 79(3): 366–76. doi.org/10.1016/j.yqres.2013.03.001

Mass, C.K., 2008. *The Weather of the Pacific Northwest.* Seattle: University of Washington Press.

Meidinger D. and J. Pojar (eds), 1991. *Ecosystems of British Columbia.* Victoria: Ministry of Forests (Special Report Series 6).

Moore, R.D. (Dan), D.L. Spittlehouse, P.H. Whitfield and K. Stahl, 2010. 'Weather and Climate.' In R.G. Pike, T.E. Redding, R.D. Moore, R.D. Winkler and K.D. Bladon (eds), *Compendium of Forest Hydrology and Geomorphology in British Columbia*. Victoria: Ministry of Forests.

Mote, P.W. and E.P. Salathé Jr, 2010. 'Future Climate in the Pacific Northwest.' *Climatic Change* 102: 29–50. doi.org/10.1007/s10584-010-9848-z

Mustoe, G.E. and W.L. Gannaway, 1995. 'Palm Fossils from Northwest Washington.' *Washington Geology* 23: 21–26.

National Ocean and Atmospheric Administration, 2015. 'What Are Atmospheric Rivers?' Viewed 31 May 2023 at: www.noaa.gov/stories/what-are-atmospheric-rivers

——, n.d. 'DataTools: 1981–2010 Normals.' Viewed 31 May 2023 at: www.climate.gov/maps-data/dataset/1981-2010-climate-normals-climographs

Pacific Climate Impacts Consortium, 2020. 'Plan2Adapt.' Viewed 31 May 2023 at: www.pacificclimate.org/analysis-tools/plan2adapt

Pearson, J. and R.J. Hebda, 2006. 'Paleoclimate of the Late Cretaceous Cranberry Arms Flora of Vancouver Island: Evidence for Latitudinal Displacement.' In J.W. Haggart, R.J. Enkin and J.W.H. Monger (eds), *Paleogeography of the North American Cordillera: Evidence For and Against Large-Scale Displacements*. Geological Association of Canada (Special Paper 46).

Peel, M.C., B.L. Finlayson and T.A. McMahon, 2007. 'Updated World Map of the Köppen-Geiger Climate Classification.' *Hydrology and Earth System Sciences Discussions, European Geosciences Union* 4(2): 439–73. doi.org/10.5194/hessd-4-439-2007

Pellatt, M.G., R.J. Hebda and R.W. Mathewes, 2001. 'High Resolution Holocene Vegetation History and Climate from Hole 1034B, ODP Leg 169S, Saanich Inlet Canada.' *Marine Geology* 174: 211–26. doi.org/10.1016/S0025-3227(00)00151-1

Philip, S., Y.S.F. Kew, G.J. van Oldenborgh, W. Yang, G.A. Vecchi, F.S. Anslow, S. Li et al., 2022. 'Rapid Attribution Analysis of the Extraordinary Heatwave on the Pacific Coast of the US and Canada in June 2021.' *Earth System Dynamics* 13: 1689–713. doi.org/10.5194/esd-13-1689-2022

Pigott, R.W. and B. Hume, 2009. *Weather of British Columbia*. Edmonton: Lone Pine Publishing.

Pinna Sustainability, 2020. 'Climate Projections for Islands Trust Area.' Viewed 31 May 2023 at: islandstrust.bc.ca/document/climate-projections-for-islands-trust-area-report-2020/

Turner, N.J. and R.J. Hebda, 2012. *Saanich Ethnobotany: Culturally Important Plants of the WSÁNEC People*. Victoria: Royal British Columbia Museum.

Whitfield, P.H., R.D. Moore, S.W. Fleming and A. Zawadzki, 2010. 'Pacific Decadal Oscillation and the Hydroclimatology of Western Canada—Review and Prospects.' *Canadian Water Resources Journal* 35: 1–28. doi.org/10.4296/cwrj3501001

Zhang, Q. and R.J. Hebda, 2005. 'Abrupt Climate Change and Variability in the Past Four Millennia of the Southern Vancouver Island, Canada.' *Geophysical Research Letters* 32, 1–4, L16708. doi.org/10.1029/2005GL022913

3

Oceanography of the Salish Sea

Sophia Johannessen

The Salish Sea connects the islands to the open ocean. It provides food, transportation and recreation, and moderates the regional climate. But the sea is changing.

Change flows in off the land, with variations in the timing and temperature of river runoff, and rushes in from the open ocean with the tide and currents. Human activities, such as fishing, shipping, wastewater discharge and shoreline protection, also change this coastal sea. The sea does not receive these changes passively. Everything that enters is subject to its unique circulation and geometry.

The Salish Sea has not always been the shape it is today. Fifteen thousand years ago, at the height of the last glaciation, the whole region was covered by 1 to 2 kilometres of ice (see Clague, this volume). Then the ice receded, and water flowed over the land. Sea level was as much as 200 metres higher than it is now. Slowly, over the next two thousand years, the land rebounded, and the islands emerged out of the ocean.

The oral history of the First Nations people of Vancouver Island records this event. The Saanich Peninsula, north of Victoria, is named for the SENĆOŦEN word W̱SÁNEĆ, meaning 'raised' or 'emerging' (Paul et al. 1995). Alternatively, the word can be translated as 'to raise rump in the air', referring to a local mountain (Mt Newton) as the rump of the peninsula

(Hudson 1970). The oral history recalls the time, 9,000–13,000 years ago, when this peninsula rose out of the ocean, giving the southwestern edge of the Salish Sea the shape it has today.

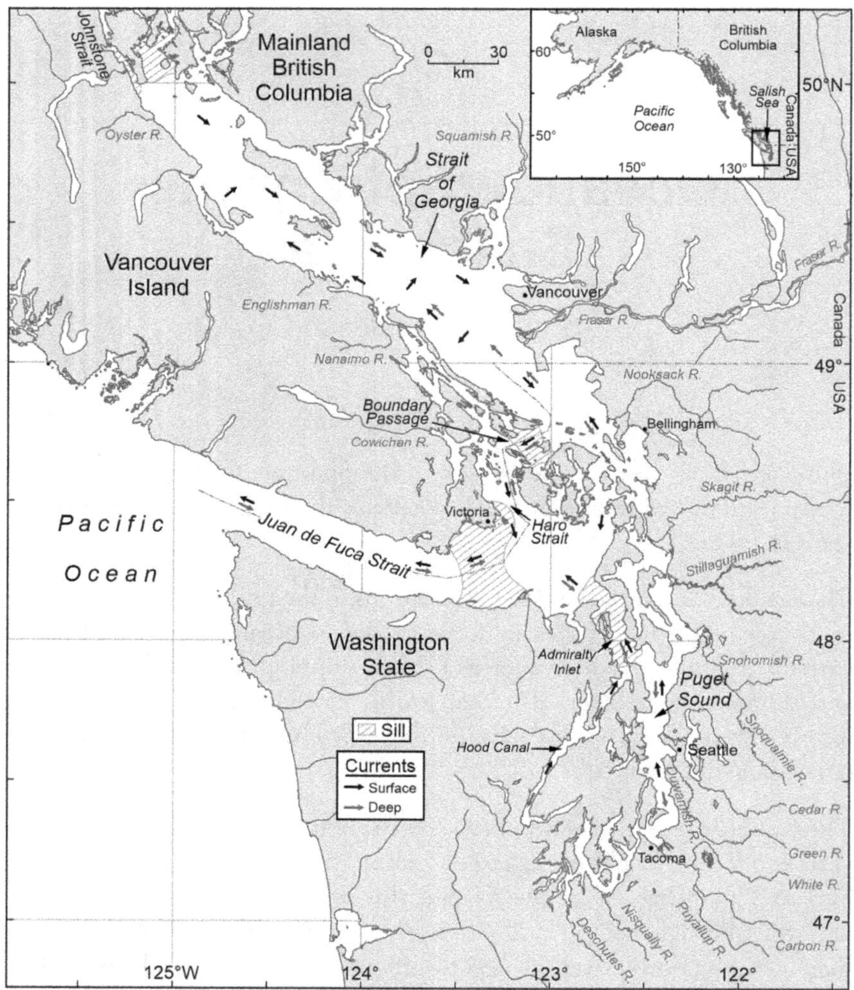

Figure 3.1 The Salish Sea, showing sills, currents and principal rivers

Note: The dashed grey line represents the Canada–US border.

Source: Sophia Johannessen.

The shape of the sea has a strong effect on the oceanography. The inner Strait of Georgia and southern parts of Puget Sound are deep basins (up to 420 and 283 m deep, respectively), bounded by shallow sills and narrow tidal passages (Thomson 1981) (Figure 3.1). Water mixes vigorously over the sills, stirring surface water, bubbles and particles of plankton and soil throughout the whole water column (Thomson et al. 2020). The intense mixing in these tidal passages has important effects on the food web and on the way in which the inner basins experience global climate change.

This chapter describes the physical, geochemical and biological oceanography of the Salish Sea, and discusses how local circulation and environmental conditions determine the effects of global and local change.

Physical Oceanography

Rivers flow into the Salish Sea, driving estuarine circulation—seaward at the surface and landward at depth (Thomson 1981). The largest is the Fraser River, which discharges into the southern Strait of Georgia, near Vancouver (Sutherland et al. 2011). Smaller rivers flow in all along the shoreline (Figure 3.1). The peak fresh water inflow occurs in late May and early June, during the freshet of the mainly snow-fed Fraser and Squamish Rivers (Sutherland et al. 2011), with a smaller peak in autumn (October/November), driven by heavy rainfall.

Fresh water flows out across the surface, toward the open ocean. It entrains seawater from underneath, which mixes with the outflow (Pawlowicz et al. 2007). A strong subsurface inflow of seawater from the open ocean through Juan de Fuca Strait balances the outflow year-round (Figures 3.1, 3.2). There is also a small exchange through Johnstone Strait, at the northern end of the Salish Sea (Khangaonkar et al. 2017).

There are shallow sills (~100 m deep) in Johnstone Strait, Haro Strait and Admiralty Inlet (Figure 3.1)—remnants of the huge glaciers that once filled the basins. The sills interrupt the smooth estuarine and tidal flow (Figure 3.2), causing intense mixing (e.g. Thomson et al. 2020; MacCready et al. 2021). The mixing is strongest during spring tides (when the sun and moon align to pull together on the earth) and weakest during neap tides (when the sun and moon work against one another, and the tidal range is small) (Masson 2002).

Estuarine circulation and tides act year-round, continually replacing surface and mid-depth water (Pawlowicz et al. 2007). However, the water in the deep basins of the Strait of Georgia and Puget Sound is only replaced episodically in late spring and summer (Masson 2002; Thomson et al. 2020), and during some neap tides in summer in Puget Sound (Deppe et al. 2018). For the rest of the year, the deep water is essentially stagnant, exchanging only slowly with the overlying water. Deep-water renewal depends on the strength of tidal mixing and the density of the water flowing in through Juan de Fuca Strait, which depends, in turn, on the strength of upwelling on the continental shelf (Thomson et al. 2020).

In the spring, the predominant winds blow from the northwest to the southeast. Along the outer coast of Vancouver Island, this causes surface water to move away from the shore and to be replaced by seawater upwelled from greater depths (Thomson 1981). The upwelled water is much denser than the surrounding water. It flows in through Juan de Fuca Strait and then, once it has passed over the entrance sills at Haro Strait (into the Strait of Georgia) or Admiralty Passage (into Puget Sound), it drops down into the deep inner basins, displacing the water that was there before (Figure 3.2) (Masson 2002; Deppe et al. 2018; Thomson et al. 2020). The properties of the inflowing ocean water are modified by the mixing over the sill (see Geochemical Oceanography section).

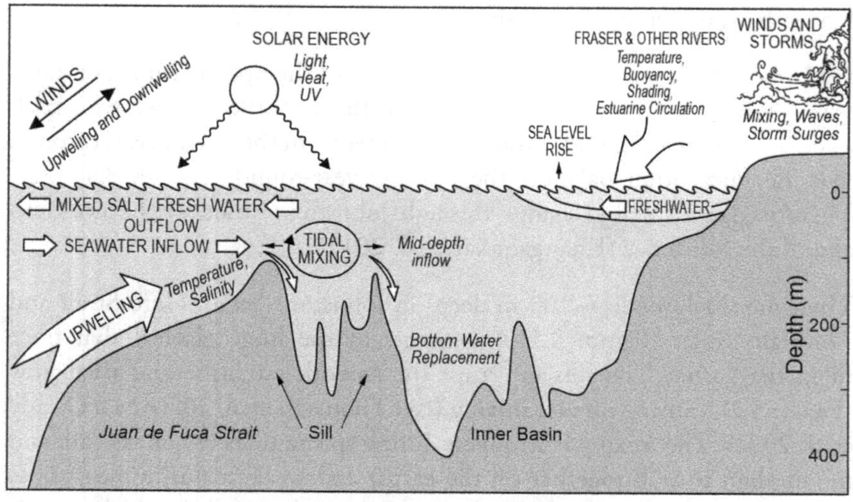

Figure 3.2 Physical circulation of the Salish Sea

Note: The figure shows inflow through Juan de Fuca Strait; mixing over entrance sills; mid-depth and deep inflow into the deep basins of the Strait of Georgia and Puget Sound.

Source: Modified from Johannessen and Macdonald (2009).

In addition to the large-scale water movements caused by estuarine circulation, tides, mixing over sills and deep-water renewal, winds also play a role in the Salish Sea. Winds produce waves and move the surface plume of the Fraser River (Pawlowicz et al. 2017; Yang et al. 2019). They also mix the surface water, breaking up the surface stratification, which helps drive summertime productivity (see Biological Oceanography section).

Chemical and Geochemical Oceanography

Nutrients and Oxygen

Phytoplankton (microscopic plants at the base of the marine food web) need nutrients to grow. Nitrate is generally the limiting nutrient in the coastal ocean (Mackas and Harrison 1997). Nitrate enters the Salish Sea from the open ocean, with the estuarine inflow. It is drawn into the surface water by estuarine circulation and winds (Collins et al. 2009: 14; Khangaonkar et al. 2018). Nutrients also enter the Salish Sea with river discharge and other surface runoff, atmospheric deposition, groundwater infiltration and local human sources, such as wastewater discharge and agricultural runoff (Sutton et al. 2013). However, all the other sources of nitrogenous nutrients are dwarfed by the subsurface inflow of nutrients from the ocean (Sutton et al. 2013) (Figure 3.3).

Nutrients are taken up by phytoplankton in surface waters and regenerated at depth, as the sinking organic matter is broken down. Consequently, nutrient concentrations are generally low at the surface and increase with depth.

Oxygen profiles have the opposite shape. The concentration of oxygen is high at the sea surface and low at depth (e.g. Masson and Cummins 2007). Oxygen diffuses into the ocean from the atmosphere and is also produced in surface waters by phytoplankton during photosynthesis. As organic matter sinks and breaks down, releasing nutrients, oxygen is consumed (Figure 3.4).

Deep-water renewal in the late spring replenishes oxygen in the deep water of the Strait of Georgia and Puget Sound (Deppe et al. 2018; Johannessen et al. 2014). Although the inflowing, upwelled water sometimes has very low concentrations of oxygen, the strong mixing over the entrance sills adds oxygen to the inflow before it flows into the deep basins.

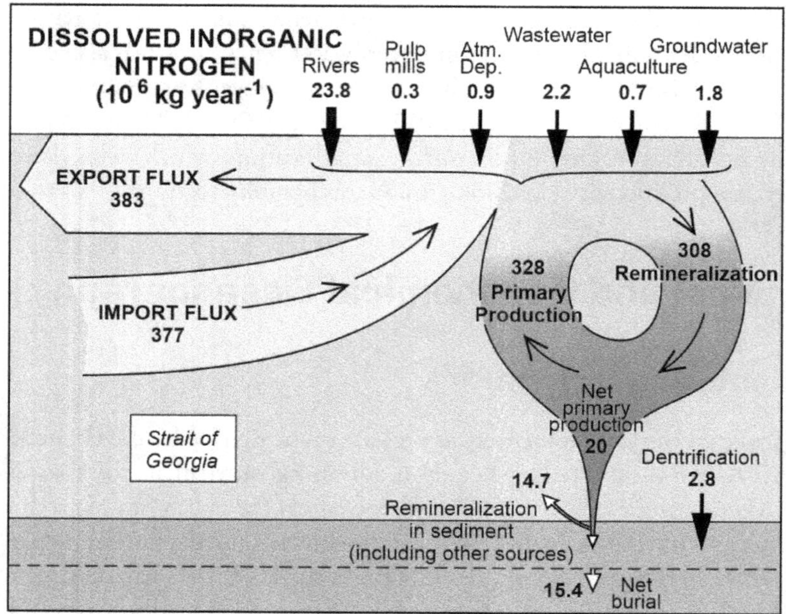

Figure 3.3 Dissolved inorganic nitrogen budget for the Strait of Georgia

Note: Import and export fluxes represent exchanges with the open ocean. Equivalent information not available for Puget Sound.

Source: Sophia Johannessen, modified from Sutton et al. (2013).

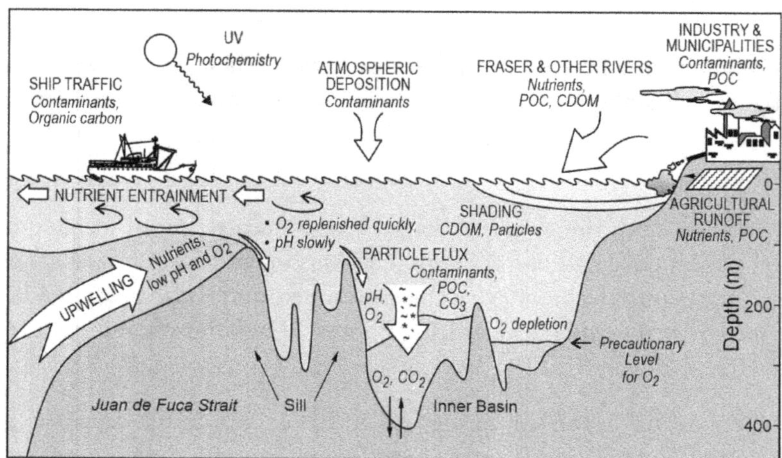

Figure 3.4 Geochemical processes in the Salish Sea

Notes: The 'upwelling' arrow on the left shows the inflow of nutrient-rich, low-oxygen water from the open ocean through Juan de Fuca Strait. Intense mixing over the entrance sills modifies the water that flows into the inner basins of the Strait of Georgia and Puget Sound.

Source: Modified from Johannessen and Macdonald (2009).

The concentration of oxygen in the deep water of Puget Sound is sometimes very low (hypoxic), particularly in Hood Canal (Bricker et al. 1999). This is likely because of the limited circulation and replacement of bottom water in that inlet (Khangaonkar et al. 2017).

Particles and Underwater Weather

The Salish Sea experiences underwater weather (Johannessen et al. 2017). Organic and inorganic particles settle through the water column. The rain of particles is affected by events in the atmosphere (rainstorms, windstorms); on land (snowmelt, river discharge, erosion); and in the surface ocean (blooms of phytoplankton, zooplankton and jellyfish).

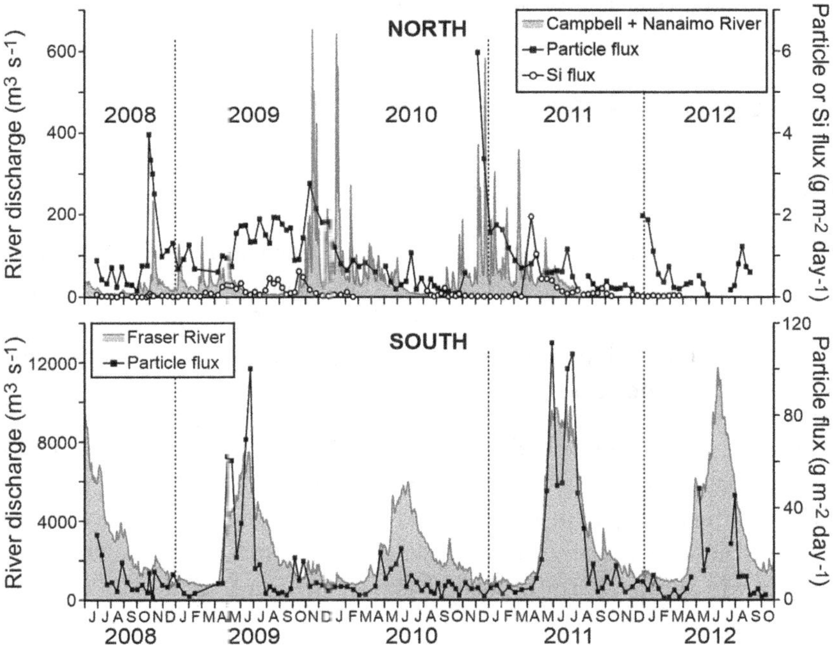

Figure 3.5 Underwater weather

Notes: Particle fluxes measured at 50 m depth using sediment traps in the northern and southern Strait of Georgia. Particle fluxes are overlain on river discharge for the two areas. Silica flux — representing diatoms — is shown for the northern station. Note different scales on the upper and lower panels.

Source: Modified from Johannessen et al. (2017).

In the southern Strait of Georgia, the flux of particles has a strong, monsoon-like seasonality dominated by the Fraser River discharge, with very high fluxes in May and June, at the time of the Fraser River freshet, and lower fluxes for the rest of the year (Figure 3.5). In the northern strait, farther from the Fraser River, there are more frequent and varied events related to local rivers and blooms. Even at its winter minimum, the particle flux in the southern strait is much higher than in the north, analogous to living in a wet climate, rather than a dry one.

The sinking particles affect dissolved oxygen, pH, carbon cycling and the quality and availability of food for bacteria, zooplankton and the benthic animals that live at the bottom of the sea. If they are not consumed in the water column, particles settle to the seafloor as sediment, burying organic matter and particle-active contaminants.

Biological Oceanography

Phytoplankton

The Salish Sea is highly productive. The annual average primary productivity in the Strait of Georgia is 280 gC m^{-2} yr^{-1} (Harrison et al. 1983; Sutton et al. 2013). Productivity is higher in Puget Sound (465 gC m^{-2} yr^{-1}) (Winter et al. 1975). The high productivity is the result of the physical and chemical environment, as described above.

The main spring bloom is dominated by diatoms (Harrison et al. 1983; Del Bel Belluz et al. 2021). These are large phytoplankton with siliceous shells that fuel a highly productive, short food chain. Later in the summer, there are blooms of other, smaller phytoplankton, including dinoflagellates, some of which cause harmful algal blooms (Harrison et al. 1983; Pospelova et al. 2010).

Phytoplankton need light and nutrients to grow. Strong stratification keeps them near the surface in the euphotic zone, where there is enough light. However, if the stratification is too strong, the phytoplankton use up all the nutrients and cannot grow any more until the nutrients are replenished.

The timing and strength of windstorms determines the timing of the spring phytoplankton bloom (Collins et al. 2009), with stronger late-winter winds delaying the onset of the spring bloom by mixing phytoplankton below the euphotic zone. However, small summer windstorms can give rise to intermittent blooms throughout the summer, by replenishing the surface water with nutrients from the underlying seawater (St John et al. 1993).

For most of the year, in Juan de Fuca Strait, Haro Strait, most of the Strait of Georgia and parts of Puget Sound, phytoplankton are limited by light, not by nutrients (Mackas and Harrison 1997; Winter et al. 1975). Light doesn't penetrate very deeply in this turbid, organic-rich coastal sea. In most of the Strait of Georgia, the light needed by phytoplankton only reaches about 20 m for most of the year (Masson and Peña 2009).

The total phytoplankton productivity of the Salish Sea does not appear to have changed in the last one hundred years (Johannessen, Macdonald et al. 2020). There are questions, however, about whether the timing or type of primary productivity might have changed (Allen and Wolfe 2013).

Zooplankton

Zooplankton are abundant in the Salish Sea. Large-bodied crustacean species like copepods, euphausiids and amphipods predominate, although there are some common gelatinous species too (Mackas et al. 2013). The total zooplankton biomass and main bloom timing have varied dramatically over the last few decades (ibid.; Perry et al. 2021). The large interannual variability seems to be related to variations in large-scale estuarine circulation and the match or mismatch between the timing of the phytoplankton and zooplankton blooms in a particular year (Mackas et al. 2013; Perry et al. 2021), although the exact mechanisms are unclear.

The timing and type of zooplankton affect the availability and nutritional quality of food for juvenile fish, invertebrates, benthic animals and seabirds (Perry et al. 2021).

Fishes

The Salish Sea supports many species of fish, including herring, hake, dogfish and pollock, as well as the iconic Pacific salmon (Fu et al. 2012). Some fish mainly eat plankton (planktivorous fishes), while others eat fish (piscivorous fishes). The two types overlap, since some planktivorous species begin to eat other fish when they grow large enough, and juveniles of many piscivorous species eat plankton. However, the distinction is useful, because planktivorous and piscivorous fishes are susceptible to different environmental pressures.

Small, planktivorous fish predominate in the Salish Sea, as appears to have been the case throughout the Holocene (Tunnicliffe et al. 2001). The most common species are Pacific herring (*Clupea pallasi*), Pacific sand lance (*Ammodytes personatus*), and smelt (*Osmeridae* spp.) (Therriault et al. 2009). Walleye pollock (*Theragra chalcogramma*), juvenile spiny dogfish (*Squalus acanthias*) and other juvenile fish also consume zooplankton. Pacific hake (*Merluccius productus*) are planktivorous when small, but the larger adults are major predators of herring (Francis and Lowry 2018). Anchovy (*Engraulis mordax*) is near the northern end of its range in the Salish Sea, but its population seems to be increasing; it is more common in warm years, which suggests that its regional population may continue to increase in the future (Duguid et al. 2019).

Small, short-lived pelagic fishes are subject to wide fluctuations in abundance. In the Strait of Georgia, the population of Pacific herring increased from 1980 to 2003 and then suddenly declined (Therriault et al. 2009). The decline was even more dramatic in Puget Sound (Francis and Lowry 2018).

Recent research has shown a strong link between the timing of herring spawning and the success of the new juveniles in the Strait of Georgia (Boldt et al. 2019), which implies that the main control on herring populations currently is the availability of food. The ideal timing for herring spawn is about 20 days before the main phytoplankton bloom. This permits the juvenile herring to emerge at a time when their zooplankton prey, especially copepods, are abundant (Boldt et al. 2019).

Competition for prey has been proposed as an explanation for the decline in herring, but Boldt et al. (2019) showed that juvenile herring actually survive better in years when juvenile salmon are also abundant, implying that the same conditions favour both species. The larger population decline in Puget Sound has been attributed to urbanisation, predation and poor water quality (Francis and Lowry 2018).

Populations of several predominantly piscivorous fish species have also declined in recent decades. Landings data show that coho (*Oncorhynchus kisutch*) and chinook salmon (*Oncorhynchus tshawytscha*) (Sobocinski et al. 2021), lingcod (*Ophiodon elongates*), Pacific cod (*Gadus macrocephalus*) and some inshore rockfishes have all declined, while the populations of the populations of the predominantly planktivorous pink (*Oncorhynchus gorbuscha*) and chum salmon (*Oncorhynchus keta*) are at historic highs (Ruggerone and Irvine 2018).

Salmon occupy a special niche in west coast marine ecosystems (see Chapter 16). Because they spend part of their lives in fresh water and part in the ocean, they are vulnerable to change in a variety of ways (Shelton et al. 2021). During their adult lives at sea, they are susceptible to large-scale climate variation that alters temperature, water-mass distribution and food supply. In fresh water, the out-migrating smolts and spawning adults are exposed to contamination, temperature rise and habitat destruction by logging, urbanisation, damming and natural events, such as the Big Bar landslide on the Fraser River in June 2019. A water temperature of less than 17° C is optimum for salmon migration, and above 18° C fish begin to manifest stress affecting survival (Crossin et al. 2008).

The cause of the long-term decline in coho, chinook, steelhead and sockeye salmonids has not been demonstrated definitively. Suggested causes have included reduced availability of food for juveniles (see Pearsall and Schmidt, this volume) and overfishing.

Models indicate that no one single factor was responsible for the decline. Predation, hatchery timing and anthropogenic activities had the strongest effects, according to the Sobocinski et al. (2021) model, while the Fu et al. (2012) model indicated that heavy fishing of Pacific herring reduced the productivity of all fish species in the Strait of Georgia.

Studies of recent survival rates have also implicated multiple stressors. Chittenden et al. (2018) reported that in the Cowichan River estuary in the Strait of Georgia, the size and timing of release of smolts put hatchery-reared chinook salmon at a disadvantage relative to wild salmon, and further, that jellyfish and harmful algal blooms in the bay could prevent juvenile salmon from leaving the estuary to find food in the wider ocean. In Puget Sound, Meador (2014) reported that chinook smolts from hatcheries in contaminated estuaries had a 47 per cent lower chance of survival than those reared in uncontaminated estuaries.

It is possible that, as a result of the large declines over the last few decades, the populations of these fish have become so low that they have lost their resiliency, and every stressor now poses an existential threat. Fortunately, many of the local stressors (habitat destruction, contaminants, fishing, etc.) are within local control and could be reduced.

Benthos

There is a hidden ecosystem associated with the mud and sand at the bottom of the Salish Sea. These communities vary with water depth. For example, bivalves, like clams, are very common in shallow water sediment, but become less common below 100 m depth, after which various types of worms become more common (Burd et al. 2008). Habitat, food supply and oxygen all affect benthic animals and determine where they can live (Burd et al. 2008).

One unusual type of benthic animal is the reef-forming hexactinellid glass sponges that were discovered in the coastal waters of British Columbia in the 1980s (Conway et al. 1991). Glass sponge reefs have since been discovered in many places along the coast, including parts of the Salish Sea. The sponge reefs provide habitat for at least 115 species of invertebrates and fish and sequester carbon at a high rate (Dunham et al. 2018). They are vulnerable to destruction by bottom-contact fisheries like trawling. Some of the Salish Sea sponge reefs have been designated as protected, with no bottom-contact fisheries permitted.

Change in the Salish Sea

The Salish Sea is changing as a result of both local and global stressors (Figure 3.6). All of these pressures are caused by humans, whether directly, as in the discharge of contaminants through wastewater, or indirectly, through fossil fuel emissions that have led to global climate change. The scale of human effects on the earth and ocean are so great that scientists have proposed that the era that we live in be called the 'Anthropocene' (from 'anthropos' = human; e.g. Lewis and Maslin 2015).

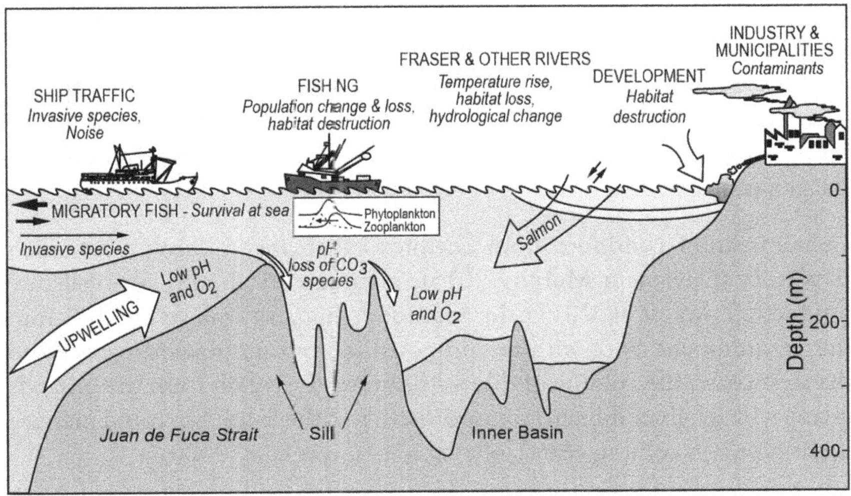

Figure 3.6 Global and local stressors on the marine ecosystem of the Salish Sea

Source: Modified from Johannessen and Macdonald (2009).

Local Stressors

Local stressors include contaminants, habitat destruction, fishing, marine traffic, invasive species and harmful algal blooms.

Contaminants

The Salish Sea has received contaminants from many sources over the last 150 years. Persistent, bioaccumulative contaminants, such as PCBs (polychlorinated biphenyls) and PBDEs (polybrominated diphenyl ethers) can enter the food web and remain in the environment for a long time. Some, like PCBs, biomagnify, increasing in concentration with every step in the food web (Ross et al. 2000). High trophic-level animals like killer whales accumulate high concentrations of these contaminants, and because they are long-lived (up to 70 years), it can take decades to reverse a trend (Hickie et al. 2007).

Discharges of many types of contaminants have decreased over time, including metals, dioxins, furans, mercury, tributyltin and PCBs (Johannessen and Macdonald 2009). Concentrations of PCBs, PBDEs and other persistent pollutants have declined in harbor seals since those contaminants were banned (Ross et al. 2013), although it will take longer for those declines to be noticeable in the longer-lived killer whales (Hickie et al. 2007). Similarly, in Howe Sound, a fjord near Vancouver, the fisheries for crab, shrimp and prawn reopened in 1995, following regulations that had resulted in a 97 per cent reduction in the discharge of dioxins and furans into the environment (Hagen et al. 1997).

Contamination continues from chemicals that are in current use, such as plastics (Davis and Murphy 2015), pharmaceuticals and personal care products (Lowe et al. 2017). In addition, ships can leak or spill oil into this environment, and, with the proposed increase in pipeline capacity to the west coast, the volume of diluted bitumen (heavy oil from the oilsands mixed with a lighter diluent) transported across the Salish Sea could increase significantly in coming years (Johannessen, Greer et al. 2020).

Wastewater

Municipal wastewater is a mixture of human waste and other chemicals (including persistent, bioaccumulative contaminants) (Johannessen et al. 2015). The human waste components (nitrogen, carbon, biological oxygen demand) have large, natural cycles, while there is no natural source of PCBs, PBDEs, etc.

High nutrient fluxes from land can cause eutrophication (excessive nutrient concentrations that lead to thick and sometimes harmful algal blooms and hypoxia; e.g. Khangaonkar et al. 2018). In Juan de Fuca Strait and most of the Strait of Georgia, natural concentrations of nutrients are already high, due to supply from the open ocean, and phytoplankton are limited by light (Mackas and Harrison 1997; Sutton et al. 2013). Eutrophication is unlikely in these areas. In the Strait of Georgia, wastewater contributes <1 per cent of the total influx of nitrogen (Johannessen et al. 2015) (Figure 3.3). However, southern Puget Sound, especially Hood Canal, has been identified as eutrophic, and excess nutrients from land have likely contributed (Bricker et al. 1999).

Throughout the Salish Sea municipal outfalls are a major route of entry for brominated flame retardants, which are common in homes and businesses (Johannessen et al. 2015), as well as for pharmaceuticals and personal care products.

Habitat Destruction

Coastal habitat is under threat. Coastal marshes are shrinking; nearly all those surveyed have lost area since the 1880s (Levings and Thom 1994). Shoreline armouring can cause beaches to become steeper and reduce the amount of wrack (washed-up seaweed and other natural materials) and associated invertebrates, which provide habitat and food, respectively, for spawning fish and migratory seabirds (Dethier et al. 2016).

Channelising and damming rivers and removing riparian vegetation stresses fish, in addition to major natural threats like the recent Big Bar landslide, which blocked fish passage through part of the Fraser River.

Other Local Stressors

Commercial fishing affects both the target species and the larger ecosystem (Myers and Worm 2003). In the Salish Sea, population crashes among commercially fished species, such as herring and salmon, have led to episodic fisheries closures. For additional discussion of changes in salmon populations see Pearsall and Schmidt (this volume).

Marine traffic was identified as one of the three main threats to the endangered southern resident killer whales, along with food availability and contaminants (National Marine Fisheries Service 2008). Traffic brings the risk of ship strikes, pollution and noise, which can interfere with animals' ability to communicate under water.

Invasive species can threaten native species by competition, habitat destruction, predation and hybridisation (Heath et al. 1995). Some species were introduced intentionally, such as the Pacific oyster, while others arrived uninvited on or inside ships and trade goods (Gillespie 2007). For example, the European green crab (*Carcinus maenas*) arrived in the Salish Sea in 2016, as larvae transported by currents, and has since established populations in several bays (Brasseale et al. 2019). Green crabs tend to damage seagrass beds, which reduces the populations of fish and other species that rely on the seagrass as habitat.

Farmed Atlantic salmon can be considered an invasive species, since they are not indigenous to the Salish Sea. There has been widespread public concern about the potential for fish farms to expose wild fish to disease, parasites and contaminants. The Canadian federal government has pledged to phase out open-net fish farms in British Columbia by 2025.

Harmful algal blooms (HABs) have occurred episodically in the Salish Sea at least since the end of the last glaciation (Mudie et al. 2002). The incidence of natural HABs can be augmented by eutrophication. Blooms have become more toxic in recent years in Puget Sound. However, the high natural concentration of nutrients in this area makes it difficult to assign a cause with confidence (Anderson et al. 2008).

Global Change

The Salish Sea is changing as a result of global climate change. Models (e.g. Khangaonkar et al. 2019) predict a number of changes by the end of the century, including warmer water, increased nutrient inflow and decreased oxygen concentration (Table 3.1).

Table 3.1 Projected climate change effects in the Salish Sea, 2000–2095

Property	Direction of change	Amount of change (Year 2095 relative to 2000)
Seawater temperature at inflow (°C)	↑	2.6° C (32%)
Air temperature (°C)	↑	3.5° C
River/runoff temperature (°C)	↑	3.2° C
Nitrate + nitrite at inflow	↑	gN m^{-3} (6%)
Nitrate + nitrite in rivers/runoff	↑	44%
Dissolved oxygen at inflow	↑	1.7 mg L^{-1} (35%)
Sea level rise	↑	l.5 m

Note: 'At inflow' refers to ocean source water at mouth of Juan de Fuca Strait. These are average modelled increases; animals will experience greater variability over shorter timescales.

Source: Khangaonkar et al. (2019).

As the climate warms, more precipitation falls as rain (largely in autumn and winter) and less as snow. This means that the large, late spring discharge of water (freshet) from the snow-fed rivers on the mainland is becoming smaller and occurring later, while runoff is increasing in autumn and winter (Morrison et al. 2002). This will affect the circulation and stratification in the Salish Sea and likely change the nutrient entrainment into the surface layer.

By the end of the century, the summer water temperatures in the Fraser River could sometimes reach 24° C (Morrison et al. 2002), which could be disastrous to salmon.

The concentration of oxygen in the deep Strait of Georgia has declined at about 0.2 mL/L/decade (89.2 mol/L/decade; Johannessen et al. 2014), due to a decline in the concentration of oxygen in the inflowing, upwelled seawater (Crawford and Peña 2013). Deep-water oxygen approaches the hypoxic threshold seasonally in the Strait of Georgia, due to the respiration of organic matter. Although the water flowing in through Juan de Fuca Strait is hypoxic, the strong mixing over the entrance sills in Haro Strait adds oxygen, so that the mixed replacement water replenishes oxygen in the deep waters of the Strait of Georgia. The deep water of Hood Canal in Puget Sound is already frequently hypoxic (Khangaonkar et al. 2018).

Globally, the ocean has taken up about a quarter of the carbon dioxide from fossil fuel burning (Watson et al. 2020). Although this has helped to slow the rate of warming of the atmosphere, it has resulted in ocean acidification, which causes numerous problems for marine animals (Doney et al. 2009). The water on the BC and Washington shelf has become more acidic in recent years (Feely et al. 2008).

The deep waters of the Salish Sea are not so well buffered against acidification as they are against hypoxia, because carbon dioxide doesn't exchange as quickly between water and air as oxygen does (Ianson et al. 2016). Consequently, acidification has been observed or modelled in the Salish Sea (Feely et al. 2010; Evans et al. 2019). In Puget Sound, acidification is thought to have been responsible for major failures in oyster aquaculture in the early 2000s (Feely et al. 2010), and for the dissolution of pteropod zooplankton shells (Bednaršek et al. 2021). Since different animals respond differently, acidification could drive changes of species assemblage in the Salish Sea in the future.

Cumulative Effects of Global and Local Change

In some cases, local and global stressors act together to produce cumulative effects. For example, sea level rise, combined with shoreline armouring, will drown the intertidal and shallow subtidal habitat of seagrasses, such as eelgrass (Figure 3.7).

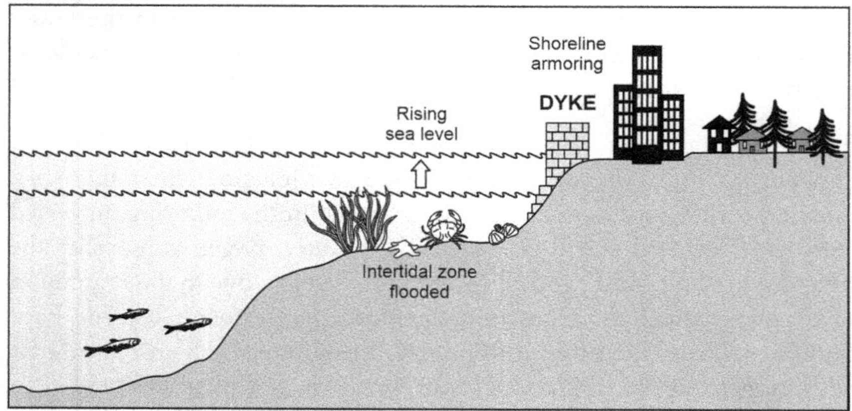

Figure 3.7 Interaction of local and global change
Note: Rising sea level meets an armoured shoreline, drowning intertidal habitat.
Source: Sophia Johannessen.

Salmon are particularly vulnerable to both global and local change, because they live in both fresh water and seawater at different times in their lives. Animals that live in the deep water are also vulnerable, because they are adapted to the relatively stable environmental conditions found at depth. Increased temperatures, deoxygenation and acidification together pose a greater threat to these animals than the sum of the three individually (Pörtner 2008); one stressor makes the animals more susceptible to changes in the others.

Conclusions

The unique character of the Salish Sea results from its position between land and the open ocean. The animals and plants of the Salish Sea feel the combined effects of climate change and local human activities. Fish and mammals that have been hunted or fished heavily have had to contend at the same time with contaminants, habitat destruction and rapidly changing environmental conditions.

An international border passes through the sea (Figure 3.1), but the seawater and all its constituents pass freely across it, as do the animals that call the sea home. Although many studies relate only to one part of the Salish Sea, oceanographers and marine biologists have come together increasingly to study the sea as a whole (e.g. Pearsall and Schmidt, this volume).

National governance can make it difficult to address stressors on a large scale, but fortunately many of the stressors are quite local. Reducing the local stressors of pollution, habitat destruction, fishing and hunting can help the ecosystem to remain resilient in the face of global-scale change (Brander 2008). This is clear in the example of Howe Sound, where marine life has returned in abundance since the cessation of toxic discharges along its shore (Hagen et al. 1997).

The Salish Sea has provided the setting and a lifeline for the people, animals and plants of the islands for thousands of years. Change is already under way. How we choose to respond is up to us.

Acknowledgements

Dr Robie Macdonald provided invaluable advice during discussions of this chapter, which draws heavily from an earlier paper (Johannessen and Macdonald 2009). Patricia Kimber prepared the figures. Two anonymous reviewers and Christopher Arnett offered comments that greatly improved the manuscript.

References

Allen, S. and M.A. Wolfe, 2013. 'Hindcast of the Timing of the Spring Phytoplankton Bloom in the Strait of Georgia, 1968–2010.' *Progress in Oceanography* 115: 6–13. doi.org/10.1016/j.pocean.2013.05.026

Anderson, D.M., J.M. Burkholder, W.P. Cochlan, P.M. Glibert, C.J. Gobler, C.A. Heil, R. Kudela et al., 2008. 'Harmful Algal Blooms and Eutrophication: Examining Linkages from Selected Coastal Regions of the United States.' *Harmful Algae* 8(1): 39–53. doi.org/10.1016/j.hal.2008.08.017

Bednaršek, N., J.A. Newton, M.W. Beck, S.R. Alin, R.A. Feely, N.R. Christman and T. Klinger, 2021. 'Severe Biological Effects Under Present-Day Estuarine Acidification in the Seasonally Variable Salish Sea.' *Science of the Total Environment* 765: 142689. doi.org/10.1016/j.scitotenv.2020.142689

Boldt, J.L., M. Thompson, C.N. Rooper, D.E. Hay, J.F. Schweigert, T.J. Quinn II, J.S. Cleary and C.M. Neville, 2019. 'Bottom-Up and Top-Down Control of Small Pelagic Forage Fish: Factors Affecting Age-0 Herring in the Strait of Georgia, British Columbia.' *Marine Ecology Progress Series* 617–18: 53–66. doi.org/10.3354/meps12485

Brander, K., 2008. 'Tackling the Old Familiar Problems of Pollution, Habitat Alteration and Overfishing Will Help with Adapting to Climate Change.' *Marine Pollution Bulletin* 56: 1957–8. doi.org/10.1016/j.marpolbul.2008.08.024

Brasseale, E., E.W. Grason, P.S. McDonald, J. Adams and P. MacCready, 2019. 'Larval Transport Modeling Support for Identifying Population Sources of European Green Crab in the Salish Sea.' *Estuaries and Coasts* 42(6): 1586–99. doi.org/10.1007/s12237-019-00586-2

Bricker, S., B. Longstaff, W. Dennison, A. Jones, K. Boicourt and C. Wicks, 1999. *National Estuarine Eutrophication Assessment: Effects of Nutrient Enrichment in the Nation's Estuaries*. Silver Spring: NOAA, National Ocean Service, Special Projects Office and National Centers for Coastal Ocean Science.

Burd, B., P. Barnes, C.A. Wright and R. Thomson, 2008. 'A Review of Subtidal Benthic Habitats and Invertebrate Biota of the Strait of Georgia, British Columbia.' *Marine Environmental Research* 66, Supplements S3–38. doi.org/10.1016/j.marenvres.2008.09.004

Chittenden, C.M., R. Sweeting, C.M. Neville, K. Young, M. Galbraith, E. Carmack, S. Vagle, M. Dempsey, J. Eert and R.J. Beamish, 2018. 'Estuarine and Marine Diets of Out-Migrating Chinook Salmon Smolts in Relation to Local Zooplankton Populations, Including Harmful Blooms.' *Estuarine, Coastal and Shelf Science* 200: 335–48. doi.org/10.1016/j.ecss.2017.11.021

Collins, A.K., S.E. Allen and R. Pawlowicz, 2009. 'The Role of Wind in Determining the Timing of the Spring Bloom in the Strait of Georgia.' *Canadian Journal of Fisheries and Aquatic Sciences* 66(9): 1597–16. doi.org/10.1139/f09-071

Conway, K., J.V. Barrie, W.C. Austin and J.L. Luternauer, 1991. 'Holocene Sponge Bioherms on the Western Canadian Continental Shelf.' *Continental Shelf Research* 11: 771–190. doi.org/10.1016/0278-4343(91)90079-L

Crawford, W.R. and M.A. Peña, 2013. 'Declining Oxygen on the British Columbia Continental Shelf.' *Atmosphere-Ocean* 51(1): 88–103. doi.org/10.1080/07055900.2012.753028

Crossin, G.T., S.G. Hinch, S.J. Cooke, D.W. Welch, D.A. Patterson, S.R.M. Jones, A.G. Lotto, et al., 2008. 'Exposure to High Temperature Influences the Behaviour, Physiology, and Survival of Sockeye Salmon During Spawning Migration.' *Canadian Journal of Zoology* 86 (2): 127–40. doi.org/10.1139/Z07-122

Davis, W. and A.G. Murphy, 2015. 'Plastic in Surface Waters of the Inside Passage and Beaches of the Salish Sea in Washington State.' *Marine Pollution Bulletin* 97 (1): 169–77. doi.org/10.1016/j.marpolbul.2015.06.019

Del Bel Belluz, J., M. Peña, J. Jackson and N. Nemcek, 2021. 'Phytoplankton Composition and Environmental Drivers in the Northern Strait of Georgia (Salish Sea), British Columbia, Canada.' *Estuaries and Coasts* 44: 1419–39. doi.org/10.1007/s12237-020-00858-2

Deppe, R.W., J. Thomson, B. Polagye and C. Krembs, 2018. 'Predicting Deep Water Intrusions to Puget Sound, WA (USA), and the Seasonal Modulation of Dissolved Oxygen.' *Estuaries and Coasts* 41(1): 114–27. doi.org/10.1007/s12237-017-0274-6

Dethier, M.N., W.W. Raymond, A.N. McBride, J.D. Toft, J.R. Cordell, A.S. Ogston, S.M. Heerhartz and H.D. Berry, 2016. 'Multiscale Impacts of Armoring on Salish Sea Shorelines: Evidence for Cumulative and Threshold Effects.' *Estuarine, Coastal and Shelf Science* 175: 106–17. doi.org/10.1016/j.ecss.2016.03.033

Doney, S., V. Fabry, R. Feely and J. Kleypas, 2009. 'Ocean Acidification: the Other CO_2 Problem.' *Annual Review of Marine Science* 1: 169–92. doi.org/10.1146/annurev.marine.010908.163834

Duguid, W.D.P., J.L. Boldt, L. Chalifour, C.M. Greene, M. Galbraith, D. Hay, D. Lowry, et al., 2019. 'Historical Fluctuations and Recent Observations of Northern Anchovy *Engraulis mordax* in the Salish Sea, Deep Sea Research Part II.' *Topical Studies in Oceanography* 159: 22–41. doi.org/10.1016/j.dsr2.2018.05.018

Dunham, A., S.K. Archer, S. Davies, L. Burke, J. Mossman, J.R. Pegg and E. Archer, 2018. 'Assessing Condition and Ecological Role of Deep-Water Biogenic Habitats: Glass Sponge Reefs in the Salish Sea.' *Marine Environmental Research* 141: 88–99. doi.org/10.1016/j.marenvres.2018.08.002

Evans, W., K. Pocock, A. Hare, C. Weekes, B. Hales, J. Jackson, H. Gurney-Smith, J.T. Mathis, S.R. Alin and R.A. Feely, 2019. 'Marine CO_2 Patterns in the Northern Salish Sea.' *Frontiers in Marine Science* 5 (536). doi.org/10.3389/fmars.2018.00536

Feely, R.A., S.R. Alin, J. Newton, C.L. Sabine, M. Warner, A. Devol, C. Krembs and C. Maloy, 2010. 'The Combined Effects of Ocean Acidification, Mixing and Respiration on pH and Carbonate Saturation in an Urbanized Estuary.' *Estuarine, Coastal and Shelf Science* 88(4): 442–9. doi.org/10.1016/j.ecss.2010.05.004

Feely, R., C. Sabine, J. Hernandez-Ayon, D.Y. Ianson and B.E. Hales, 2008. 'Evidence for Upwelling of Corrosive 'Acidified' Water onto the Continental Shelf.' *Science* 320: 1490–2. doi.org/10.1126/science.1155676

Francis, T. and D. Lowry, 2018. 'Assessment and Management of Pacific Herring in the Salish Sea: Conserving and Recovering a Culturally Significant and Ecologically Critical Component of the Food Web.' Orcas Island: The SeaDoc Society.

Fu, C., S. Yunne-Jai, R.I. Perry, J.A. King and H. Liu, 2012. 'Exploring Climate and Fishing Impacts in an Ecosystem Model of the Strait of Georgia, British Columbia.' In G.H. Kruse et al. (eds), *Global Progress in Ecosystem-Based Fisheries Management*. Fairbanks: Sea Grant/University of Alaska. doi.org/10.4027/gpebfm.2012.04

Gillespie, G., 2007. 'Distribution of Non-Indigenous Intertidal Species on the Pacific Coast of Canada.' *Nippon Suisan Gakkaishi* 73: 1133–7. doi.org/10.2331/suisan.73.1133

Hagen, M.E., A.G. Colodey, W.D. Knapp and S.C. Samis, 1997. 'Environmental Response to Decreased Dioxin and Furan Loadings from British Columbia Coastal Pulp Mills.' *Chemosphere* 34(5): 1221–9. doi.org/10.1016/S0045-6535(97)00420-7

Harrison, P.J., J.D. Fulton, F.J.R. Taylor and T.R. Parsons, 1983. 'Review of the Biological Oceanography of the Strait of Georgia: Pelagic Environment.' *Canadian Journal of Fisheries and Aquatic Sciences* 40: 1064–94. doi.org/10.1139/f83-129

Heath, D.D., P.D. Rawson and T.J. Hilbish, 1995. 'PCR-Based Nuclear Markers Identify Alien Blue Mussel (*Mytilus* spp.) Genotypes on the West Coast of Canada.' *Canadian Journal of Fisheries and Aquatic Sciences* 52 (12): 2621–7. doi.org/10.1139/f95-851

Hickie, B.E., P.S. Ross, R.W. Macdonald and J.K. Ford, 2007. 'Killer Whales (*Orcinus orca*) Face Protracted Health Risks Associated with Lifetime Exposure to PCBs.' *Environmental Science & Technology* 41 (18): 6613–9. doi.org/10.1021/es0702519

Hudson, D. R., 1970. 'Some Geographical Terms of the Saanich Indians of Southern Vancouver Island.' Victoria: University of Victoria, Department of Linguistics. Available from Institute of Ocean Sciences, Fisheries and Oceans Canada, Sidney, BC.

Ianson, D., S.E. Allen, B.L. Moore-Maley, S.C. Johannessen and R.W. Macdonald, 2016. 'Vulnerability of a Semienclosed Estuarine Sea to Ocean Acidification in Contrast with Hypoxia.' *Geophysical Research Letters* 43(11): 5793–801. doi.org/10.1002/2016GL068996

Johannessen, S.C., C.W. Greer, C.G. Hannah, T.L. King, K. Lee, R. Pawlowicz and C.A. Wright, 2020. 'Fate of Diluted Bitumen Spilled in the Coastal Waters of British Columbia, Canada.' *Marine Pollution Bulletin* 150: 110691. doi.org/10.1016/j.marpolbul.2019.110691

Johannessen, S.C. and R.W. Macdonald, 2009. 'Effects of Local and Global Change on an Inland Sea: The Strait of Georgia, British Columbia, Canada.' *Climate Research* 40: 1–21. doi.org/10.3354/cr00819

Johannessen, S.C., R.W. Macdonald, B. Burd, A. van Roodselaar and S. Bertold, 2015. 'Local Environmental Conditions Determine the Footprint of Municipal Effluent in Coastal Waters: A Case Study in the Strait of Georgia, British Columbia.' *Science of the Total Environment* 508: 228–39. doi.org/10.1016/j.scitotenv.2014.11.096

Johannessen, S.C., R.W. Macdonald and J.E. Strivens, 2020. 'Has Primary Production Declined in the Salish Sea?' *Canadian Journal of Fisheries and Aquatic Sciences* 78(3): 312–21. doi.org/10.1139/cjfas-2020-0115

Johannessen, S.C., R.W. Macdonald, C.A. Wright and D.J. Spear, 2017. 'Short-Term Variability in Particle Flux: Storms, Blooms and River Discharge in a Coastal Sea.' *Continental Shelf Research* 143: 29. doi.org/10.1016/j.csr.2017.05.016

Johannessen, S.C., D. Masson and R.W. Macdonald, 2014. 'Oxygen in the Deep Strait of Georgia, 1951–2009: The Roles of Mixing, Deep-Water Renewal and Remineralization of Organic Carbon.' *Limnology and Oceanography* 59(1): 211–22. doi.org/10.4319/lo.2014.59.1.0211

Khangaonkar, T., A. Nugraha, W. Xu and K. Balaguru, 2019. 'Salish Sea Response to Global Climate Change, Sea Level Rise, and Future Nutrient Loads.' *Journal of Geophysical Research: Oceans* 124(6): 3876–904. doi.org/10.1029/2018JC014670

Khangaonkar, T., A. Nugraha, W. Xu, W. Long, L. Bianucci, A. Ahmed, T. Mohamedali and G. Pelletier, 2018. 'Analysis of Hypoxia and Sensitivity to Nutrient Pollution in Salish Sea.' *Journal of Geophysical Research: Oceans* 123 (7): 4735–61. doi.org/10.1029/2017JC013650

Khangaonkar, W.L., and W. Xu, 2017. 'Assessment of Circulation and Inter-Basin Transport in the Salish Sea Including Johnstone Strait and Discovery Islands Pathways.' *Ocean Modelling* 109: 11–32. doi.org/10.1016/j.ocemod.2016.11.004

Levings, C.D. and R.M. Thom, 1994. 'Habitat Changes in Georgia Basin: Implications for Resource Management and Restoration.' In R.C.H. Wilson, R.J. Beamish, F. Aitkens and J. Bell (ed.), *Review of the Marine Environment and Biota of Strait of Georgia, Puget Sound and Juan de Fuca Strait. Proceedings of the BC/Washington Symposium on the Marine Environment, January 13 & 14,* 330–51. Sidney, British Columbia: Fisheries and Oceans Canada.

Lewis, S.L. and M.A. Maslin, 2015. 'Defining the Anthropocene.' *Nature* 519 (7542): 171–80. doi.org/10.1038/nature14258

Lowe, C., S. Lyons and J. Krogh, 2017. 'Pharmaceuticals and Personal Care Products in Municipal Wastewater and the Marine Receiving Environment Near Victoria Canada.' *Frontiers in Marine Science* 4. doi.org/10.3389/fmars.2017.00415

MacCready, P., R.M. McCabe, S.A. Siedlecki, M. Lorenz, S.N. Giddings, J. Bos, S. Albertson, N.S. Banas and S. Garnier, 2021. 'Estuarine Circulation, Mixing, and Residence Times in the Salish Sea.' *Journal of Geophysical Research: Oceans* 126 (2): e2020JC016738. doi.org/10.1029/2020JC016738

Mackas, D., M. Galbraith, D. Faust, D. Masson, K. Young, W. Shaw, S. Romaine, M. Trudel, J. Dower, et al., 2013. 'Zooplankton Time Series from the Strait of Georgia: Results From Year-Round Sampling at Deep Water Locations, 1990–2010.' *Progress in Oceanography* 115: 129–59. doi.org/10.1016/j.pocean.2013.05.019

Mackas, D.L. and P.J. Harrison, 1997. 'Nitrogenous Nutrient Sources and Sinks in the Juan De Fuca Strait/Strait of Georgia/Puget Sound Estuarine System: Assessing the Potential for Eutrophication.' *Estuarine, Coastal and Shelf Science* 44: 1–21. doi.org/10.1006/ecss.1996.0110

Masson, D., 2002. 'Deep Water Renewal in the Strait of Georgia.' *Estuarine, Coastal and Shelf Science* 54: 115–26. doi.org/10.1006/ecss.2001.0833

Masson, D. and P. Cummins, 2007. 'Temperature Trends and Interannual Variability in the Strait of Georgia, British Columbia.' *Continental Shelf Research* 27: 634–49. doi.org/10.1016/j.csr.2006.10.009

Masson, D. and A. Peña, 2009. 'Chlorophyll Distribution in a Temperate Estuary: The Strait of Georgia and Juan De Fuca Strait.' *Estuarine Coastal and Shelf Science* 82: 19–28. doi.org/10.1016/j.ecss.2008.12.022

Meador, J.P. and D.H. MacLatchy, 2014. 'Do Chemically Contaminated River Estuaries in Puget Sound (Washington, USA) Affect the Survival Rate of Hatchery-Reared Chinook Salmon?' *Canadian Journal of Fisheries and Aquatic Sciences* 71 (1): 162–80. doi.org/10.1139/cjfas-2013-0130

Morrison, J., M.C. Quick and M.G. Foreman, 2002. 'Climate Change in the Fraser River Watershed: Flow and Temperature Projections.' *Journal of Hydrology* 263: 230–44. doi.org/10.1016/S0022-1694(02)00065-3

Mudie, P.J., A. Rochon and E. Levac, 2002. 'Palynological Records of Red Tide-Producing Species in Canada: Past Trends and Implications for the Future.' *Palaeo* 180: 159–86. doi.org/10.1016/S0031-0182(01)00427-8

Myers, R. and B. Worm, 2003. 'Rapid Worldwide Depletion of Predatory Fish Communities.' *Nature* 423: 280–3. doi.org/10.1038/nature01610

National Marine Fisheries Service, 2008. 'Recovery Plan for Southern Resident Killer Whales (*Orcinus orca*).' Seattle, Washington: National Marine Fisheries Service, Northwest Region, United States.

Paul, P.K., P.C. Paul, E. Carmack and R.W. Macdonald, 1995. 'The Care-Takers: The Re-emergence of the Saanich Indian Map.' Sidney: WSÁNEĆ Nation and Fisheries and Oceans Canada.

Pawlowicz, R., R. Di Costanzo, M. Halverson, E. Devred and S.C. Johannessen, 2017. 'Advection, Surface Area and Sediment Load of the Fraser River Plume Under Variable Wind and River Forcing.' *Atmosphere-Ocean* 55 (4–5): 293-313. doi.org/10.1080/07055900.2017.1389689

Pawlowicz, R., O.R. Riche and M. Halverson, 2007. 'The Circulation and Residence Time of the Strait of Georgia Using a Simple Mixing-Box Approach.' *Atmosphere-Ocean* 45(4): 173–93. doi.org/10.3137/ao.450401

Perry, R.I., K. Young, M. Galbraith, P. Chandler, A. Velez-Espino and S. Baillie, 2021. 'Zooplankton Variability in the Strait of Georgia, Canada, and Relationships with the Marine Survivals of Chinook and Coho Salmon.' *PLoS ONE* 16 (1): e0245941. doi.org/10.1371/journal.pone.0245941

Pörtner, H.-O., 2008. 'Ecosystem Effects of Ocean Acidification in Times of Ocean Warming: A Physiologist's View.' *Marine Ecology Progress Series* 373: 203–17. doi.org/10.3354/meps07768

Pospelova, V., S. Esenkulova, S.C Johannessen, M.C. O'Brien and R.W. Macdonald, 2010. 'Organic-Walled Dinoflagellate Cyst Production, Composition and Flux From 1996 to 1998 in the Central Strait of Georgia (BC, Canada): A Sediment Trap Study.' *Marine Micropaleontology* 75(1): 17–37. doi.org/10.1016/j.marmicro.2010.02.003

Ross, P.S., G.M. Ellis, M.G. Ikonomou, L.G. Barrett-Lennard and R.F. Addison, 2000. 'High PCB Concentrations in Free-Ranging Pacific Killer Whales, *Orcinus orca*: Effects of Age, Sex and Dietary Preference.' *Marine Pollution Bulletin* 40(6): 504–15. doi.org/10.1016/S0025-326X(99)00233-7

Ross, P.S., M. Noël, D. Lambourn, N. Dangerfield, J. Calambokidis and S. Jeffries, 2013. 'Declining Concentrations of Persistent PCBs, PBDEs, PCDEs and PCNs in Harbor Seals (*Phoca vitulina*) From the Salish Sea.' *Progress in Oceanography* 115: 160–70. doi.org/10.1016/j.pocean.2013.05.027

Ruggerone, G. and J. Irvine, 2018, 'Numbers and Biomass of Natural- and Hatchery-Origin Pink Salmon, Chum Salmon, and Sockeye Salmon in the North Pacific Ocean, 1925–2015.' *Marine and Coastal Fisheries* 10: 152–68. doi.org/10.1002/mcf2.10023

Shelton, A.O., G.H. Sullaway, E.J. Ward, B.E. Feist, K.A. Somers, V.J. Tuttle, J.T. Watson and W.H. Satterthwaite, 2021. 'Redistribution of Salmon Populations in the Northeast Pacific Ocean in Response to Climate.' *Fish and Fisheries* 22(3): 503–17. doi.org/10.1111/faf.12530

Sobocinski, K.L., C.M. Greene, J.H. Anderson, N.W. Kendall, M.W. Schmidt, M.S. Zimmerman, I.M. Kemp, S. Kim and C.P. Ruff, 2021. 'A Hypothesis-Driven Statistical Approach for Identifying Ecosystem Indicators of Coho and Chinook Salmon Marine Survival.' *Ecological Indicators* 124: 107403. doi.org/10.1016/j.ecolind.2021.107403

St John, M.A., S.G. Marinone, J. Stronach, P.J. Harrison, J. Fyfe and R.J. Beamish, 1993. 'A Horizontally Resolving Physical-Biological Model of Nitrate Concentration and Primary Productivity in the Strait of Georgia.' *Canadian Journal of Fisheries and Aquatic Sciences* 50: 1456–66. doi.org/10.1139/f93-166

Sutherland, D.A., P. MacCready, N.S. Banas and L.F. Smedstad, 2011. 'A Model Study of the Salish Sea Estuarine Circulation.' *Journal of Physical Oceanography* 41(6): 1125–43. doi.org/10.1175/2011jpo4540.1

Sutton, J.N., S.C. Johannessen and R.W. Macdonald, 2013. 'A Nitrogen Budget for the Strait of Georgia, British Columbia.' *Biogeosciences* 10: 7179–94. doi.org/10.5194/bg-10-7179-2013

Therriault, T., D. Hay and J. Schweigert, 2009. 'Biological Overview and Trends in Pelagic Forage Fish Abundance in the Salish Sea (Strait of Georgia, British Columbia).' *Marine Ornithology* 37: 3–8.

Thomson, R.E., 1981. *Oceanography of the British Columbia Coast, Canadian Special Publication of Fisheries and Aquatic Science.* DFO.

Thomson, R.E., E.A. Kulikov, D.J. Spear, S.C. Johannessen and W.P. Wills, 2020. 'A Role for Gravity Currents in Cross-Sill Estuarine Exchange and Subsurface Inflow to the Southern Strait of Georgia.' *Journal of Geophysical Research: Oceans* 125(4). doi.org/10.1029/2019JC015374

Tunnicliffe, V., J.M. O'Connell and M.R. McQuoid, 2001. 'A Holocene Record of Marine Fish Remains from the Northeastern Pacific.' *Marine Geology* 174(1): 197–210. doi.org/10.1016/S0025-3227(00)00150-X

Watson, A.J., U. Schuster, J.D. Shutler, T. Holding, I.G.C. Ashton, P. Landschützer, D.K. Woolf and L. Goddijn-Murphy, 2020. 'Revised Estimates of Ocean-Atmosphere CO_2 Flux Are Consistent with Ocean Carbon Inventory.' *Nature Communications* 11(1): 4422. doi.org/10.1038/s41467-020-18203-3

Winter, D.F., K. Banse and G.C. Anderson, 1975. 'The Dynamics of Phytoplankton Blooms in Puget Sound: A Fjord in the Northwestern United States.' *Marine Biology* 29(2): 139–76. doi.org/10.1007/BF00388986

Yang, Z., G.G. Medina, W.-C. Wu, T. Wang, L. Leung, L. Castrucci and G. Mauger, 2019. 'Modeling Analysis of the Swell and Wind-Sea Climate in the Salish Sea.' *Estuarine, Coastal and Shelf Science* 224: 289–300. doi.org/10.1016/j.ecss.2019.04.043

4

Geology and Geomorphology

Steven Earle and John Clague

The subject of this chapter is the geology and geomorphology of the islands within the northern Salish Sea. These islands are the product of geological events that extend back in time more than 400 million years, although their form (geomorphology) has been strongly influenced by repeated continental and alpine glaciation over the past 3 million years. We describe the landscape legacy left by these geologic events and briefly discuss how this rich landscape has nurtured humans from the time they first arrived on the South Coast more than 14,000 years ago to the present.

Generalised Geology

The Salish Sea has remarkable geological diversity. Many similar-sized regions of the world have only one type of rock, of one specific age, but the Salish Sea includes geological materials of almost every type known, formed by a myriad of different processes both here and elsewhere, and spanning over 400 million years of earth history. Thanks to the power of coastal erosion, the rocks are generally well exposed and amenable to investigation, and thanks to numerous universities, colleges, government-run geological institutions and enthusiastic geologists on both sides of the border, the geology of the region is quite well understood.

The generalised geology of the Salish Sea is shown in Figure 4.1. The area has been divided into eight major geological entities based on the origins and ages of the materials exposed on the land surface.

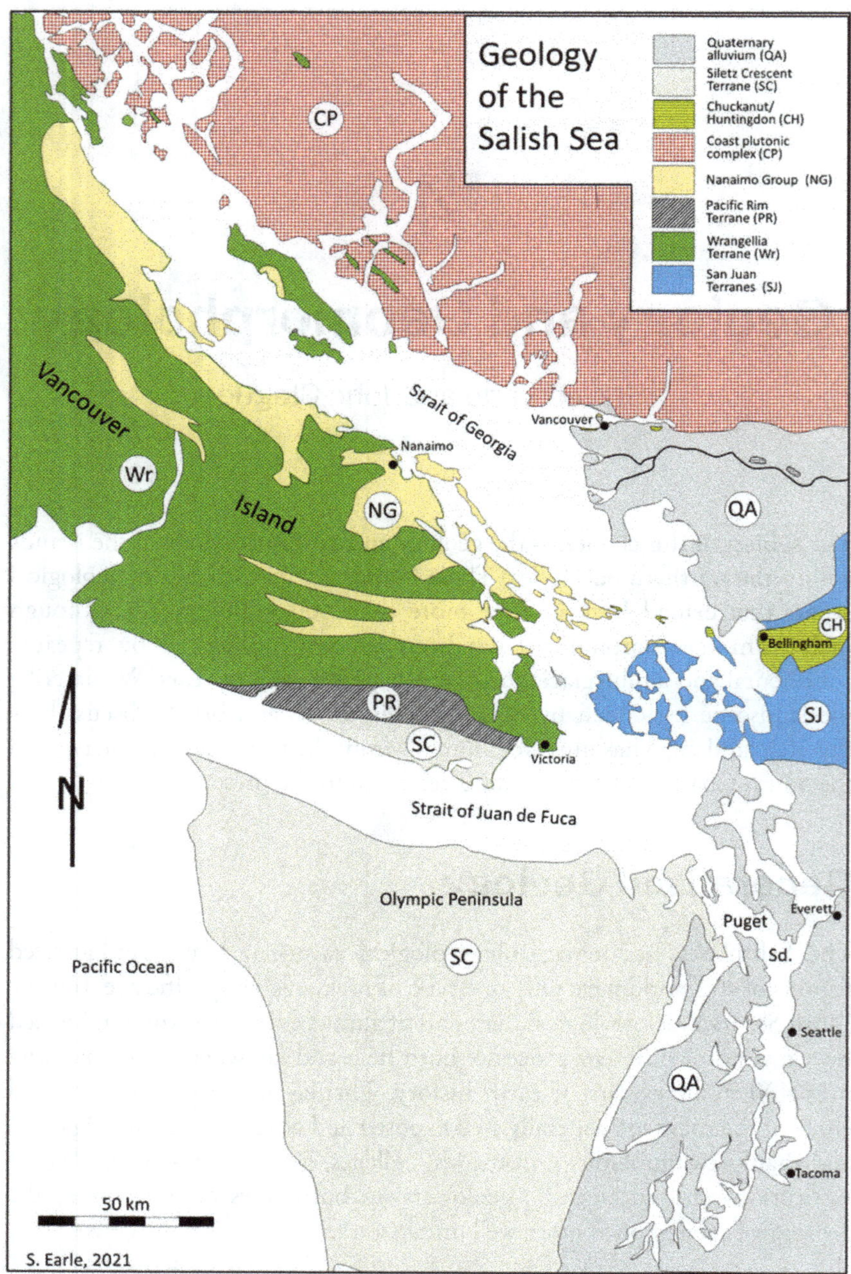

Figure 4.1 Major geological features of the Salish Sea region
Source: Steven Earle.

Most of these are described in more detail below, but to quickly summarise, they are as follows, from oldest to youngest:

- the San Juan Terranes, comprising most of the San Juan Islands, parts of northwestern Washington around Bellingham and extending into southern British Columbia (BC) around Chilliwack,
- the Wrangellia Terrane, comprising most of Vancouver Island and parts of the islands within the Strait of Georgia, and extending onto the mainland coast north of Vancouver,
- the Coast Plutonic[1] Complex, which makes up most of the coastal mainland north of Vancouver but also extends into the inland part of northwestern Washington,
- the Pacific Rim Terrane, which extends across southern Vancouver Island from Port Renfrew to Victoria,
- the Nanaimo Group sedimentary rocks, which underly the eastern edge of Vancouver Island and most of the Gulf Islands,
- the Chuckanut Formation (and its equivalent Huntingdon Formation in Canada), which underlies much of the region around Bellingham, and extends north underneath the Fraser Delta, reappearing in Vancouver and points to the east within the Fraser Valley,
- the Crescent Terrane, which underlies most of the Olympic Peninsula and the southern tip of Vancouver Island, and
- Quaternary alluvium of glacial, fluvial and other origins that cover the bedrock in much of the area south of Vancouver, most of the islands of Puget Sound and much of the mainland both north and south of Seattle.

Salish Sea Geological History: Exotic Origins and Local Deposition

There are two main types of geological entities in the area of the Salish Sea, those that formed elsewhere and arrived here through the movement of tectonic plates and those that formed here. Although the rocks of the region date back as far as 425 million years, there was very little land in this area prior to about 110 Ma (million years ago)—only relatively deep ocean water on top of oceanic crust, perhaps dotted with small volcanic islands. At that time the 'hard' western shore of North America was as much as 250 km to the east.

1 Also referred to as Coast Crystalline or Coast Belt.

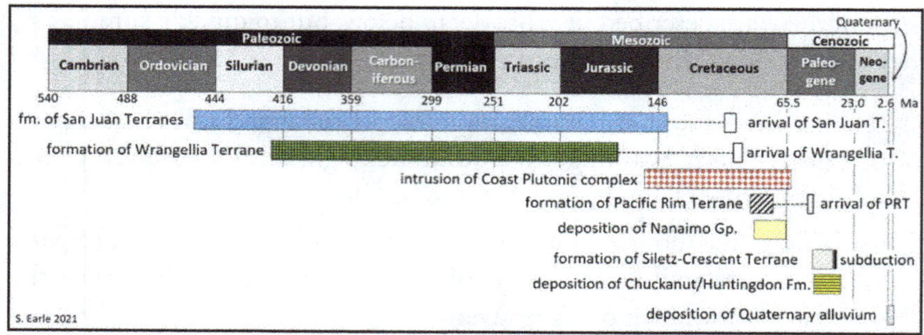

Figure 4.2 Graphical summary of geological evolution in the Salish Sea region

Note: Dates of period boundaries are shown, in millions of years.

Source: Steven Earle.

The oldest rocks of the Salish Sea, of middle-Silurian age (ca. 425 Ma), are part of the San Juan Terranes,[2] which comprise the San Juan Islands and parts of the adjacent coast of Washington (Figure 4.2). The various San Juan Terranes include a wide variety of rock types spanning almost 300 million years, the youngest being early Cretaceous (ca. 140 Ma). Some of the San Juan rocks are assumed to have formed far away from their current location while others appear to have formed on the margin of North America, but well south of here (Brown et al. 2007). The San Juan Terranes moved north to their current location during the Cretaceous, along the north–south trending Straight Creek–Fraser Fault system, situated about 100 km east of the Salish Sea, arriving here between 115 and 85 Ma (Brown et al. 2007).

The Wrangellia Terrane, which makes up most of Vancouver Island but extends onto the mainland north of Vancouver, includes rocks that are as old as late Silurian (420 Ma) and as young as early Jurassic (180 Ma). There is evidence to show that most of the rocks of Wrangellia formed well offshore from North America and south of the equator, and that they arrived here at between 90 and 100 Ma, moving towards the coast as part of a plate that was subducting beneath the North America Plate (Monger et al. 1982). In fact, Wrangellia extends well east of Vancouver Island; it includes parts of Salt Spring and Quadra Islands, almost all of Lasqueti and Texada Islands and numerous smaller islands in the Strait of Georgia and much of the floor of the strait as well. Small fragments of Wrangellia Terrane rocks can be found at least 50 km into the Coast Range northwest of Vancouver (Figure 4.1).

2 A terrane is a crust fragment formed on a tectonic plate and accreted to another plate.

From the late Jurassic to late Cretaceous, overlapping the arrival of the Wrangellia Terrane, oceanic crust was subducting beneath the western edge of North America, producing volcanism and associated igneous intrusive rocks at depth in the crust (Gehrels et al. 2009). The resulting volcanic rocks are now almost entirely eroded, and the Coast Plutonic Complex, which stretches from northwestern Washington to southeastern Alaska, is the main remnant of this igneous activity. The small remaining slices of Wrangellia within the Coast Range (Figure 4.1) illustrate how the Plutonic Complex has intruded along—and effectively obliterated—the boundary between Wrangellia and the rest of North America.

The metamorphic rocks of the Leech River Complex of the Pacific Rim Terrane are interpreted to have been originally deposited as clay- to sand-sized sediments in a marine environment close to edge of North America during the early Cretaceous (Fairchild and Cowan 1982; Groome 2000; Groome et al. 2003). The terrane then moved north and was partially subducted beneath the western and southern margins of Vancouver Island at around 55 Ma. This succeeded in pushing the crust of Vancouver Island closer to the mainland. Metamorphism of these rocks to slate and schist is interpreted to have taken place between 100 and 45 Ma (Groome et al. 2003).

The accretion of the San Juan and Wrangellia Terranes resulted in significant uplift of the crust in the Salish Sea region, and the rate of erosion increased dramatically within the newly formed mountain ranges. The resulting sediments accumulated mostly in the oceanic basin between 'Vancouver Island' and the mainland from around 90 Ma to approximately 65 Ma, forming the Nanaimo Group sedimentary rocks (Mustard 1994). At that time most of the Nanaimo Group sediments and sedimentary rocks were still submerged, but they were later uplifted during the accretion of the Pacific Rim Terrane (ca. 55 Ma) and the Siletz-Crescent Terrane (ca. 42 Ma) as Vancouver Island was pushed closer to the mainland, to become part of eastern Vancouver Island and islands in the Salish Sea.

Formation of seafloor crust continued offshore from this region through much of the Mesozoic, and some of that crust has since become part of the continent as the Siletz-Crescent Terrane (McRory and Wilson 2013), which comprises ocean-crust basalt and gabbro ranging in age from approximately 56 to 40 Ma. At around 42 Ma the northern edge of the Siletz-Crescent Terrane was partly subducted beneath the continental crust of Vancouver Island, pushing Wrangellia eastward and closing the gap between the island

and the mainland even further. Subsequent subduction of the Juan de Fuca Plate beneath this accreted oceanic crust has sufficiently elevated the Siletz-Crescent oceanic crust to form land, with mountains up to 2,400 m above sea level.

Accumulation of sediments in the region resumed at about 55 Ma and continued to at least 35 Ma with the deposition of the Chuckanut Formation in the US, and the equivalent Huntingdon Formation in Canada (Mustard and Rouse 1994; Gilley 2003; Mustoe et al. 2007). Unlike the Nanaimo Group rocks, most of which were deposited in ocean water, these rocks accumulated on land in a fluvial environment.

Geology and the Terrain

The Salish Sea exists because of plate tectonics. The Strait of Georgia and Puget Sound are part of a subduction-related forearc basin in that they occupy a geographical depression that lies between the Cascadia subduction zone (offshore from Vancouver Island and Washington State) and the Cascadia volcanic arc (the line of subduction-related volcanoes, including Mounts Garibaldi, Baker, Rainier and others (Figure 4.3)). This depression continues south into the Willamette Valley of Oregon. A forearc basin develops because the subduction of dense oceanic crust beneath continental crust creates additional gravitational pull on the crust as a whole. The Salish Sea is also partly a foreland basin, related to the depression of the crust by the Coast and Cascadia Ranges.

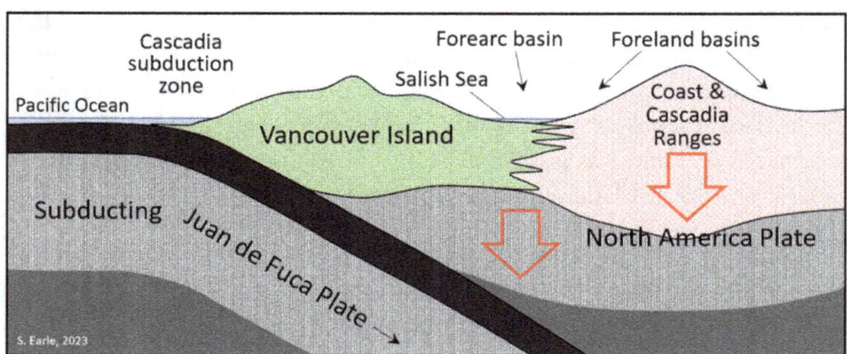

Figure 4.3 Plate tectonic scenario

Note: The figure shows the gravitational forces (red arrows) that have contributed to the formation of the Salish Sea basin.

Source: Steven Earle.

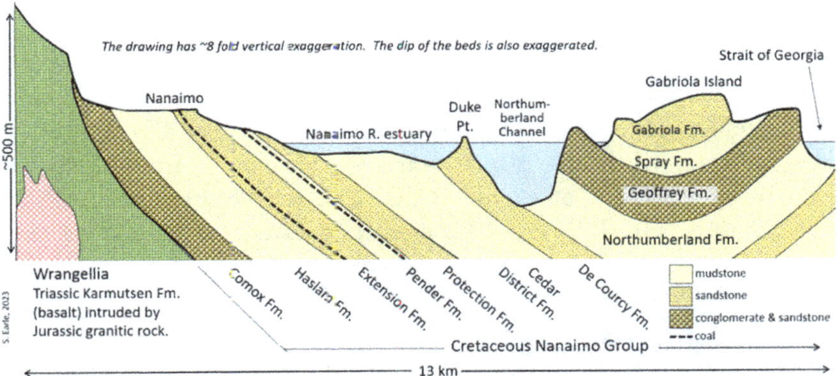

Figure 4.4 East–west geological cross-section through Wrangellia and Nanaimo rock groups

Source: Steven Earle.

As noted already, the Salish Sea is home to a wide range of rock types, and these rocks have varying resistance to the mechanical erosion inflicted by water (rain, streams and waves) and by glacial ice. That varying resistance is responsible for many of the key topographical features that contribute to the Salish Sea region, including its liveable places, agricultural land, mountains, fascinating islands, channels, bays and harbours. For example, the eastern 10 to 25 km of central Vancouver Island (and some parts farther inland as well) (Figure 4.1), are underlain by the relatively weak rocks of the Nanaimo Group. Streams and glaciers have eroded these rocks to create a relatively low-relief region that is suitable for agriculture, settlements and cities. This principle is illustrated on Figure 4.4, which shows how the city of Nanaimo has been built on the relatively flat land underlain by the soft Nanaimo Group rocks. Immediately to the west of Nanaimo (left on the drawing) the harder crystalline rocks of Wrangellia form hills and mountains that are generally too steep for habitation.

Although the Nanaimo Group rocks are consistently softer than the older crystalline rocks of the region (Wrangellia and San Juan Terranes and Coast plutonic rocks), these sedimentary layers have their own lithological variations that have contributed to the diverse topography of the region. As can be clearly seen in Figure 4.4, water bodies are underlain by soft mudstone, while points and islands, and ridges and cliffs on those islands, are underlain by harder sandstone and conglomerate.

Three Types of Islands

Northern Salish Sea islands are of three general types, each corresponding to the geologic materials that form them (Table 4.1, Figure 4.5). Two of the three types are 'hard rock' islands, which are resistant to erosion, whereas the third type consists of sediments that have not been transformed into rock and thus are easily and quickly eroded by waves, streams and glaciers.

Table 4.1 Composition of islands in the northern Salish Sea

North Strait of Georgia			South Strait of Georgia		
	Island	Primary component		Island	Primary component
1	Quadra	Wrangellia, Coast Belt, Quadra Sand	14	Gabriola	Nanaimo Group
2	Read	Coast Belt	15	Mudge	Nanaimo Group
3	Cortez	Coast Belt	16	De Courcy	Nanaimo Group
4	Marina	Quadra Sand	17	Valdes	Nanaimo Group
5	Hernando	Quadra Sand	18	Galiano	Nanaimo Group
6	Harwood	Quadra Sand	19	Saltspring	Wrangellia, Nanaimo Group
7	Savary Island	Quadra Sand	20	Mayne	Nanaimo Group
8	Texada	Wrangellia, Nanaimo Group	21	Saturna	Nanaimo Group
9	North Thormanby	Quadra Sand	22	Prevost	Nanaimo Group
10	Thormanby	Coast Belt	23	North Pender	Nanaimo Group
11	Denman	Nanaimo Group, Quadra Sand	24	South Pender	Nanaimo Group
12	Hornby	Nanaimo Group	25	Sidney	Quadra Sand, Wrangellia
13	Lasqueti	Wrangellia, Nanaimo Group	26	James	Quadra Sand

Note: Numbers correspond to islands shown on Figure 4.5.

Source: John Clague's tabulation.

The first of the three general types of islands are those formed of 'hard' Coast Crystalline, Wrangellia and San Juan metamorphic and igneous rocks up to perhaps 200 million years old. Islands fringing the northernmost Strait of Georgia, including Quadra, Cortez and Read islands, are of this type, as is Lasqueti Island in the southern Strait of Georgia.

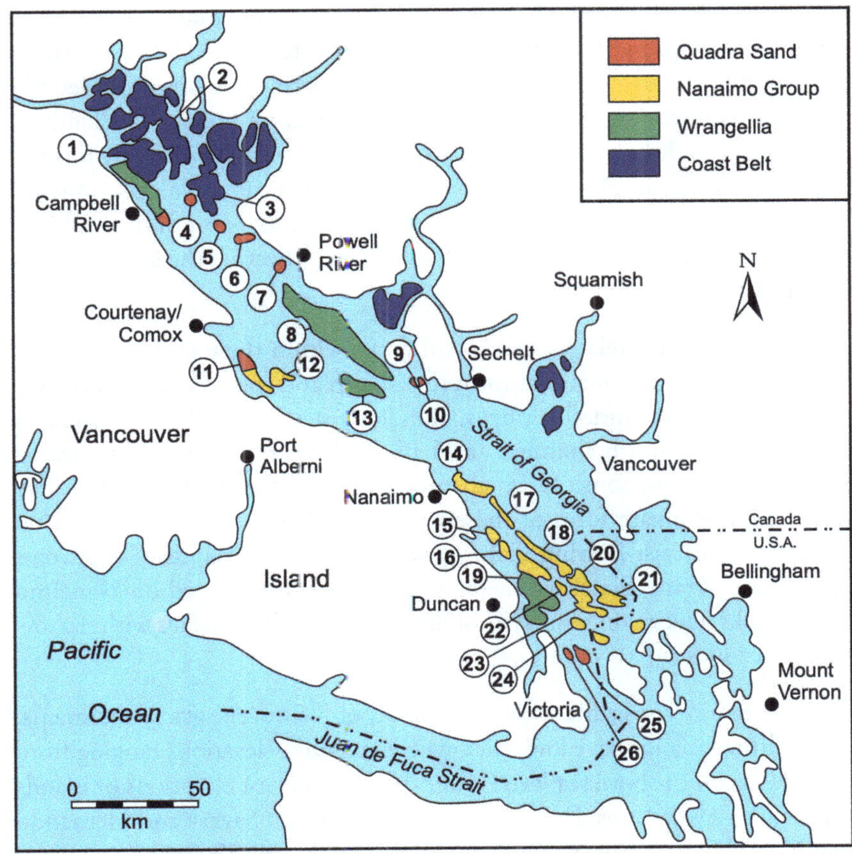

Figure 4.5 Rock types of islands in Strait of Georgia

Source: John Clague.

Texada Island is another member of this group and has an elongate northwest–southeast form controlled by its geologic structure and Pleistocene glacial erosion. Much of rugged spine of Vancouver Island is underlain by Wrangellia volcanic rocks. The San Juan Islands south of the Strait of Georgia consist largely of Mesozoic and Palaeozoic San Juan Terrane rocks.

The shorelines of islands formed of Coast Crystalline, Wrangellia and San Juan rocks are rocky and rough, with small beaches dominated by gravel and boulders (Clague and Bornhold 1980). The intertidal zone along these shorelines is irregular and narrow, dropping off into deep water over a short distance.

The second general type are islands formed of sedimentary rocks of the Nanaimo Group, mainly sandstone, conglomerate and shale. These rocks underlie the lowlands of Vancouver Island bordering the Strait of Georgia from near Duncan on the south to Campbell River on the north. Islands in the southern and central Strait of Georgia are also entirely or largely formed of these rocks. Examples are Denman, Hornby, Gabriola, Valdes, Galiano, Mayne, Saturna, and North and South Pender islands. Most of these islands are elongate in a northwest–southeast direction, reflecting the strike of beds of Nanaimo Group rocks.

Geomorphic details relate to folds and faults within these rocks and to the types of sedimentary rock that form the islands and their shores. Sandstone and conglomerate underlie ridges, hills and steep coastal cliffs, whereas shale is found beneath channels and bays. Ridges are commonly separated by elongate valleys, some of which support small lakes and wetlands. The pronounced elongate form of these islands has been enhanced by erosion caused by southeast-flowing Pleistocene glaciers. Like islands of the Coast Crystalline, Wrangellia and San Juan groups, islands formed on Nanaimo Group rocks have mainly rocky shorelines and gravelly beaches with narrow intertidal zones.

Third are islands consisting entirely or mainly of Pleistocene glacial sediments. Most of these islands have low relief, with maximum elevations ranging from about 50 to 80 m above sea level, lower than the highest elevations of islands of the other two groups. Examples include Harwood, Savary and Hernando islands in the northern Strait of Georgia, and North Thormanby, Sidney and James islands in the southern Strait of Georgia. Thick Pleistocene glacial sediments are also common along the shores of eastern Vancouver Island as far north as Campbell River. The islands of this third group are developed in Quadra Sand capped by Fraser Glaciation till. Quadra Sand comprises loose sand and gravel, which is easily eroded by currents and waves.

Some of the eroded sediment is transported by currents along the shoreline and deposited in prominent spits, such as Sidney Spit. The sandy beaches on islands in the Strait of Georgia and along the coast of eastern Vancouver Island owe their existence to erosion of Quadra Sand by waves. These beaches are much wider than beaches of the other two island groups. The backshore areas of some of these beaches are vulnerable to erosion (Figure 4.6), and in many instances have been protected with engineered structures. Similarly, several islands in northern Puget Sound (e.g. Camano and southern Whidbey islands) are underlain, in part, by Esperance Sand and have sandy beaches and eroding shorelines that require protection.

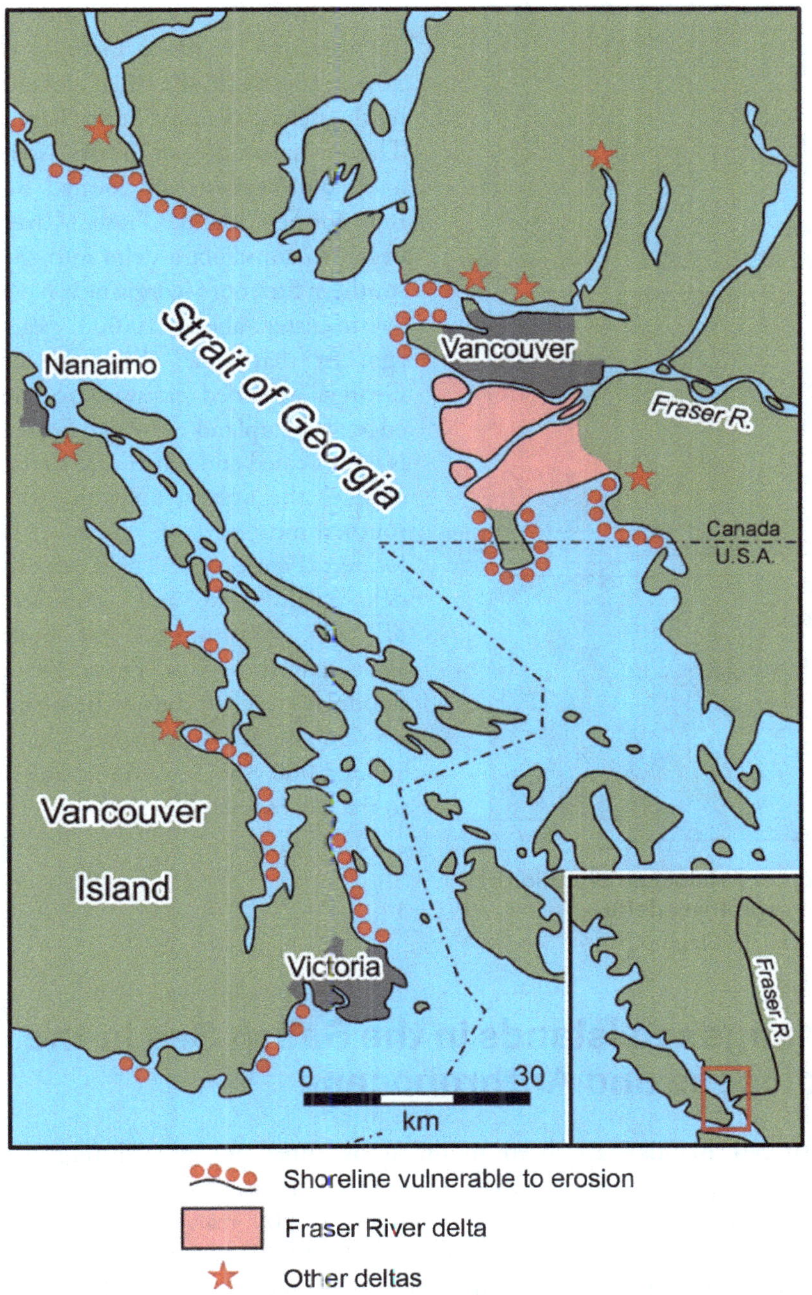

Figure 4.6 Generalised distribution of vulnerable shorelines in northern Strait of Georgia

Source: John Clague.

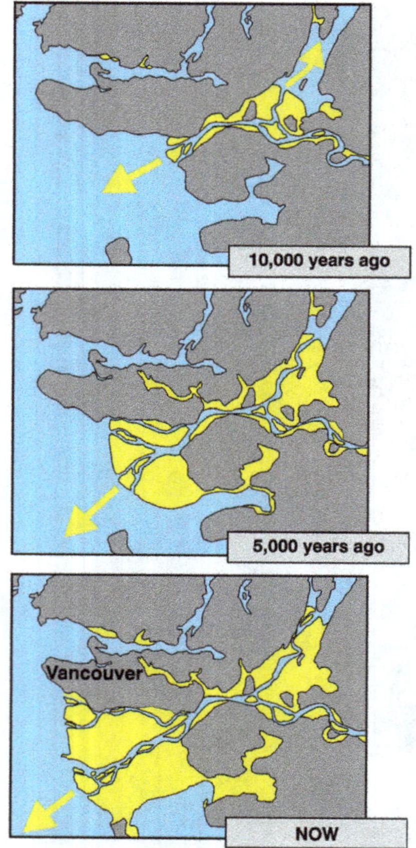

Figure 4.7 Holocene evolution of the Fraser River delta
Source: John Clague.

One former Pleistocene island in the southern Strait of Georgia is now connected to the BC mainland by the deltaic plain of Fraser River. The communities of Tsawwassen and Point Roberts are located on this former island. Fraser River began to construct a delta into the southern Strait of Georgia near New Westminster about 10,000 years ago. At that time, the Strait of Georgia extended eastward to the edge of the upland on which Surrey is now located and northward to the edge of the upland on which the southernmost part of Vancouver is located (Figure 4.7). The extensive delta plain west and southwest of New Westminster was built since that time. The Fraser Delta floodplain reached the northern and western shores of the former island about 3,000 years ago, connecting it to the BC mainland.

Changes to Islands in the Salish Sea in the Holocene and Anthropocene

Salish Sea islands have been modified by three processes during the Holocene Epoch: sea level change, waves and currents, and human activity. Sea level at the beginning of the Holocene, 11,700 years ago, was lower than it is today. Glacio-isostatic uplift of the Pacific coast was complete by that time, and sea level began to 'track' the global, or eustatic, curve. Large bodies of glacier ice remained over central and northern Canada and parts of Scandinavia at the start of the Holocene, thus eustatic sea level was lower than it is today. There is some uncertainty about the level of the sea in the Strait of Georgia at this time, although geologic data in the western Fraser Lowland suggest a value of about –15 to –20 m with respect to the present datum (Mathews et al. 1970). In any case, shorelines of all islands in the Strait of Georgia at the beginning of the Holocene were seaward of their present locations; amounts in each case depended on offshore bathymetry. Shorelines of Savary and Hernando islands, for example, were up to several hundred metres seaward of their present locations (Figure 4.8), partly because sea level was lower than today, partly because shallow marine platforms border these islands, and partly because waves subsequently eroded the Pleistocene sediments that form the islands.

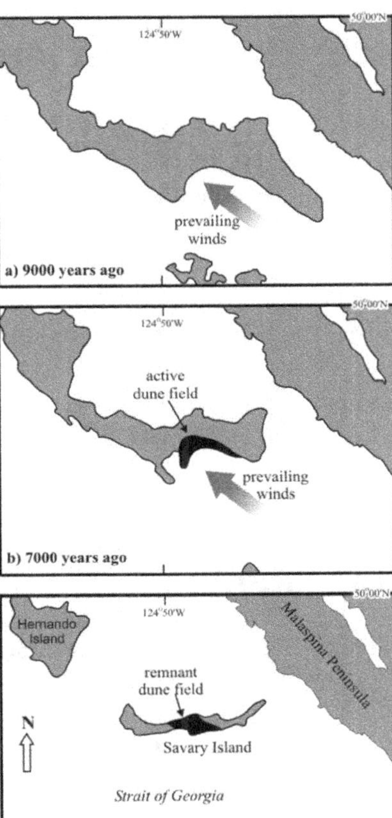

Figure 4.8 Reduction in the area of Hernando and Savary islands

Note: This reduction is due to sea level rise and erosion by waves driven by prevailing southeasterly winds. Some of the eroded sediment became airborne and accumulated as a field of sand dunes on Savary Island.

Source: John Clague.

Sea level only briefly remained at its Holocene low position. It slowly rose to near the present datum over the first six millennia of the postglacial period in response to the final melting of North Hemisphere ice sheets and warming of oceans (Mathews et al. 1970). The rise in sea level eroded Pleistocene islands such as Savary and Hernando, and inundated formerly exposed fringes of all Salish Sea islands.

Sea level has varied no more than 1–2 m over the past 5,000–6,000 years and the 'hard' islands in the Salish Sea have changed little in form over that time. In contrast, shorelines of 'soft' Pleistocene islands, as well as those along the east coast of Vancouver Island and the BC mainland that are backed by Pleistocene sediments have retreated due to erosion by waves. These shorelines are especially susceptible to erosion during fall and winter storms when strong winds create large waves and when low atmospheric pressures associated with Pacific storms slightly elevate the sea surface (Clague and Bornhold 1980). The waves carry sediment offshore into deeper water, and currents locally transport sediment along the shoreline to locations where it accumulates in spits and offshore bars. Notable examples of large spits formed in this manner are Goose Spit near Comox and Sidney Spit at the north end of Sidney Island.

Geological Resources of the Salish Sea Islands

Geology also has important economic and social implications for the Salish Sea region. Coal in the Nanaimo Group rocks was critical to the development of central Vancouver Island from 1850 to 1950. There are also thin coal beds in the Chuckanut Formation, but those have not been mined extensively at any time. Metal has been important on Quadra Island (copper) and Texada Island (gold and iron), and base-metal and gold mining continue to be important on Vancouver Island. Limestone quarries have operated on many of the Salish Sea islands for well over a century, on Vancouver Island (e.g. Quatsino Formation), and also on San Juan Island (Deadman Bay Formation) and Texada Island (Quatsino Formation). The sandstones of the Nanaimo Group and Chuckanut Formations have been widely used in the past for building stone and also for pulp-industry mill stones.

Extensive glacial and fluvial deposits of gravel and sand are distributed throughout the region and are very important as sources of aggregate for construction of buildings and roads.

In many of the rural parts of the Salish Sea region residents rely on private sources for domestic water, and in most cases this is groundwater supplied from private wells. In the older crystalline rocks of the San Juan and Wrangellia Terranes, porosity and permeability are controlled mostly by fracturing (Eddy 1975; Kenny et al. 2006). In general, fractured-rock aquifers are less predictable as water sources and less secure from surface contamination than porous-rock aquifers. Intergranular porosity is better developed in the Nanaimo Group and Chuckanut Formation conglomerates and sandstones (Allen and Suchy 2001; Sullivan 2005), although the marine-deposited sandstones of the Nanaimo Group have relatively poor permeability compared with typical sandstones (Earle and Krogh 2006).

Earthquakes and Tsunami

The Salish Sea area is tectonically active. The Juan de Fuca Plate is subducting beneath the North America Plate at a rate of close to 4 cm/year and the two plates are locked against each other in the area offshore from Vancouver Island and the Washington coast. In other words, although the plates themselves are still moving, they are not sliding past each other along this part of the boundary and the result is that the rocks are deforming elastically. The energy represented by this deformation is released frequently as small to medium-sized quakes and some larger ones (e.g. the 1946 Vancouver Island earthquake, and the 2001 Nisqually earthquake), and infrequently as very large megathrust earthquakes like the one of approximate magnitude 9 in 1700.

It is widely accepted that we can expect a large to very large (magnitude 7–9) earthquake in the Salish Sea region in the future, although we do not understand the seismic situation here well enough to be able to predict how soon that might be. The most effective ways to mitigate the effects of a future quake are to ensure that all buildings, but especially public buildings like schools and hospitals, and all other types of infrastructure (bridges, dams etc.) are designed to withstand the maximum expected shaking, and also to ensure that our population is well informed about the risks and about what they can do to take care of themselves both during and after an earthquake.

Both Washington State and BC have funded programs for the seismic upgrade of schools (e.g. Walsh and Schelling 2011) and hospitals, as well as roads, bridges, dams and other infrastructure (e.g. Kennedy and Huffman 2005). And both jurisdictions also have a variety of earthquake emergency measures in place at the federal, state/province, community and neighbourhood levels.

Reliable earthquake prediction does not currently exist (NEPEC 2016), and unreliable prediction is clearly worse than none at all. On the other hand, early detection of earthquakes that have already happened is possible and does provide some protection for the public. A megathrust quake on the subduction zone offshore is potentially the most damaging earthquake scenario for this region, and early detection systems that would provide between one and two minutes of warning to the Salish Sea population that such an event has already happened are currently being constructed on the seafloor and on land on both sides of the border.[3] Such warnings can provide time for utilities to close gas pipeline valves, disable specific electrical circuits, stop or slow trains, clear bridges or tunnels of traffic, and allow people in institutions and at home to take immediate steps to protect themselves.

The 1700 earthquake is known to have produced a significant tsunami along the western margin of Vancouver Island and parts of the Washington Coast, as has been related in numerous First Nation oral histories (Arima et al. 1991; Ludwin et al. 2005). The damaging effects of the tsunami from this and future large offshore quakes are assumed to be limited within the Salish Sea. A tsunami model created for a Juan de Fuca megathrust earthquake by Cherniawsky and Fine (2014) shows that while wave heights are likely to exceed 3 m on many exposed areas of the Vancouver Island and Washington coasts, most shorelines within the Salish Sea will likely be exposed to wave heights of less than 1 m. Some exceptions include the ends of inlets, such as those around Victoria, on Orcas Island and within Puget Sound.

Three quiet, but still 'active' volcanoes loom over the Salish Sea, from north to south: Mt Garibaldi, Mt Baker and Mt Rainier (Figure 4.2). All of these volcanoes have the potential to erupt with only a few months warning, and in all three cases there would be implications for the Salish Sea.[4]

3 See for example: shakealert.org/, www.oceannetworks.ca/services/earthquake-early-warning/.

4 While there are volcanic rocks on some of the Salish Sea islands, the youngest of these are late Jurassic in age (at least as old as 145 million years) and most are sea-floor volcanic rocks, not the products of terrestrial volcanism. Even if there had been a 'volcano' associated with any of these rocks (which is unlikely), erosion would have removed all trace of a volcanic edifice.

Closure

Eroded remnants of the Paleozoic Wrangellia and San Juan Terranes, and the Mesozoic and Cenozoic Coast Crystalline, Nanaimo Group and Chuckanut rocks are preserved today as islands in the Salish Sea. Some of the islands partly or entirely comprise much younger Pleistocene glacial sediments. All Salish Sea islands have been shaped by Pleistocene glaciers that flowed over them between about 25,000 and 17,000 years ago, and the topography of the region is strongly controlled by differences in the resistance to erosion of strong crystalline rocks, weaker sandstone and conglomerate and weaker-still mudstone. At the end of the last Pleistocene glaciation, relative sea level was up to 200 m higher than today due to glacioisostatic depression of the crust by glacier ice covering much of BC. Glaciomarine sediments accumulated on submerged coastal lowlands at this time, and streams and rivers built deltas out into the sea. Sea level quickly fell to slightly below its present level as glacio-isostatic rebound returned the crust to its pre-glacial level. During the past 5,000 years, sea level has remained stable in the Strait of Georgia, providing a base level into which deltas such as the Fraser River delta have been built and large spits formed.

The hard rock islands have changed little during the Holocene Epoch. In contrast, the shores of islands cored by Quadra Sand and other Pleistocene sediments have retreated considerably under the assault of waves and currents. Erosion of these 'soft' sediments has produced the sandy and gravelly beaches nested among the rocky shorelines of the Strait of Georgia.

The Salish Sea islands have significant geological resources, including coal, base and precious metals, building stone, limestone and aggregate. Many residents of the region, especially on the smaller islands, rely on either bedrock or unconsolidated aquifers for water.

The region is tectonically active, and subduction of the Juan de Fuca Plate beneath the North America Plate is responsible for frequent small to medium earthquakes, and potentially active volcanoes. There is a risk of a large earthquake within decades or centuries, including an offshore megathrust quake that could produce a damaging tsunami.

References

Allen, D.M. and M. Suchy, 2001. 'Geochemical Evolution of Groundwater on Saturna Island, British Columbia.' *Canadian Journal of Earth Sciences* 38: 1059–80. doi.org/10.1139/e01-007

Arima, E.Y., L. Clamhouse, E. Edgar, C. Jones and D. St Claire, 1991. *Between Ports Alberni and Renfrew: Notes on West Coast Peoples*. Ottawa: Canadian Museum of Civilization, Canadian Ethnology Service (Mercury Series Paper 121). doi.org/10.2307/j.ctt22zmdnn

Brown, E.H., B.A. Housen and E.R. Schermer, 2007. 'Tectonic Evolution of the San Juan Islands Thrust System.' In P. Stelling and D.S. Tucker (eds), *Floods, Faults, and Fire: Geological Field Trips in Washington State and Southwest British Columbia*. Geological Society of America (Field Guide 9). doi.org/10.1130/2007.fld009(08)

Cherniawsky, J. and I. Fine, 2014, 'Models of Tsunami Waves at the Institute of Ocean Sciences,' Viewed 23 July 2023 at: web.archive.org/web/20161111051033/www.oceannetworks.ca/sites/default/files/pdf/tsunami_workshop/JC-IF-Tsunami-talk-PortAlberni2014.pdf

Clague, J.J. and B.D. Bornhold, 1980. 'Morphology and Littoral Processes of the Pacific Coast of Canada.' In S.B. McCann (ed.), *The Coastline of Canada*. Ottawa: Geological Survey of Canada (Paper 80–10).

Earle, S. and E. Krogh, 2006. 'Elevated Fluoride Levels in a Sandstone and Mudstone Aquifer System, Eastern Vancouver Island, Canada.' Paper presented at International Association of Hydrogeologists conference, Vancouver, October.

Eddy, P.A., 1975. 'Quaternary Geology and Groundwater Resources of San Juan County, Washington.' In R.H. Russel (ed.), *Geology and Water Resources of the San Juan Islands*. Seattle: Department of Ecology, Office of Technical Services (Water Supply Bulletin 46).

Fairchild, L. and D. Cowan, 1982. 'Structure, Petrology, and Tectonic History of the Leech River Complex Northwest of Victoria, Vancouver Island.' *Canadian Journal of Earth Sciences* 19: 1817–35. doi.org/10.1139/e82-161

Gehrels, G. et al., 2009. 'U-Th-Pb Geochronology of the Coast Mountains Batholith in North-Coastal British Columbia: Constraints on Age and Tectonic Evolution.' *GSA Bulletin* 121: 1341–61. doi.org/10.1130/B26404.1

Gilley, B.H.T., 2003. Facies Architecture and Stratigraphy of the Paleogene Huntingdon Formation at Abbotsford, British Columbia. Simon Fraser University (MSc Thesis).

Groome, W.G., 2000. Magmatism and Metamorphism in the Leech River Complex, Southern Vancouver Island, British Columbia, Canada—Implications for Eocene Tectonics of the Pacific Northwest. Simon Fraser University (MSc Thesis).

Groome, W.G., D.J. Thorkelson, R.M. Friedman, J.K. Mortensen, N.W.D. Massey, D.D. Marshall and P.W. Layer, 2003. 'Magmatic and Tectonic History of the Leech River Complex, Vancouver Island, British Columbia: Evidence for Ridge–Trench Intersection and Accretion of the Crescent Terrane.' In V.B. Sisson, S.M. Roeske and T.L. Pavlis (eds), *Geology of a Transpressional Orogen Developed During Ridge–Trench Interaction Along the North Pacific Margin*. Boulder: Geological Society of America (Special Paper 371). doi.org/10.1130/0-8137-2371-X.327

Kennedy, D. and S. Huffman, 2005. 'Seismic Retrofit Design Criteria, British Columbia Ministry of Transportation.' Viewed August 2018 at: www2.gov.bc.ca/assets/gov/driving-and-transportation/transportation-infrastructure/engineering-standards-and-guidelines/bridge/volume-4/seismic-retrofit-design-criteria.pdf

Kenny, S., M. Wei and K. Telmer, 2006. 'Factors Controlling Well Yield in a Fractured Metamorphic Bedrock Aquifer, District of Highlands, Vancouver Island, British Columbia.' Paper presented at International Association of Hydrogeologists conference, Vancouver, October.

Ludwin, R., R. Dennis, D. Carver, A. McMillan, R. Losey, J. Clague and C. Jonientz-Trisler et al., 2005. 'Dating the 1700 Cascadia Earthquake: Great Coastal Earthquakes in Native Stories.' *Seismological Research Letters* 76/2. doi.org/10.1785/gssrl.76.2.140

Mathews, W.H., J.G. Fyles and H.W. Nasmith, 1970. 'Postglacial Crustal Movements in Southwestern British Columbia and Adjacent Washington State.' *Canadian Journal of Earth Sciences* 7: 690–702. doi.org/10.1139/e70-068

McCrory, P.A. and D.S. Wilson, 2013. 'A Kinematic Model for the Formation of the Siletz-Crescent Terrane by Capture of Coherent Fragments of the Farallon and Resurrection Plates.' *Tectonics* 32: 718–36. doi.org/10.1002/tect.20045

Monger, J.W.H., R.A. Price and D.J. Tempelman-Kluit, 1982. 'Tectonic Accretion and the Origin of the Two Major Metamorphic and Plutonic Welts in the Canadian Cordillera.' *Geology* 10: 70–5. doi.org/10.1130/0091-7613(1982)10<70:TAATOO>2.0.CO;2

Mustard, P.S., 1994. 'The Upper Cretaceous Nanaimo Group, Georgia Basin.' In J.W.H. Monger (ed.), *Geology and Geological Hazards of the Vancouver Region, Southwestern British Columbia (Volume 481).* Geological Survey of Canada.

Mustard, P.S. and G.E. Rouse, 1994, 'Stratigraphy and Evolution of Tertiary Georgia Basin and Subjacent Upper Cretaceous Sedimentary Rocks, Southwestern British Columbia and Northwestern Washington State.' In J.W.H. Monger (ed.), *Geology and Geological Hazards of the Vancouver Region, Southwestern British Columbia (Volume 481)*. Geological Survey of Canada. doi.org/10.4095/203247

Mustoe, G.E., R.M. Dillhoff and T.A. Dillhoff, 2007. 'Geology and Paleontology of the Early Tertiary Chuckanut Formation.' In P. Stelling, and D.S. Tucker (eds), *Floods, Faults, and Fire: Geological Field Trips in Washington State and Southwest British Columbia* (Geological Society of America Field Guide 9). doi.org/10.1130/2007.fld009(06)

NEPEC (National Earthquake Prediction Evaluation Council), 2016. 'Evaluation of Earthquake Predictions: Recommendations to the USGS Earthquake Hazards Program from the National Earthquake Prediction Evaluation Council (NEPEC).' Viewed 23 June 2023 at: www.usgs.gov/media/files/recommendations-nepec-evaluating-earthquake-predictions

Sullivan, W.M., 2005. The Hydrogeology of North Lummi Island, Washington. Western Washington University (MA Thesis).

Walsh, T. and J. Schelling, 2011. 'Washington State School Seismic Safety Pilot Project—Providing Safe Schools for Our Students, Report to Washington State Seismic Safety Committee.' Viewed August 2018 at: www.dnr.wa.gov/publications/ger_ofr2011-7_school_pilot_project.pdf

5

Postglacial Dispersal and Ecosystem Assembly in the San Juan and Gulf Islands

Russel L. Barsh and Madrona Murphy

The composition of contemporary San Juan and Gulf Islands landscapes is a product of geophysical substrates such as topography and soils; weather patterns and microclimates; processes of colonisation and dispersal across the islands by species established on the nearby mainland; and of course, human interventions. These factors have changed over time and continue to change. Erosion alters topography and soils. Climate change has subjected the islands to warming and cooling cycles for millennia. The assembly of ecosystems within the archipelago has affected the ability of later-arriving species to establish, thrive and expand their range. Indigenous cultures have been displaced by European cultures that are, themselves, changing in the ways that they perceive and manipulate the environment.

At the same time, relatively small islands everywhere share characteristics that influence the way their landscapes assemble and change. They all have a relatively high ratio of shoreline to inland, resulting in broadly marine-influenced ecosystems as well as a great deal of disturbance created by tides, storms, tsunamis and sea level changes. Water-dispersed, salt-tolerant coastal vegetation is an obvious example, but less obvious perhaps is the tendency of islands' terrestrial ecosystems to be dominated by marine-adapted animals, ranging from midges that lay their eggs in tide-pools to birds and mammals that forage along shorelines and dive for fish and shellfish. Likewise, islands are poor habitats for animals that require extensive foraging areas, such as large ungulates and big cats. Islands often lack the diversity of species of nearby mainland areas due to their small size as well as their isolation.

As individual islands, the San Juan and Gulf Islands vary not only greatly in size but in the degree of their isolation. There are islands large enough to have supported, historically, herds of elk and wolf packs, but also hundreds of islands in the archipelago that are less than a hectare, and many barely large enough to pitch a tent. The effect of island size on species diversity should be greatest at the low end of the scale. No white-footed mice were found on islands smaller than 25 hectares by Redfield (1976), for example. Smaller islands are also more easily overwhelmed by small generalist invaders such as rats, and more quickly transformed by human activity such as camping and foraging. Behavioural adaptation to island resources, such as river otters (*Lutra canadensis pacifica*) learning to dive for herring and swim to smaller islands for seabird chicks (Speich and Pitman 1984), may be a matter of decades, as opposed to millennia for speciation. The San Juan and Gulf Islands are small enough to require behavioural adaptations by many animals, but are too young, and too much exposed to influences from nearby continental ecosystems, including human activity, to have much likelihood of producing novel endemic species. However, as Joseph Bennett and Peter Arcese (2013) have suggested, the San Juan and Gulf Islands are isolated enough to serve as refuges for some plant species that have largely disappeared from more interconnected and developed parts of Washington and British Columbia.

Geophysical Context

The structural geology of the San Juan Islands differs from Vancouver Island, the Gulf Islands and mainland Washington in ways that profoundly influence ecosystems (see Chapter 4). In the Gulf Islands and the northwesternmost San Juan Islands (Sucia, Waldron, Stuart), precipitation can infiltrate and travel through porous bedrock; whereas in most of the San Juan Islands, more precipitation runs off non-porous bedrock into the sea. Absence of a winter snow pack results in seasonal streamflows, moreover. The largest streams in the San Juan group dwindle to 0.01 m^3/s or less by late summer (Barsh 2010). Fish either exit streams in early summer or retreat to shaded, isolated stream pools for several months. Most natural glacial-kettle lakes with sphagnum bogs have been excavated for peat or to create artificial lakes.

Shallow fens and vernal pools, more widespread at the time of contact, have also mostly been drained for agriculture or excavated and diked as ponds. Unlike beaver dams, evidence of which abound in parts of the San Juan and Gulf Islands, artificial reservoirs do not leak continuously, and lose more water to evaporation. Contemporary freshwater habitats tend to be seasonal or artificial.

Weather interacts with geology to produce the characteristic appearance of many smaller islands: wildflower meadows with scattered oak or arbutus on the south side facing the sun and prevailing winds, and densely treed with Douglas firs and other conifers on the cooler north side (Rigg 1913). This is more pronounced in the San Juan group, where annual precipitation at low elevations is significantly reduced by the rain shadow of the Olympic Mountains. The rain shadow effect also extends to the Victoria Capital district, which historically had landscapes more like San Juan Island than the Gulf Islands—in particular, Garry oak-dominated meadows rather than meadows with copses of Douglas fir or shore pines.

Precipitation increases markedly with elevation in the San Juan and Gulf Islands. At the highest point in the archipelago atop Mount Constitution (731 m), annual precipitation exceeds 1,200 mm—greater than Seattle (960 mm), or Vancouver (1,110 mm)—but steep terrain, thin soils and non-porous metamorphic rock favour drought-tolerant plants. Some of the archipelago's rarest plants are only found close to the summit of Mount Constitution—or at sea level on the arid south coast of Lopez. Mistletoe (*Arceuthobium* spp.)[1] thrives on the summit of Mount Constitution but nowhere else in the San Juan and Gulf Islands. Sphagnum fens and a unique sphagnum floating bog are also characteristic of the higher elevations of this mountain, as are some characteristically alpine species such as showy Jacob's-ladder (*Polemonium pulcherrimum*) and rosy pussytoes (*Antennaria rosea*). Conditions at Mount Sullivan and other peaks on Saltspring and Saturna are comparable.

It should not be surprising that native wildflowers in the San Juan and Gulf Islands (including traditional Indigenous food plants) are adapted to summer drought. Most leaf out in winter when water is plentiful, and go dormant by June (Barsh and Murphy 2016). Drought has a greater impact on

1 Collections at the Washington State (Burke) Herbarium were identified as *tsugense* (i.e. the 'hemlock' mistletoe) but on Mount Constitution this hemiparasite only grows on shore pines (*Pinus contorta*), so is potentially a novel species.

woody species that must draw water higher to reach their leaves. Anticipated future changes—extreme precipitation events that increase runoff as well as warmer, drier summers—are likely to favour arbutus, Douglas fir and shore pine at the expense of western red cedar and Garry oak (Hebda 1997; Bachelet et al. 2011; Pellatt et al. 2012), recreating ecosystems that prevailed under warmer early postglacial conditions.

Young, steep, rocky and limited by the distribution of water, the islands are fine-scale habitat mosaics that are rich in *ecotones*—transitions between different plant communities—resulting in many unusual combinations of species not found in larger but less complex mainland landscapes. Six square kilometres of Mount Constitution offer red cedar and shore pine woodlands; sphagnum bogs; fens; natural lakes with rare aquatic plants; extensive xeric wildflower meadows; and even manzanita thickets; excluding areas that have been cleared or otherwise modified for recreational uses. Aspect, slope, elevation and substrate variation produce relatively small microclimates. A fine-scale mosaic of conditions sets the stage for pockets of unusual plant species and mobile, generalist animals.

Changing Postglacial Communities

Many ecosystems disappeared beneath periodic advances of the continental ice sheet between approximately 110,000 and 12,000 years ago. Since then, a period of gradual warming and deglaciation, followed some 4,000 years ago by a cooling trend, have resulted in a succession of postglacial landscapes in the San Juan and Gulf Islands. During the same period, dramatic changes in relative sea levels have altered conditions for the dispersal of plants and animals from the mainland to the islands, influencing the order in which species arrived and their assembly into landscapes and ecosystems.

Sea levels low enough to permit dry land colonisation of the islands from the mainland may have existed during the early stages of warming and glacial retreat, some 14,000 to 11,000 years ago, evidenced by the remains of large herbivores in island bogs (Kenady et al. 2011). However, the Salish Sea basin was soon flooded with glacial melt water, dammed by terminal moraines to the south and by the remaining ice to the west. 'Lake Bretz' provided a relatively brief opportunity for freshwater fish and amphibians to disperse simultaneously to re-emerging rivers and streams on the mainland and the islands. Beneficiaries probably include the islands' reticulate sculpins (*Cottus perplexus*), salamanders (*Ambystoma* spp.), and signal crayfish (*Pacifastacus leniusculus*), which we recently rediscovered on San Juan Island. Some plants probably also crossed Lake Bretz to the islands, including a rare water lobelia (*Lobelia dortmanna*), only found in a few remaining high-elevation lakes. When the ice jam in the Strait of Juan de Fuca finally decayed roughly 9,000 to 8,000 years ago, seawater flooded the Salish Sea basin and further separated the islands from each other and from the mainland.

High elevations such as Mount Constitution were probably exposed by ice thinning at an early stage of de-glaciation and flooding of the Salish Sea basin. Propagules of lichens, mosses and possibly vascular plants may have reached the islands from exposed portions of the Cascade, Olympic and Vancouver Island Ranges carried by winds, or the feet or faeces of birds, while much of the lowlands were still buried in ice. About one third of the lichen species found today in the islands are circumboreal; they would have been living on the perimeter of the continental ice sheet, and followed its northward retreat (Noble 1982). As Judith Harpel (1997) observed, most of the islands' mosses are also circumboreal, but include southern elements that presumably colonised the Salish Sea during the warm conditions that prevailed until about 4,000 years ago. As the climate cooled, these 'Mediterranean' species retreated to isolated refugia in the San Juan and Gulf Islands' rain shadow, including well drained higher elevations of the Southern Gulf Islands such as Mount Tuam, which are characterised by semi-arid type plant communities.

The succession of vascular plant communities in the islands has been recovered from the analysis of pollen deposited in lake sediments, nearshore marine sediments and sphagnum peat. Different sites produce somewhat different chronologies, reflecting differences in microclimates as well as conditions of pollen deposition. A lake sediment core from Pender Island reveals a shift about 8,000 years ago from an herbaceous landscape with scattered Douglas firs, to dense mixed woodlands still dominated by Douglas firs. About 4,000 to 3,500 years ago, when the regional climate began to cool, there was a shift towards a higher ratio of conifers to deciduous trees (Lucas and Lacourse 2013). The overall trend has been towards more conifers, less grass and fewer hardwoods, with oaks declining since about 6,000 years ago.

A more recently analysed peat core from Mount Constitution on nearby Orcas Island (Leopold et al. 2016) reflects the greater rain shadow effect on the San Juan group, with hardwoods spreading somewhat earlier and persisting somewhat longer. The Orcas core confirms that a variety of conifers were already established in exposed, elevated portions of the San Juan and Gulf Islands between 13,700 and 12,000 years ago, followed by the spread of ferns and alders. Perhaps within the warming and melting alpine conditions prevailing in that period, there were seasonally ice-free corridors connecting the islands to mainland forests to the south, where the ice had retreated a millennium earlier. As warming continued, Mediterranean-climate shrubs such as hairy manzanita (*Arctostaphylos columbiana*) colonised the Salish Sea from Oregon, only to retreat again when the climate trend reversed. Relic patches of manzanita remain on some high-elevation dry slopes on Mount Constitution, Eagle Cliff (Cypress Island) and Chuckanut Mountain on the Washington mainland, as well as the heights of Mount Erskine and Mount Maxwell on Saltspring Island and Bodega Hill on Galiano Island: a pattern similar to the mosses and lichens.

Although largely composed of regionally common species, contemporary plant communities in the islands frequently represent unusual combinations of these species that are absent on the mainland (Chappell 2006): for example, open canopy Douglas fir–grand fir woodlands with an understorey of native tall grass (*Festuca occidentalis*); or a madrone–Douglas fir forest with Alaska onion grass (*Melica subulata*) understorey unique to Lime Kiln State Park on San Juan Island. Widespread species reached the emerging islands one by one over a long period of time and 'assembled' in a different order than on the mainland, resulting in new configurations.

Rare, Endemic and Disjunct Plant Species

Plant communities in the islands today combine species found in wet coastal Pacific habitats with species more commonly found in the dry foothills on the east side of the Coast Range, such as soopolallie (*Shepherdia canadensis*). This nitrogen-fixing shrub is common and widespread in south central British Columbia (BC) and eastern Washington, while on the west side of the mountains it is concentrated in the central Salish Sea rain shadow: Lopez Island, San Juan Island, Sucia and Matia Islands, as well as nearby Whidbey and Fidalgo Islands at the east end of the archipelago, with patches at Victoria, Galiano and Valdez Islands in the Gulf group, and the Saanich Peninsula.

Some plant species that are abundant on the mainland of the Salish Sea may either remain absent from the islands or are found in circumstances suggesting relatively recent colonisation. One example is kinnikinnick (*Arctostaphylos uva-ursi*). Widespread on the mainland of Washington and

BC from sea level to foothills, it is found only on Patos Island, Lummi Rocks and tiny Castle Island in the San Juan group as well as Gabriola Island at the northern extreme of the Gulf group—potentially the results of four independent colonisation events. Gardeners have no difficulty getting this species to establish and thrive on other islands, therefore its native distribution was due to limitations on dispersal rather than habitat availability.

The contemporary distribution of Pacific huckleberry (*Vaccinium parvifolium*) is similar. This shrub is common and widespread throughout western Oregon, Washington and BC, including the Victoria Capitol Region, but rare in the San Juan and Gulf Islands. Collections are reported from Saltspring Island and from Matia Island in the San Juan group. The authors have seen isolated patches on Decatur, Cypress and Patos islands and a few individuals on Lopez Island. These occurrences are on the edges of the archipelago, closest to the mainland or Vancouver Island, suggesting a combination of infrequent bird drops with poor growing conditions where seeds fell. Indeed, the plants on Decatur and Lopez are growing from Douglas fir logging stumps, suggesting a relatively recent deposit of bird faeces.

A different story is suggested by the two *Camassia* species whose blue-purple flowers are iconic elements of San Juan and Gulf Islands coastal meadows. Both species were eaten by Indigenous peoples. *Camassia quamash*, or blue camas, is found throughout the Pacific Northwest and continues to be wild harvested by Native American communities in Washington, Idaho and Montana. *Camassia leichtlinii*, or great camas, the larger of the two, is largely restricted to the Salish Sea and the Willamette Valley of Oregon; and is presumed to be the cultivar described by early explorers and settlers (Barsh and Murphy 2016). In the San Juan group, where there is physical evidence of pre-contact gardens, all but 2 of 157 collections made in 2009–2012 were *C. leichtlinii*. While nuclear genetic markers suggest that *C. quamash* spread northward from multiple Ice Age refugia in Oregon (Tomimatsu at al. 2009), *C. leichtlinii* may have been domesticated (or strongly selected) in the postglacial era and widely traded. It forms a single regional genetic cluster. Human agency may have been responsible for the spread of great camas but not blue camas in the San Juan and Gulf Islands.

Only one plant species unique to the San Juan and Gulf Islands has been identified—*Castelleja victoriae*, an owl clover first identified in 2007—but a number of Salish Sea regional endemics are especially well represented in the islands. California buttercup (*Ranunculus californicus*) is abundant on the west side of San Juan Island, the south coast of Lopez Island, and Victoria, for example, but is found nowhere else in Washington State or BC. Golden paintbrush (*Castilleja levisecta*), originally found only in arid coastal meadows of Lopez and San Juan Islands and the Victoria Capitol Region is on the Endangered Species List in the United States and is red-listed in Canada. Junipers from the Salish Sea, recently identified as a cryptic species, *Juniperus maritima* or seaside juniper (Adams 2007), are widespread throughout the San Juan and Gulf Islands. Of course, some island endemics may yet be discovered among the desmids that Moore (1930) could not identify in isolated island bogs; or the unusually diverse lichen communities that Rhoades (2009) has been exploring on coastal rocky outcrops.

Animal Colonisation and Dispersal

Compared to vascular plants, animals have experienced greater restrictions on colonisation and dispersal in the San Juan and Gulf Islands. There are roughly half as many native species of mammals, reptiles and amphibians in the San Juan group as are found in mainland western Washington, as little as 10 kilometres distant by water. Overcoming the sea barrier can be achieved by various means. Animals that swim and can tolerate extended immersion in cold saline water have an excellent chance of reaching the islands, and the added advantage of island-hopping to avoid food shortages and place more habitat resources at their disposal. In the Salish Sea this includes native beavers, muskrats, mustelids (short-tailed weasel, river otter, mink), island-hopping Columbia deer (a peculiarity of the Salish Sea Islands) and, at least historically, grey wolves. The only large terrestrial mammal that did not disperse freely is Roosevelt elk (*Cervus elaphus roosevelti*), found on Orcas and Lopez Islands until early European settlers exterminated them. It is conceivable that they were relics of early migrations to the islands 11,000 to 13,000 years ago.

Very small mammals (excluding bats) are more likely to have arrived by being swept to sea on floating mats of woody debris launched by severe floods in mainland rivers. Rafts could make the crossing to the islands in a matter of hours with favourable tides and winds. One of the authors has seen rafts of cottonwood in the Oldman River of southern Alberta transporting bears, porcupines, coyotes and smaller animals for hours during spring floods. Beneficiaries of rafting in the San Juan and Gulf Islands may have included shrews (*Sorex vagrans*), voles (*Microtus townsendii*), white-footed mice (*Peromyscus leucopus*), raccoons and squirrels.

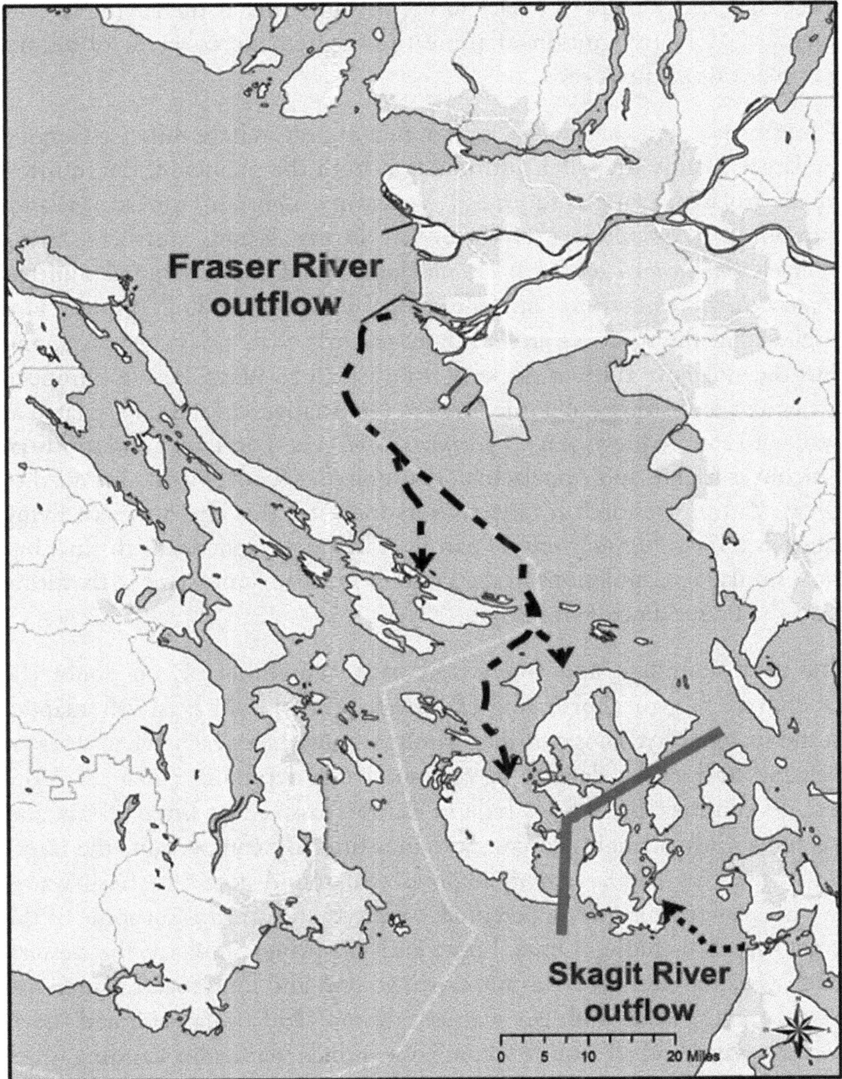

Figure 5.1 Major postglacial wildlife rafting routes and the 'Kennerly Line'
Source: Russel L. Barsh and Madrona Murphy.

The principal source of logjam rafts would have been the Fraser River, which would mainly deposit debris and passengers on north Orcas Island, north San Juan Island and the outer Gulf Islands. The smaller Skagit River would mainly have influenced colonisation of Lopez, Decatur and Blakeley islands, resulting in a division of the San Juan group into two separately assembling ecosystems. We refer to the dividing line as the Kennerly Line (Figure 5.1) in recognition of the earliest systematic collector of animal specimens from the islands.

While shrews, voles and white-footed mice appear to have dispersed among the larger islands after their introduction from the mainland, the squirrels remain isolated: Douglas squirrels (*Tamiasciurus douglasii*) on Orcas Island, chipmunks (*Neotamias townsendii*) on Lopez Island, northern flying squirrels (*Glaucomys sabrinus*) on San Juan Island and native red squirrels (*Tamiasciurus hudsonicus*) on Saltspring Island in the Gulf group. This implies that the squirrels arrived more recently than voles or shrews, and thus did not have time to disperse from island to island before European settlement resulted in habitat reduction. Non-native squirrels are isolated as well, with eastern grey squirrels (*Sciurus carolinensis*) on Lopez and southern Vancouver Island, and Appalachian fox squirrels (*Sciurus niger*) restricted to Orcas. Current research by the authors indicates that the northern flying squirrels (*Glaucomys sabrinus*) of San Juan Island are genetically distinct but most similar to populations in the Fraser watershed, consistent with rafting from the Fraser down Haro Strait.

The most abundant and widespread native amphibian is no doubt the Pacific tree frog or chorus frog (*Pseudacris regilla*), which is well adapted to the seasonal hydrology of the islands: it reproduces early when water is plentiful, and its fast-developing eggs survive in depressions as small as tire ruts. In summer, these frogs hide in damp moss or leaf litter by day and are active only at night, to conserve moisture. By comparison, the larger red-legged frog (*Rana aurora*) requires year-round access to fresh water, and is restricted to the few perennial wetlands and streams on some of the larger islands including Orcas, Lopez and Saltspring. Similarly, the western toad (*Anaxyrus boreas*), terrestrial northwestern and long-toed salamanders (*Ambystoma gracile* and *A. macrodactylum*), and rough-skinned newt (*Taricha granulosa*) all require secure home ponds that retain standing water in summer—more common on larger islands with greater topographic relief. Beavers once augmented the few natural perennial ponds on the larger islands, but today most summer standing water is found in dug ponds that are often deep, poorly vegetated (or vegetated chiefly with invasive reed canary grass), and frequently disturbed by human activity.

Newts are much more abundant and widespread in the San Juan and Gulf Islands than terrestrial salamanders, which may be due in part to their toxicity; only garter snakes routinely dine on them (Johnson et al. 2013;

Stokes et al. 2015). Newts are patchily abundant on Lopez, Orcas, San Juan, Gabriola, Galiano and South Pender islands, as well as southern Vancouver Island. Toads, by contrast, were historically restricted to the drier San Juan group. There are records from Blakeley, Cypress, Lopez, Orcas and San Juan Islands but none in the Gulf group. Toads disappeared from their last refuges on relatively undeveloped Blakeley and Lopez islands in the 1990s. Loss of naturally occurring home ponds to development may have been a significant factor in their disappearance, exacerbated by predation by raccoons, garter snakes and invasive bullfrogs (*Lithobates catesbeianus*).

Unlike newts and toads, the islands' terrestrial salamanders remain isolated. *A. gracilis* appears to be restricted to a few small drainages on Lopez and San Juan Islands, while *A. macrodactylum* is restricted to several wetlands on Orcas Island, nearby Shaw Island and Mount Tuam on Saltspring Island. All of these populations occupy hilly terrain at least 150 meters above current sea level, suggesting arrival in early postglacial times when the islands were connected to the mainland by a freshwater lake. Their failure to disperse more widely suggests habitat limitations, as does the fact that the Lopez and San Juan populations are mainly neotenic, remaining in their home pond while retaining larval 'tadpole' characteristics.

Three garter snakes are found throughout the islands: *Thamnophis elegans vagans*, *Thamnophis ordinoides* and *Thamnophis sirtalis* (Guest 1974; Engelstoft and Ovaska 2000) These snakes swim to pursue aquatic prey or escape predators. The 'common' garter snake (*T. ordinoides*) seems best adapted to living in proximity to humans, and is frequently found around homes and gardens; while the 'wandering' garter snake (*T. elegans vagans*) is most adapted to pursuing aquatic prey, and is often seen on small, uninhabited islands. Like the garter snakes, the northern alligator lizard (*Elgaria coerulea*) is viviparous and well adapted to a cool, moist environment. It is more widespread than previously reported and can be found on San Juan, Orcas, Shaw, Saltspring and Saturna islands but is surprisingly absent on Lopez, the driest of the San Juan and Gulf Islands. This is presumably due to the vagaries of dispersal by rafting.

By comparison, the oviparous sharp-tailed snake, *Contia tenuis*, is rare and restricted to higher elevations of Orcas, San Juan, Saltspring, Galiano, North and South Pender, and the Saanich highlands. Its patchy distribution and preference for steep, well-drained habitats suggests that it is a relic of the warmer and drier regional climate that prevailed 5,000–7,000 years ago (Oldham et al. 2021). A genetic study of the San Juan Island population is currently under way.

It is unclear whether the western pond turtle (*Actinemys marmorata*) or western painted turtle (*Chrysemys picta*) were native to the islands. There were reports of pond turtles on San Juan and Blakeley islands as recently as the 1990s, and painted turtles are present in small numbers today on Lopez, Orcas, San Juan, Saltspring and South Pender. Exotic red-eared sliders (*Trachemys scripta elegans*) are the most widespread turtle, however. Oviparous like the sharp-tailed snake, turtles bury and abandon eggs, making them vulnerable to predation of eggs and hatchlings by great blue herons, minks and otters. They may have been more successful and widespread prior to the cooling trend that brought the islands their cedars and salmon some 4,000 years ago.

Flight is a more consistent means of dispersal than rafting, and bats are sufficiently strong fliers to have colonised the islands as soon as there was sufficient insect prey to support them. Indeed, at least 9 of the 11 bat species known from the Salish Sea mainland are also found throughout the islands (Barsh 2015). As a result, bats rather than rodents are the most diverse mammalian group in the San Juan and Gulf Islands. The absence of rabies in San Juan and Gulf Islands bats suggests that they rarely fly to mainland Washington, where the incidence of rabies in bats is low but persistent. Nightly recordings of thousands of bats at 'listening posts' around the San Juan group, initiated by the authors in 2014, have thus far found no differences in bat diversity between the larger islands. The most abundant species are two small mouse-eared bats, *Myotis californicus* and *Myotis yumanensis*, that appear to forage routinely over beaches and saltwater and to visit smaller islands, changing their routes every few nights.

Townsend's big-eared bat (*Corynorhinus townsendii*) is significantly more abundant in the San Juan group than elsewhere in Washington State or BC, for reasons that remain unclear. Like other bat species in the San Juan Islands, Townsend's bats remain sporadically active throughout the winter, feeding primarily on woodland-associated moths, which may improve their chances for survival.

Birds should have little difficulty moving between the islands and the mainland, and indeed there are many examples of species that utilise the islands seasonally—not only long-distance migrators such as ducks, geese and swans, but also many songbirds such as crossbills, cedar waxwings, western tanagers and Swainson's thrush. The distribution of birds in the San Juan and Gulf Islands should reflect habitat availability rather than colonisation history. Indeed, some birds have either disappeared or suffered significant range reductions in the San Juan and Gulf Islands since 1900, presumably due to development. This includes western bluebirds, extirpated by the 1950s, and Steller's jays, once common in the San Juan group but now largely restricted to Orcas Island (Miller 1935; Goodge 1950; Retfalvi 1963). Widespread use of bird feeders has meanwhile converted seasonal visitors such as Anna's hummingbird into year-round residents.

Birds can be significant vectors for the dispersal of invertebrates, especially crustaceans, molluscs and insects with aquatic eggs and larvae. Mud caked on the feet of waterfowl can contain eggs that survive until the carrier lands in another body of water (Figuerola et al. 2004; Frisch et al. 2007). This may explain the widespread distribution in the San Juan and Gulf Islands of small freshwater species such as fairy shrimp (*Eubranchipus oregonus*) and fingernail clams (*Sphaerium patella*), as well as terrestrial molluscs such as *Monadenia fidelis*, the common side-band snail, which the authors have observed frequently on small uninhabited islands. Birds sometimes airlift larger animals. The authors once observed a red-tailed hawk drop a white-footed mouse from an altitude of 20 m into leaf litter on San Juan Island. The mouse survived.

The authors' observations of pollinators in the San Juan group suggest that at least two thirds of the species of bees and hoverflies reported from nearby mainland areas are also found in the islands, including both native and introduced Eurasian species, and bumblebees were seen in flight between larger and smaller islands. Not all insects are capable of long-distance flight. The initial colonisation of the San Juan and Gulf Islands by beetles, for example, was most likely by rafting. Their subsequent dispersal within the archipelago may have been at least partly under their own power. This implies lower diversity in the islands than the mainland but little difference between islands. Few surveys of terrestrial arthropods are available, unfortunately, and they are limited to larger, extensively disturbed, ferry-connected islands (e.g. Worley 1932; Strange et al. 2016). James Bergdahl's collection of Carabid ground beetle species from the San Juan and Gulf Islands—thousands of specimens representing 115 species—has yet to be published.

Humans and other vertebrates may also have played a large role of the dispersal of insects and arachnids. The authors' current study of ticks in the San Juan group has thus far identified two native and one introduced species, associated with native wildlife such as deer as well as humans and their pets. It is common for island farms to import truckloads of hay from the mainland to feed cattle and horses, and a number of Lepidoptera pupate in chrysalises tethered to grass stems. Plants and soil can be conveyors of arthropod eggs and larvae. These vectors of dispersal have existed as long as humans have travelled from the mainland to the islands, especially in the company of dogs and unprocessed food plants.

In the light of nine millennia of postglacial human activity in the San Juan and Gulf Islands, it may come as a surprise that any mammals are still restricted to one or two of the islands. The explanation may simply be a matter of time. Belgian hares introduced a century ago to San Juan Island and Long Island (Couch 1929; Hall 1977), and more recently to Lopez Island, have joined native rodents in the diets of bald eagles and other native raptors (Retfalvi 1970), but have not yet spread more widely. European foxes were introduced repeatedly to San Juan Island and at least once on Cypress Island for their fur and as a rabbit control (Hall 1977), but for a century have not colonised other islands. Appalachian fox squirrels, introduced at Olga (Orcas Island) in 2002, have spread throughout that island but not yet appeared elsewhere. Eastern grey squirrels were introduced on Lopez about 50 years ago, and remain restricted to that island.

At the same time, native mammals that arrived hundreds to thousands of years ago may represent genetically unique populations that in time could change enough behaviourally (if not also morphologically) to become novel species. Genetic differentiation across an archipelago at a fine scale has been shown in wolves and flying squirrels of the BC coast and southeast Alaska (Bidlack and Cook 2002; Stronen et al. 2014). Dalquest (1940) proposed a distinct subspecies of Townsend's vole in the San Juan Islands on the basis of morphometric analysis of eight individuals, mainly from Shaw Island, but this was never tested genetically. Nor has the purported San Juan Island subspecies of the large marble butterfly (*Euchloe ausonides*), which is abundant in eastern Washington. On the other hand, two small populations of coastal cutthroat trout (*Oncorhynchus clarkii clarkii*) have been identified in the San Juan group, one isolated for millennia above a natural barrier (Glasgow et al. 2016); and the authors recently confirmed the genetic distinctives of San Juan Island's northern flying squirrels, suggesting the need for more studies of this nature.

Conclusions

Over the course of the last 12,000 years, plants and animals have established in the San Juan and Gulf Islands by varied means. Earliest pioneers were lichens and mosses that settled on a few ice-free mountain peaks. When the ice began to retreat, flooding the San Juan group with melt water, plants, fish and amphibians migrated to the islands from rivers and streams on the continental mainland, while birds helped bring seeds and freshwater invertebrates, and bats established their role as the islands' most diverse mammals. Swimming and rafting continued the process of ecosystem assembly once saltwater replaced fresh water in the Salish Sea basin, delivering rodents, other mammals and additional reptiles to the islands. Humans returned to the islands at this stage, roughly 8,000 years ago, becoming the main driver of further changes.

In light of the rather young geological age of the islands as they exist today, their limited surface area and their close proximity to continental ecosystems, it is not surprising that unique, endemic animal species have not been found. On the other hand, the islands have isolated some species from forces that have nearly extirpated them from mainland area, such as contagious disease (rabies, thus far), and human disturbance. The abundance in the islands of otherwise rare Townsend's big-eared bats was noted earlier in this chapter. Another example is the Propertius duskywing butterfly (*Erynnis propertius*), one of the few animals that rely exclusively on Garry oaks, found in small stands of oaks on Waldron, San Juan, Orcas, Saltspring, Thetis and Galiano islands, and relic oak prairies on southeast Vancouver Island (Hellmann et al. 2008). The role of the San Juan and Gulf Islands as refugia in an urbanised region should not be overlooked.

Acknowledgements

The authors gratefully acknowledge the unpublished data and insights contributed by associates at Kwiáht including Joe Behnke, Nathan Hodges and Christian Oldham; the leadership of Peter Arcese and his students at the University of British Columbia; the pioneering efforts of the Islands Trust, Gulf Islands National Park and the US National Park Service to inventory, map and protect the islands' diversity; and the personal influence and inspiration of the late Eugene Kozloff. Previously unpublished distribution data summarised in this chapter were collected with support of the National Fish and Wildlife Foundation, Charlotte Martin Foundation, Mills Davis Foundation and Orcas Island Community Foundation.

References

Adams, R.P., 2007. '*Juniperus maritima*, the Seaside Juniper, a New Species from Puget Sound, North America.' *Phytologia* 89: 263–83.

Bachelet, D., B.R. Johnson, S.D. Bridgham, P.V. Dunn, H.E. Anderson and B.M. Rogers, 2011. 'Climate Change Impacts on Western Pacific Northwest Prairies and Savannas.' *Northwest Science* 85: 411–29. doi.org/10.3955/046.085.0224

Barsh, R.L., 2010. 'Structural Hydrology and Limiting Summer Conditions of San Juan County Fish-Bearing Streams.' Lopez Island: Kwiáht.

——, 2015. 'Bats, Most Diverse and Cosmopolitan Mammals of the San Juan Islands, WA, Are Residents and Active Year-Round.' Lopez Island: Kwiáht.

Barsh, R.L. and M. Murphy, 2016. 'Coast Salish Camas Cultivation.' HistoryLink.org Essay 11220. Viewed 31 May 2023 at: www.historylink.org/File/11220

Bennett, J.R. and P. Arcese, 2013. 'Human Influence and Classical Biogeographic Predictors of Rare Species Occurrence.' *Conservation Biology* 27: 1–5. doi.org/10.1111/cobi.12015

Bidlack, A.L. and J.A. Cook, 2002. 'A Nuclear Perspective on Endemism in Northern Flying Squirrels (*Glaucomys sabrinus*) of the Alexander Archipelago, Alaska.' *Conservation Genetics* 3: 247–59.

Booth, D.B., K.G. Troost, J.J. Clague and R.B. Waitt, 2003. 'The Cordilleran Ice Sheet.' *Developments in Quaternary Science* 1: 17–43. doi.org/10.1016/S1571-0866(03)01002-9

British Columbia Ministry of Forests, 2022. 'Biogeoclimatic Zones of British Columbia.' Viewed 8 June 2023 at: www.for.gov.bc.ca/hre/becweb/resources/maps/ProvinceWideMaps.html

Chappell, C.B., 2006. 'Upland Plant Associations of the Puget Trough Ecoregion, Washington.' Olympia: Washington Department of Natural Resources (Natural Heritage Report 2006-01).

Couch, K., 1929. 'Introduced European Rabbits in the San Juan Islands, Washington.' *Journal of Mammalogy* 10: 334–6. doi.org/10.2307/1374122

Dalquest, W.W., 1940. 'New Meadow Mouse From the San Juan Islands, Washington.' *The Murrelet* 21: 7–8. doi.org/10.2307/3535209

Engelstoft, C. and K.E. Ovaska, 2000. 'Artificial Cover-Objects as a Method for Sampling Snakes (*Contia tenuis* and *Thamnophis* spp.) in British Columbia.' *Northwestern Naturalist* 81: 35–43. doi.org/10.2307/3536898

Figuerola, J., A.J. Green and T.C. Michot, 2004. 'Invertebrate Eggs Can Fly: Evidence of Waterfowl-Mediated Gene Flow in Aquatic Invertebrates.' *American Naturalist* 165: 274–80. doi.org/10.1086/427092

Frisch, D., A.J. Green and J. Figuerola, 2007. 'High Dispersal Capacity of a Broad Spectrum of Aquatic Invertebrates Via Waterbirds.' *Aquatic Sciences* 69: 568–74. doi.org/10.1007/s00027-007-0915-0

Glasgow, J., J. de Groot, R. Barsh, M. O'Connell and N. Gayeski, 2016. 'Evaluating the Genetic Composition and Status of Coastal Cutthroat Trout (*Oncorhynchus clarkii clarkii*) in the San Juan Islands, Washington.' Duvall: Wild Fish Conservancy.

Goodge, W., 1950. 'Some Notes on the Birds of the San Juan Islands.' *The Murrelet* 31: 27–28. doi.org/10.2307/3535807

Guest, C.,1974. 'The Vertebrate Fauna of Three Ecological Reserves on the Gulf Islands of British Columbia.' Victoria, BC: University of Victoria, Department of Biology. Viewed 31 May 2023 at: ecoreserves.bc.ca/wp-content/uploads/2012/02/Guest_1974_Vertebrates_Saturna_Mt_Tuam_Mt_Maxwell.pdf

Hall, L.S., 1977. 'Feral Rabbits on San Juan Island, Washington.' *Northwest Science* 51: 293–7.

Harpel, J.S., 1997. The Phytogeography and Ecology of the Mosses Within the San Juan Islands, Washington State. Vancouver: University of British Columbia (PhD thesis).

Hebda, R., 1997. 'Impact of Climate Change on Biogeoclimatic Zones of British Columbia and the Yukon.' In E. Taylor and B. Taylor (eds), *Responding to Global Climate Change in British Columbia and Yukon*. Vancouver: Ministry of Environment.

Hellmann, J.J., S.L. Pelini, K.M. Prior and J.D.K. Dzurisin, 2008. 'The Response of Two Butterfly Species to Climatic Variation at the Edge of Their Range and the Implications for Poleward Range Shifts.' *Oecologia* 157: 583–92. doi.org/10.1007/s00442-008-1112-0

Johnson, J., B.G. Gall and E.D. Brodie, Jr, 2013. 'Predator Avoidance in Lab-Reared Juvenile Rough-Skinned Newts, *Taricha granulosa*.' *Northwestern Naturalist* 94: 103–9. doi.org/10.1898/12-20.1

Kenady, S.M., M.C. Wilson, R.F. Schalk and R.R. Mierendorf, 2011. 'Late Pleistocene Butchered *Bison antiquus* From Ayer Pond, Orcas Island, Pacific Northwest: Age Confirmation and Taphonomy.' *Quaternary International* 233: 130–141. doi.org/10.1016/j.quaint.2010.04.013

Leopold, E.B., P.W. Dunwiddie, C. Whitlock, R. Nickmann and W.A. Watts, 2016. 'Postglacial Vegetation History of Orcas Island, northwestern Washington.' *Quaternary Research* 85: 380–90. doi.org/10.1016/j.yqres.2016.02.004

Lucas, J.D. and T. Lacourse, 2013. 'Holocene Vegetation History and Fire Regimes of *Pseudotsuga menziesii* Forests in the Gulf Islands National Park Reserve, southwestern British Columbia, Canada.' *Quaternary Research* 79: 366–76. doi.org/10.1016/j.yqres.2013.03.001

Miller, R.C., E.D. Lumley and F.S. Hall, 1935. 'Birds of the San Juan Islands, Washington.' *The Murrelet* 16: 51–65. doi.org/10.2307/3533918

Moore, C.S., 1930. 'Some Desmids of the San Juan Islands.' Seattle: University of Washington.

Noble, W.J., 1982. The Lichens of the Coastal Douglas-Fir Dry Subzone of British Columbia. University of British Columbia (PhD thesis).

Oldham, C., R. Barsh and M. Murphy, 2021. '*Contia Tenuis* (Common Sharp-Tailed Snake) Reproduction.' *Herpetological Review* 52: 418–9.

Pellatt, M.G., S.J. Goring, K.M. Bodtker and A.J. Cannon, 2012. 'Using a Down-Scaled Bioclimate Envelope Model to Determine Long-Term Temporal Connectivity of Garry Oak (*Quercus garryana*) Habitat in Western North America: Implications for Protected Area Planning.' *Environmental Management* 49: 802–15. doi.org/10.1007/s00267-012-9815-8

Redfield, J.A., 1976. 'Distribution, Abundance, Size, and Genetic Variation of *Peromyscus maniculatus* on the Gulf Islands of British Columbia.' *Canadian Journal of Zoology* 54: 463–74. doi.org/10.1139/z76-053

Retfalvi, L., 1963. 'Notes on the Birds of San Juan Islands, Washington.' *The Murrelet* 44: 12–13. doi.org/10.2307/3536274

——, 1970. 'Food of Nesting Bald Eagles on San Juan Island, Washington.' *The Condor* 72(3): 358–61. doi.org/10.2307/1366014

Rhoades, F., 2009. *Lichens of South Lopez Island, San Juan County, Washington State*. Seattle, WA: Washington Native Plant Society (Occasional Publication No. 9).

Rigg, G.B., 1913. 'Forest Distribution in the San Juan Islands: A Preliminary Note.' *The Plant World* 16: 177–82.

Speich, S.M. and R.L. Pitman, 1984. 'River Otter Occurrence and Predation on Nesting Marine Birds in the Washington Islands Wilderness.' *The Murrelet* 65: 25–7. doi.org/10.2307/3534208

Stokes, A.N., A.M. Ray, M.W. Buktenica, B.G. Gall, E. Paulson, D. Paulson, S. French, et al., 2015. 'Otter Predation on *Taricha granulosa* and Variation in Tetrodotoxin Levels with Elevation.' *Northwestern Naturalist* 96: 13–21. doi.org/10.1898/NWN13-19.1

Strange, J.P., J.B. Koch, W.S. Shepherd, B. Hopkins, J. Long, E. Lichtenberg and C. Looney, 2016. 'Bumblebees Community Composition and Population Genetic Diversity in the North Cascades and Coast Network.' Viewed 31 May 2023 at: depts.washington.edu/pnwcesu/reports/P13AC00713_Final_Report.pdf

Stronen, A.V., E.L. Navid, M.S. Quinn, P.C. Paquet, H.M. Bryan and C.T. Darimont, 2014. 'Population Genetic Structure of Gray Wolves (*Canis lupus*) in a Marine Archipelago Suggests Island-Mainland Differentiation Consistent with Dietary Niche.' *BMC Ecology* 14 (1): 11. doi.org/10.1186/1472-6785-14-11

Tomimatsu, H., S.R. Kephart and M. Vellend, 2009. 'Phylogeography of *Camassia quamash* in Western North America: Postglacial Colonization and Transport by Indigenous Peoples.' *Molecular Ecology* 18: 3918–28. doi.org/10.1111/j.1365-294X.2009.04341.x

Worley, L.G., 1932. 'The Spiders of Washington: With Special Reference to Those of the San Juan Islands.' *University of Washington Publications in Biology* 1 (1).

Part Two: History

6

Origin Stories of the Central Salish Archipelago

Chris Arnett

> If archaeologists or scientists of other Western disciplines are to better understand and communicate with Indigenous peoples, they must be open to the idea that Western science is not the only method by which knowledge can be created; other peoples have successfully created knowledge with their own methods. (Harris 2006: 34)

There are two origin stories for the Canadian Salish Archipelago—one Indigenous, the other non-Indigenous. Both attempt to explain the world. The chapters in this book reflect non-Indigenous stories of place and represent ways of looking at the world imported from elsewhere. Non-Indigenous ways of knowing have dominated our understanding of the Central Coast Salish past for over a century. These ways of knowing often use evolutionary-based models of culture ('what people do') that depict a trajectory of social and cultural development described by non-Indigenous people (Matson and Coupland 1995). According to this view, the nineteenth-century Central Coast Salish who inhabited these islands had evolved into a class society of hereditary nobles, commoners and slaves, (not unlike the contemporaneous non-Indigenous culture). This model, with its emphasis on conflict and inequality, is often used to interpret the material culture back thousands of years (Coupland et al. 2016).

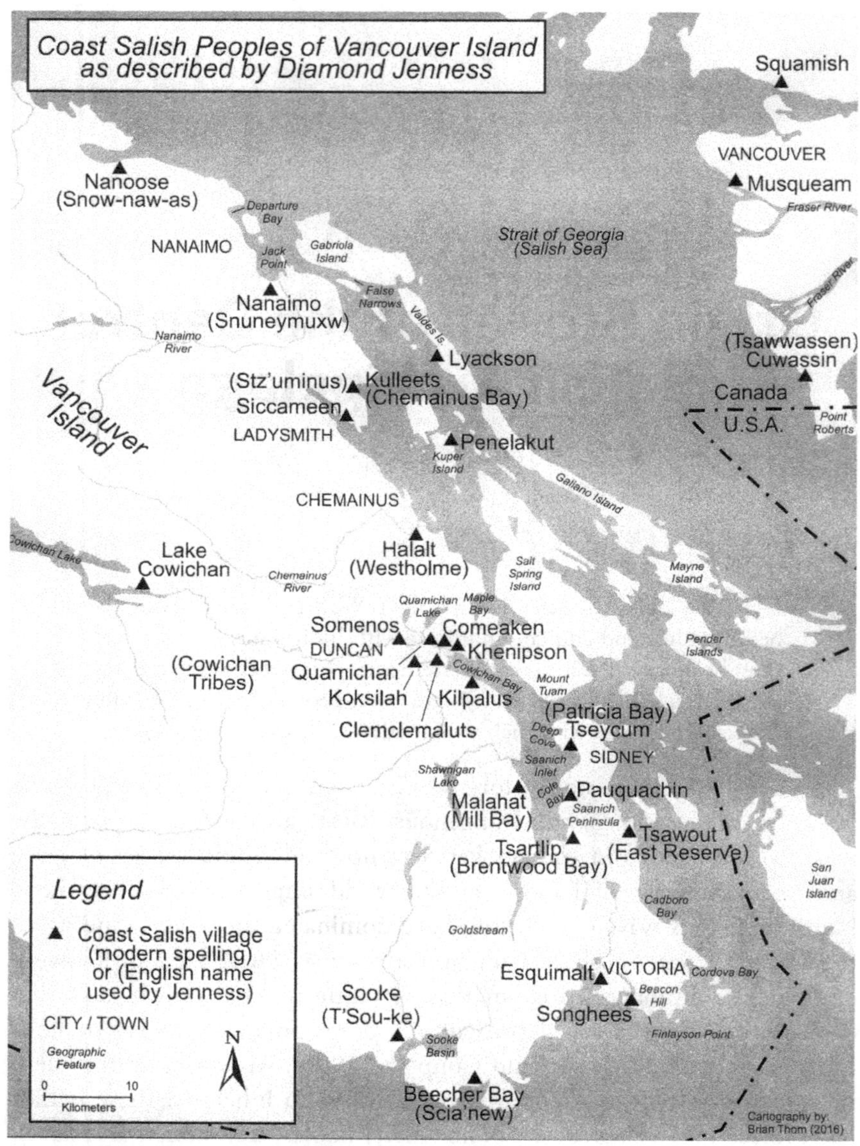
Coast Salish Peoples of Vancouver Island
as described by Diamond Jenness
Squamish
VANCOUVER
Musqueam
Fraser River
Nanoose
(Snow-naw-as)
Departure Bay
NANAIMO
Jack Point
Gabriola Island
Strait of Georgia
(Salish Sea)
Nanaimo
(Snuneymuxw)
False Narrows
Valdes Is.
Nanaimo River
Vancouver Island
Lyackson
(Tsawwassen)
Cuwassin
(Stz'uminus)
Kulleets
(Chemainus Bay)
Siccameen
LADYSMITH
Canada
U.S.A.
Point Roberts
Penelakut
Kuper Island
Galiano Island
CHEMAINUS
Cowichan Lake
Lake
Cowichan
Chemainus River
Halalt
(Westholme)
Salt Spring Island
Mayne Island
Quamichan Lake
Maple Bay
Somenos
Comeaken
DUNCAN
Khenipson
(Cowichan
Tribes)
Quamichan
Cowichan Bay
Koksilah
Kilpalus
Mount Tuam
Pender Islands
Clemclemaluts
(Patricia Bay)
Tseycum
Deep Cove
SIDNEY
Saanich Inlet
Shawnigan Lake
Malahat
(Mill Bay)
Cole Bay
Pauquachin
Saanich Peninsula
Tsawout
Tsartlip
(Brentwood Bay)
(East Reserve)
San Juan Island
Cadboro Bay
Legend
Coast Salish village
(modern spelling)
or (English name
used by Jenness)
CITY / TOWN
Geographic Feature
0
10
Kilometers
N
Goldstream
Esquimalt
VICTORIA
Cordova Bay
Beacon Hill
Sooke
(T'Sou-ke)
Songhees
Finlayson Point
Sooke Basin
Beecher Bay
(Scia'new)
Cartography by Brian Thom (2016)

Figure 6.1 Coast Salish peoples of Vancouver Island

Source: Courtesy Barnett Richling.

The W̱SÁNEĆ, Quw'utsun Mustimuhw and Snuneymuxw sources cited here suggest that nineteenth-century Central Coast Salish culture was significantly affected by non-Indigenous influence (disease) and the appropriation of non-Indigenous values (see also Moon 1978 and Moss 2011).[1] Why is this important? Tuhiwai Smith (1999: 34, 62) argues that constructed colonial histories, and I would add archaeology, can do violence to public perceptions of Indigenous people. Contrary to being antagonistic to non-Indigenous ways of knowing (see McGhee 2008), Indigenous ways of knowing can both challenge and support archaeological interpretations. In these days of 'Truth and Reconciliation', how might archaeological sites reflect Indigenous values? Indigenous theory offers a more critical analysis of archaeological data, history and the environment without prior assumptions.

In this chapter, the work of Nuu-chah-nulth scholar Richard Atleo (1999, 2004, 2011) on the significance of origin stories provides a theoretical framework to revisit three early twentieth-century narratives by David Latasse (W̱SÁNEĆ), John Humphreys (Quw'utsun), and Albert Wesley (Snuneymuxw) that illuminate Indigenous ways of knowing the past in the Central Salish Archipelago and Southern Vancouver Island. This follows the pioneering work of Duncan McLaren (2006) who used Central Coast Salish Indigenous narratives to suggest historical sequences. In doing so I hope to further demonstrate how Indigenous accounts can both complement and challenge non-Indigenous ways of knowing with regard to history, archaeology and other sciences.

Origin Stories/Indigenous Theory

Richard Atleo (Umeek) a hereditary chief and scholar of the Ahousaht Nuu-chah-nulth (McMillan this volume) has articulated how Indigenous values, or Indigenous theory, are explained by origin stories validated by the vision quest (Atleo 1999, 2004, 2011). This creates a social order which includes not only humans but all other entities, physical and spiritual, with strict protocols governing interaction to ensure successful social reproduction and sustainability based on the notion of respect.

1 This chapter uses the orthography for SENĆOŦEN (with its distinctive use of all-caps and diacritics) and Hul'q'umi'num' (with its distinctive use of roman characters which cluster multiple letters to make a single sound) when writing from Indigenous speakers, and the international phonetic alphabet (IPA) for quoting from academic or historical anthropological sources. Though the spelling may differ, the underlying words are the same.

As allegories of behaviour geared towards social and environmental sustainability, origin stories emphasise the interrelationship of all things spiritual and physical. This view is deceptively simple in theory but difficult in practice. Prior knowledge requires intergenerational teachings of place and direct experience by the *ʔuusumč́* (*oosumich*), or vision quest, a research method used by Nuu-chah-nulth and many other North American Indigenous people, to test the theory that everything is one (Atleo 2004: 71–94, 2011: 144–45). The vision quest with its self-isolation and introspection allows individuals to experience physiologically the truth of the teachings. The cognate SENĆOŦEN term is QEĆÁSET, 'to get spiritual power, become spiritually strong' (Montler 2018: 413).

The equivalent Hul'q'umi'num' term *kwa'yuthut* means literally 'to step away, move away' (physically away from the home and spiritually away from the physical world) through ritual exercises and bathing in certain locales (Hukari and Peter 1995: 32). This was not exclusive. Everyone did it. Learning was a lifelong process. As W̱SÁNEĆ elder Chris Paul (Montler 2018: 413) explained: 'the children of long ago trained themselves as they grew'. Places and place names in the landscape with associated origin stories gave guidance physically (knowing the landscape and hearing the stories) and spiritually (knowing oneself and Creation). The adaptive success of the theory and the method is evident in the long record of successful adaptation by people to the environment of the Salish Archipelago (Carlson and Hobler 1993: 50; Angelbeck and McMillan, this volume).

The catalyst for the common origin of all things in the Salish Archipelago is a supernatural entry called Xeel's in Hul'q'umi'num', or XÁ,EL,S in SENĆOŦEN, referred to in English as the Transformer, the Changer or the Creator (Hukari and Peter 1995: 27; Montler 2018: 797). The phonetic equivalent *χéʔəl's* indicates his primary role in the Indigenous cosmology. '*χéʔəl''s*', said the late Ruby Peter, 'is the one who creates and changes things'. As Ruby Peter (personal communication, 2014) explained to me, the word *χéʔəl's* is a contraction from the root *χéla* ('something rare/precious') and the activity marker *əl's*, thus someone 'creating something rare/sacred'. *χéla* is related to *χ́éχe* 'sacred'. All of his work, Creation, is precious and sacred. The word XÁ,EL,S has other shades of meaning related to his activities, as the late Snuneymuxw elder Ellen White (2006: 6) explains in a book she wrote about him:

> The ancestors called our Creator Xeel's. Xeel's made things new so he could help make life easier for the people and animals he called his children. Xi' means to appear suddenly before you. Xew's is to make things new. That is why they called him Xeel's.

This being, under different names, manifests himself in all of the Creation stories of Indigenous people in British Columbia and is responsible for establishing the present-day order, in effect the biodiversity, topography and culture of the world. Since he created everything that exists, everything has a common origin and is thus related. 'Biological differentiation', Atleo (2004: 86) writes, 'is understood as the result of transformation from a common source of being'. Through XÁ,EL,S everything is related, including the very land itself, as captured in this poignant account by David Latasse and his wife (Jenness 1934/5: 9) who describe the creation the Salish Archipelago.

> A number of Indians were travelling in their canoes some of them had joined there canoes together in twos by means of boards and piled all their luggage on the boards. Suddenly X̣e.ls appeared, sprinkled them with something and changed them to stone – the islands that now lie off SE Vancouver Island.

To this day Saanich people refer to the San Juans and Canadian Gulf islands as TELEṮÁĆES, 'relative of the deep' (Fritz 2012: 190–1). The term 'relative' could in theory be applied to anything, animate or not, a way to visualise the world in terms of relationships and connections as opposed to disconnections. In this social order all relationships involve protocols. As Atleo (2011: 92) explains:

> Protocols are agreements between life forms. Their purpose is to ensure that life forms exercise mutual; recognition, mutual responsibility and mutual respect, their responsibility to other life forms. Non-human life forms will respond in kind when mutual respect are practiced, then there is balance, harmony is achieved and the goal of environmental and economic sustainability may be realized.

This is not aboriginal essentialism (see McGhee 2008) but a way of looking at the world with the idea of respect, equality and generosity as natural laws (Atleo 2011) as opposed to a culture of disrespect, inequality and greed.

One of XÁ,EL,S' most important works was to drop human beings on top of the hills along the Fraser River, at Point Roberts (then an island), on the mountains of East Vancouver Island, and elsewhere (Jenness 1955: 10–15). The descendants of these ancestors include speakers of Northern Straits Salish (SENĆOŦEN, Ləkʷəŋən), North Lushootseed (*xʷləmiʔčósən*,[2] *xʷləšúcid*[3]) and Island Halkomelem (Hul'q'umi'num'). These first ancestors did marvellous things, established protocols with the sockeye and other species that provided a living, and created reef nets, fish weirs, deer nets, winter dancing and the potlatch. The institution of the potlatch was introduced by syalutsa', an ancestor 'who made this custom into a strict law' (Bouchard and Kennedy 2002: 142). Until recent times the potlatch was the mechanism by which 'wealth', the generosity of nature, was distributed among a community of equals (Suttles 1987: 30–31) before historic changes to demographics, the social structure and the economy began to take place in the Salish Archipelago in historic times (Masco 1995).

Sometime during 1934–35 the Canadian anthropologist Diamond Jenness interviewed a well known Saanich Chief known by his English name, David Latasse, and his wife Mary, who lived at Tsartlip on the Saanich Peninsula (Richling 2016). David Latasse's frequent mention in the press as a spokesman for Saanich people and his detailed work with Jenness are evidence of his knowledge. He narrated the following summary of southern Vancouver Island history which gives the common descent from X̣e.ls (XÁ,EL,S); the social order; the connections between peoples; and the origin of the class system:

> X̣e.ls, the great Creator and Transformer, created human beings in various places from some kind of earth, but he fashioned the stomach of the woman from bands of cherry bark so that they might expand during pregnancy. At Duncan he created a man called Hayletha [*syalutsa'*], at Sooke another man called Tayakamat, with a wife and daughter, and a maid for the daughter. The girl and her maid walked from Sooke up to Duncan, where they spied on Hayletha as he talked with a female image he made from rotten cedar; for Hayletha was lonely, and whenever he left his home to hunt or fish he would leave his distaff and the wool he spun on it in the hands of the image, as though it would spin it for him. After he had gone, the girls slipped into the house, spun the wool, and hid in some bushes before he returned. Hayletha was delighted at the industry

2 Lummi.

3 Lushootseed.

of his image, and the next time he went out left more wool for it to spin; but this time the girls spun the wool and burned the image. He was sorely puzzled when he found only its ashes; but called out at last 'I don't know who you are but come out and let me see you.' The girls came out from their hiding place, and Hayletha married Tayakamat's daughter. From their children sprang the groups of Indians around Duncan, while the Sooke Indians are descendants of Tayakamat's other children. Later X̣e.ls created at Malahat a man named Hwan'am and his wife, and from this last couple came the Malahat and Saanich Indians. In the earliest times all alike ranked as nobles, but after they began to raid and enslave one another some of the nobles married their slaves, and the offspring of the mixed marriages became the commoners. (Richling 2016: 44)

Figure 6.2 X̣e.ls and some of his works, according to David and Mary Latasse

Source: Diagram by Chris Arnett, background image derived from Google Maps.

The above account is only a sketch of a more detailed narrative (see Bouchard and Kennedy 2002: 135–44) but it explains the common origin, equality and kinship of the speakers of the different languages of the Salish Archipelago (through XÁ,EL,S, the Creator of all things), and a marked change from when 'all alike ranked as nobles' to 'began to raid and enslave one another'. A long period of intact cultural teachings was terminated by non-Indigenous teachings. One could argue (as many do) that raiding and slavery were ancient practices (Ruyle 1973; Leland 1997), but archaeological evidence is lacking (Leland 1997: 202–5). An alternative explanation is that the value system that existed prior to contact was modified by a non-Indigenous system based on inequality, introduced to and appropriated by Indigenous participants. This might explain the intense period of warfare, stockaded villages and slave raiding beginning around 1790 (Angelbeck 2008) but little attention has been paid to non-Indigenous agency in the process.

The Latasses' broad historical brush includes the story of the first man (inventor of the potlatch and other institutions narrated in greater detail elsewhere), but more importantly identifies a situation foreign to the Indigenous teachings: the commodified slavery of a European and American capitalist economy, leading to the creation of a class of 'commoners'. The fact that the Hul'q'umi'num' word *skwy'uyuth'* means 'slave' and 'prisoner of war' could have originated at this time (Hukari and Peter 1995: 78).

No slaves were identified among the Nuu-chah-nulth by the first Spanish explorers, who certainly would have recognised any. Moreover, Nuu-chah-nulth oral traditions associate Europeans, particularly Spanish, with the introduction of slavery (Leland 1997: 213). The appropriation of a non-Indigenous value system of capitalism and the commodification of beings including people led directly to the tiered class system encountered and described by early non-Indigenous arrivals who came from slave cultures themselves.

John Humphreys was born in 1865, the son of Englishman John Humphreys and Tistnamot (Amelia) a Hul'q'umi'num'-speaking woman from Kwamutsun. In his accounts of Quw'utsun Mustimuhw history written in the early twentieth century (n.d.: unpaginated) he describes an old social order characterised by family units equal in status but different in size:

> Each village had its own chiefs. Not one but several, for each family had its own chief. The villages were divided into families, each family consisting of a man, his sons and grandsons and as there was very little disease in those days the families increased rapidly and grew to great strength. If a house became too small to hold the whole family an addition was made to it, and sometimes a family occupied half a dozen houses or more. The chiefs often held meetings to make laws for hunting and fishing, as this was their means of living and consequently great care had to be taken in making these laws. Each family had its own hunting and fishing ground, and the tract of land each one claimed was proportionate to the size of the family. If it was a large family it had a large tract of land and vice versa. Sometimes one family had to its credit many miles of land for its hunting and fishing purposes. If one Indian found another man poaching on his land he immediately shot and killed him as it was within the law to do so. There were very few laws but what there were strictly kept. As the punishment for breaking them was very strict. Also, every man could get his own property and consequently did not have to trespass on his neighbours land.

This is another remarkable document by an informed elder describing Indigenous social structure, self-governance, land tenure and resource management. Family groups are equals and differentiated only by size. In other words, a large house may mean a large family rather than a separate class of nobles, until nineteenth-century Indigenous elites began to appropriate non-Indigenous housing values. Humphreys makes no mention of slavery or classes or disease, which suggests that the account describes pre-colonised, sovereign Indigenous people prior to plague, demographic collapse and a new economy. He subtitled his writing 'As told by the Indians Themselves'. As Latasse and Humphreys indicate, the Central Coast Salish were once composed of families of equally ascribed status as human beings with 'all alike as nobles' where 'every man could get his own property and consequently did not have to trespass on his neighbours land'. It was not so much for fear of punishment that 'the few laws were strictly kept'; rather that such behaviour was, given the origin stories and protocols, unthinkable.

Albert Wesley (Jenness 1934/5: 276) of Snuneymuxw ancestry provides a third Indigenous historical narrative to explain the origins of the *stacem,* a class of people created among Central Coast Salish as the result of historical events precisely associated with the arrival of non-Indigenous people and their pathogens:

> During the great winter [in the time of Wesley's great grandfather, c. 1800, and probably earlier] so many people died that numerous families became entirely extinct, while of others there remained only babies who were found sucking their dead mothers' breast. The survivors rescued these and brought them up.; but since they had no parents or kinfolk (and in some cases their parents were not known). They were called *stacem* (a word meaning low people) and not allowed to marry people of good birth. They built small houses for themselves near the Big House that sheltered groups of closely related families, and they were called on to help those families whenever their services were required. In return they received protection. They were not slaves however, not *sqauias* [*skwuy'uyuth*], for they could not be bought and sold; they were as much a part of the village as the people they served. There were *stacem* at Nanoose, at Sechelt, Kuper Island and Coquitlam. At Sechelt and Nanoose there were none before the great winter … The children of a *sqwaieth* [*skwuy'uyuth*] and of a man of good family could not be sold, but were of low caste, like *stacem*. There were really only two classes – the high class and the low class, over and above a few slaves.

This is a very important account both for its reference to the historical appearance of *stacem* as an actual class (as opposed to a system) sometime in the late eighteenth or early nineteenth century, as the result of 'the great winter'. 'Great winters' as natural disasters are mentioned in Salish historical sequences before smallpox, and Duncan McLaren (2006: 201) suggests that, as chronological markers, they could represent the onset of the Little Ice Age some 1,200 years ago. However, the timing of Wesley's narrative 'in the time of his great grandfather' (who was probably born c. 1800 or earlier) 'with numerous families becoming extinct', combined with other Indigenous-derived accounts from southern Vancouver Island (Deans 1900) and Salt Spring Island (Harris 1997: 9–10), suggests that Wesley is describing the winter smallpox epidemic of 1782 and its aftermath.

This may have led to the creation of a class of people, the remnants of devastated families taken in by other families, and a process under way just as non-Indigenous people began to appear in greater numbers. Suttles (1987: 11–13) described these people from the perspective of his mid-twentieth-century informants who described *stacem* as 'poor people' without the proper knowledge and advice formerly passed down through families since ripped asunder by smallpox. Far from an institution, these people were the result of historical events.

Wesley notes the difference between the commoners and the slaves and that the child of a slave and a noble was a commoner, exactly as described above by Latasse above. His comment 'over and above a few slaves' mirrors the 'inverted pear' of Wayne Suttles model with most people 'high born' (the original 'all nobles alike'), a smaller group of *stacem*, and an even smaller group of slaves (Suttles 1987). Slaves were never numerous among Central Coast Salish. Suttles' model is absolutely applicable to the mid-nineteenth-century Salish Archipelago. Prior to that time period, we have only Indigenous sources.

The above three historical narratives by esteemed elders portray two distinct time periods: one when Indigenous teachings were intact, and another when they began to erode. Even though non-Indigenous people are not mentioned specifically in any of the accounts, the association of disease, raiding and slavery suggest that the creation of the class system observed in the nineteenth century was influenced by non-Indigenous contact. The Salish Archipelago was a centre of significant demographic collapse in the late eighteenth century (Harris 1997). Epidemics began around 1782 and estimates of fatalities are in the 90 per cent range. The impact on institutions and people was catastrophic, as Wesley indicates. Archaeological modelling of Central Coast Salish culture has failed to see the direct connection between the epidemic, the utter collapse of the pre-contact Coast Salish culture, and its material signature.

Theorising *Hwunitum*

The Hul'q'umi'num' word for a non-Indigenous person of European ancestry is *hwunitum'* 'arriving out of nowhere', a name that reflects their sudden appearance in a long-established Central Coast Salish world (Florence James, personal communication, ~2001). Over the past century and more, archaeology posits a history of this area based on the interpretation of material culture; stone and bone tools, seated figure bowls, beaded necklaces, post moulds, trench embankments, portable art, rock art and occupations areas (shell middens), and clam gardens (Angelbeck, this volume). This evidence is often interpreted in light of nineteenth- and twentieth-century data or cultural ecology to explain the physical remains in terms of inequalities, of a class society riven with conflict. H. Martin Wobst (2006) demonstrated a number of years ago how academic archaeology often interprets the past in terms of what is familiar (say, contemporary class conflict) without fully considering cultural perspectives that might offer other avenues for interpretation.

The ancient protocols between the sockeye and Indigenous people allowed a sustainable fishery for millennia (Mitchell 1970; Fritz 2012; Claxton 2015). The beginnings of what archaeologists call the Developed Northwest Coast Pattern (as exemplified by nineteenth-century Central Coast Salish culture) is correlated with the earliest evidence of salmon storage and assumptions about human behaviour. Thus, cultural development involves more than simple environmental determinism and must include the interaction of resources, resource technology and social structure, and likely involved coercion and violence (Matson and Coupland 1995: 154).

The assumed 'coercion and violence' is most often perceived in terms of material inequalities, which are assumed to be evidence of social inequalities along with the nineteenth-century ethnographic. While Grier (2007) cautions against uncritical use of the ethnographic record to interpret the past, the assumptions of a Central Coast Salish culture of 'material wealth-based inequality', is projected ever deeper into the past (Coupland et al. 2016). Instead of inequality, how might equality be represented in the archaeological record prior to colonisation?

Histories narrated by Indigenous elders suggest a different interpretation of history, supported by the archaeological and the historical record. Contrary to these Indigenous accounts, Matson and Coupland (1995: 306) state, without argument, that 'hereditary social inequality was not a product of historical contact but rather the long-term evolutionary trends begun long ago'. Contingency-based, agent-centred historical models are used by archaeologists to interpret the distant material past but are rarely used to account for the recent. Taken together, the narratives of three early twentieth-century Coast Salish historians, David Latasse, John Humphreys and Albert Wesley offer an Indigenous perspective on this past nineteenth-century social structure and the difference between the two.

Conclusion

Origin stories suggest that the class system was foreign to Indigenous society prior to adaptation as the result of contact with non-Indigenous sources. This is not to say that conflict never occurred in pre-colonial times. The origin stories and other oral traditions and histories are full of examples to show that the balance between the positive and negative forces of existence required constant vigilance, struggle and perseverance (Atleo 1999, 2004, 2011).

Local Indigenous theory preserved in place names, stories and oral traditions brings an important Indigenous and historical perspective to non-Indigenous methods. Heather Harris (2006: 35), quoted at the beginning of this chapter, reiterates the significance of Indigenous theory as a method:

> Accepting Indigenous methods of creating and preserving knowledge and adding that knowledge to that obtained by archaeological methods is bound to expand archaeological horizons. Ideas gleaned from Indigenous knowledge and thought are likely to expand and transform archaeological theory and indicate new directions for archaeological investigation and new ideas for archaeological interpretation.

Indigenous theory provides a different interpretation about the nature of existence than others that are evolutionary based. One sees interconnectedness and equality, the other, disconnection and inequality. As Atleo (2004: iv) explains:

> The notion that all things are one stems directly from assumptions found in Nuu-chah-nulth [and Central Coast Salish] Origin Stories that predate the conscious historical notion of civilization and scientific progress, making them very important data to understanding any historical processes.

In the late eighteenth and early nineteenth century, to survive in the new world economy, Indigenous people appropriated non-Indigenous ideas and values (commodities, social structure). Relatives (people, animals, plants, etc) were commodified. Pre-encounter Indigenous social structure was fundamentally different from the non-Indigenous world-view because it included non-humans, animate and inanimate, visible and invisible, extending the social beyond human beings to include everything; land, animals, birds, fish etc. The impediments of non-Indigenous culture are vast but not insurmountable if Indigenous ways of knowing, based on responsibility and sustainability, can be applied. Despite catastrophic collapse, Indigenous people proved incredibly resilient and as more non-Indigenous people come to terms with the history, an Indigenous resurgence is already under way.

References

Angelbeck, W.O., 2008. They Recognize No Superior Chief: Power, Practice, Anarchism and Warfare in the Coast Salish Past. University of British Columbia (PhD thesis).

Atleo, E.R., 1999. 'Message from the Guest Editor: A Long-term Perspective of First Nations Educational Experience.' *Learning Quarterly* 3/1: 2–5.

——, 2004. *Tsawalk: A Nuu-chah-nulth Worldview*. Vancouver: UBC Press.

——, 2011. *Principles of Tsawalk: An Indigenous Approach to Global Crisis*. Vancouver: UBC Press.

Bouchard, R. and D. Kennedy (eds), 2002. *Indian Myths and Legends from the North Pacific Coast of America. A Translation of Franz Boas's 1895 Edition of Inianische Sagen von der Nord-Pacischen Kuste Amerikas*. Vancouver: Talonbooks.

Carlson, R. and P.M. Hobler, 1993. 'The Pender Canal Excavations and the Development of Coast Salish Culture.' *BC Studies* 99: 25–52.

Claxton, N. XEMTOLT, 2015. To Fish as Formerly: A Resurgent Journey Back to the Saanich Reef Net Fishery. University of Victoria (PhD thesis).

Coupland, G., D. Britton, T. Clark, J.S. Cybulski, G. Frederick, A. Holland, B. Letham and G. Williams, 2016. 'A Wealth of Beads: Evidence for Material Wealth-Based Inequality in the Salish Sea Region, 4000–3500 CAL B.P.' *American Antiquity* 81(2): 294–315. doi.org/10.7183/0002-7316.81.2.294

Deans, J., 1900. 'Traditional History of Vancouver Island: Being a Tale of the Terrible Sill-kous in its Connection to Mee-Acan, or Beacon Hill.' *Victoria Daily Times*, 22 December.

Fritz, J., 2012. The SWELSWÁLET of the WSÁNEĆ Nation: Narratives of a Nation (Re)Building Process. University of Victoria (PhD thesis).

Grier, C., 2007. 'Consuming the Recent for Constructing the Ancient: The Role of Ethnography in Coast Salish Archaeological Interpretation.' In B.G. Miller (ed.), *Be of Good Mind: Essays on the Coast Salish*. Vancouver: UBC Press.

Harris, C., 1997. *The Resettlement of British Columbia: Essays on Colonialism and Geographical Change*. Vancouver: UBC Press.

Harris, H., 2006. 'Indigenous Ways of Knowing as Theoretical and Methodological Foundations for Archaeological Research.' In C. Smith and H.M. Wobst (eds), *Indigenous Archaeologies: Decolonizing Theory and Practice*. London: Routledge.

Hukari, T. and R. Peter, 1995. *Hul'qumi'num' Dictionary*. Duncan, British Columbia: Cowichan Tribes.

Humphreys, J., n.d. 'History of the Cowichan Indians as They Tell it Themselves.' MS. F/3/H88. Victoria: BC Archives.

Jenness, D., 1934/5. 'Saanich and Other Coast Salish Notes and Myths.' VII-G-9M, Box 39, fl. Victoria: BC Archives.

——, 1955. 'The Faith of a Coast Salish Indian.' Victoria: The British Columbia Provincial Museum and Records Service (Anthropology in British Columbia Memoir 3).

Leland, D., 1997. *Aboriginal Slavery on the Northwest Coast of North America.* Berkeley and Los Angeles: University of California Press.

Masco, J.,1995. '"It is a Strict Law that Bids us Dance": Cosmologies, Colonialism, Death, and Ritual Authority in the Kwakwaka'wakw Potlatch, 1849 to 1922.' *Comparative Studies in Society and History* 37: 41–75. doi.org/10.1017/S00104 17500019526

Matson, R.G. and D. Coupland, 1995. *The Prehistory of the Northwest Coast.* Orlando, Academic Press.

McGhee, R., 2008. 'Aboriginalism and the Problems of Indigenous Archaeology.' *American Antiquity* 73(4): 579–97. doi.org/10.1017/S0002731600047314

McLaren, D., 2006. 'Uncovering Historical Sequences in Central Coast Salish Narratives.' In R. Carlson (ed.), *Archaeology of Coastal British Columbia: Essays in Honour of Professor Philip M. Hobler*. Burnaby: SFU Archaeology Press.

Mitchell, D.H., 1971. 'Archaeology of the Gulf of Georgia Area: A Natural Region and its Cultural Types.' *Syesis* 4/1.Victoria: The British Columbia Provincial Museum.

Montler, T., 2018. *SENĆOTEN: A Dictionary of the Saanich Language*. Seattle: University of Washington Press.

Moon, B.J., 1978. 'Vanished Companions: The Changing Relationship of the West Coast People to the Animal World.' In B. Efrat and W.J. Langlois (eds), *Nutka: Captain Cook and the Spanish Explorers on the Coast.* Victoria: BC Provincial Archives.

Moss, M., 2011. *Northwest Coast: Archaeology as Deep History.* Washington, DC: Society of American Archaeology.

Richling, B. (ed.), 2016. *The WSANEC and their Neighbours: Diamond Jenness on the Coast Salish of Vancouver Island, 1935.* Oakville: Rock's Mills Press.

Ruyle, E., 1973. 'Slavery, Surplus and Social Stratification on the Northwest Coast: The Ethnogenetics of an Incipient Stratification System.' *Current Anthropology* 14: 603–31. doi.org/10.1086/201394

Smith, T., 1999. *Decolonizing Methodologies: Research and Indigenous Peoples.* London: Zed Books.

Suttles, W., 1987. *Coast Salish Essays*. Seattle and London: University of Washington Press.

White, E., 2006. *Legends and Teachings of Xeel's, the Creator*. Vancouver: Pacific Educational Press.

Wobst, H.M., 2006. 'Power to the (Indigenous) Past and Present! Or: The Theory and Method Behind Archaeological Theory and Method.' In C. Smith and H.M. Wobst (eds), *Indigenous Archaeologies: Decolonizing Theory and Practice.* London: Routledge.

7

The Longevity of Coast Salish Presence: An Archaeological History of the Salish Sea

Bill Angelbeck

Coast Salish peoples have lived within the Salish Sea for thousands of years, since a Time of Transformation long ago, according to their oral traditions (see Chapter 6, this volume). Archaeological evidence indeed extends back to arrivals from Siberia well over 14,000 years ago, crossing overland or boating along the shoreline.[1] People have lived in the region for generations upon generations, as maintained in oral histories. Over those millennia, cultures changed again and again, expressed in the technologies created or adopted, the types of housing, engagement with plants and animals, socio-political organisation and symbols displayed internally or to others.

Salishan cultures, and others of the Northwest, defy the distinctions between mobile foragers and settled agricultural societies. People lived in large villages and towns built of plank houses based largely upon an economic foundation of hunting, gathering and fishing. In contrast, hunters and

1 For contrast, the time of European presence in the region since the 1790s amounts to about 1.5 per cent of the time that people have lived here. The entire histories of established cities in the region—Seattle (1851), Victoria (1862), Vancouver (1886)—comparatively amounts to only 1 per cent of the Salishan history of occupation in the region. This is based minimally on the oldest site in the region, discussed below, which is only our earliest evidence. Undoubtedly earlier sites will likely be found, which will make these fractions even slimmer. This gives entirely different take on the notion from the 'Occupy' movement, with their slogan of 'We are the 99%'.

foragers in other parts of the world are highly mobile, shifting their camps as they move throughout their territories to harvest and hunt plants and animals as befitting the season.

The lush environs of the Salish Sea allowed large villages to be occupied for many months over the winter while foraying to other parts of their territories the rest of the year. Coast Salish communities participated in activities that go well beyond foraging, such as the intensive cultivation of wapato (Hoffman et al. 2016) and walled intertidal clam gardens (Lepofsky et al. 2015).

Unusual among foraging peoples, the bulk of people in Coast Salish society were considered noble or high class. They engaged in elaborate ceremonies, often with potlatching: occasions for earning names or titles, and a form of governance. House posts, monumental funerary art, and masks allowed artists to maintain traditional forms while adapting and experimenting with contemporary contexts and mediums. This heritage provided a connection between the ancestral past and generations of the present and future.

Here, I present an overview of the archaeological knowledge about the Salish Sea, highlighting major cultural periods of history.[2] This is a brief presentation covering a vast period which cannot do justice to the depth of this history and its variability. We see continuities and changes through time in environments, economies, technologies, social relations, political organisations, customary practices and arts. No doubt there are continuities and changes in value systems, spiritual beliefs and philosophies as well. I aim to show that strands from this long history are woven into contemporary forms of Salishan identity.

This chronological overview of archaeological evidence and interpretation of 14,000+ years of Indigenous presence highlights the vast longevity of occupation by Salishan peoples in the region, in contrast to the 200+ years since the presence of colonisers. It is a point of agreement—shared between archaeo-historical and Indigenous historical narratives—that Salishan peoples have occupied these coasts and islands for a time representing over 700 generations. These forms of evidence allow us to bear witness to the longevity of Salishan presence.

---

2 Throughout, for general statements, I draw upon some of the main overviews of Northwest Coast culture history, including: RG Matson and Gary Coupland's (1995) *The Prehistory of the Northwest Coast*, Kenneth Ames and Herbert Maschner's (1999) *Peoples of the Northwest Coast* and Madonna Moss's (2011) *The Northwest Coast: Archaeology as Deep History*. For examples of overviews of particular cultures organised by Indigenous groups themselves, see Carlson's (2001) *A Stó:lō Coast–Salish Historical Atlas* or Miller et al.'s (2020) *Cultural and Historical Atlas of the Upper Skagit Indian Tribe*.

Earliest Peoples of the Salish Sea

The first peoples to inhabit the Salish Sea arrived long ago, at the end of the Ice Age. Much of the planet's water was frozen in glacial sheets in the northern hemisphere. Lower sea levels allowed a land bridge to form between Siberia and Alaska, opening a route into North America amid ice sheets. Some argue that early exploration took place in watercraft along the coast, inhabiting refuges of kelp, which would have attracted numerous fishes and predators such as seals and sea lions (Fladmark 1979; Erlandson et al. 2007). First people encountered a continent filled with megafauna—mammoths, mastodons, giant sloths, large bison and more. The archaeological sites from these times suggest people living in small bands that were highly mobile across the landscape, predominantly making their living by hunting megafauna.

The earliest evidence in the Salish Sea region is from the Manis site on the Olympic Peninsula, dating to 13,800 cal BP (Waters et al. 2007) (Figure 7.1).[3] Hunters killed a mastodon—a bone spearhead still embedded in its skeleton. On Orcas Island, there is evidence of bison hunting about 13,500 cal BP (Kenady et al. 2007). While there is no presence of tools, there are distinctive fractures of bone and percussion scars that indicate human butchering, rather than toothmarks of carnivores.

The first defined culture of the area is Clovis, with the majority of sites dating to 13,150 to 12,850 cal BP (Waters and Stafford 2007; Fiedel 2014: 14). Clovis spearheads (Figure 7.2) have been found throughout the Salish Sea, including eight fluted spearheads found throughout Puget Sound and the Lower Fraser region, as at Stave Lake (Croes et al. 2008; McLaren et al. 2020). Clovis and other early groups during this early Paleoamerican period predominantly lived in small bands. They hunted megafauna, especially mammoth, which required communal hunts (Frison 1987).

---

3 Archaeologists typically cite dates as BP, which refers to radiocarbon years 'before present', with 'present' being AD 1950, the approximate time radiocarbon analysis began. This standardises the archaeological dates without need to adjust the date each year, allowing for better consistency across publications. 'Cal BP' further indicates that the radiocarbon dates have been calibrated to Western calendar years before present, as the raw radiocarbon results need some adjustment in order to match our calendar years.

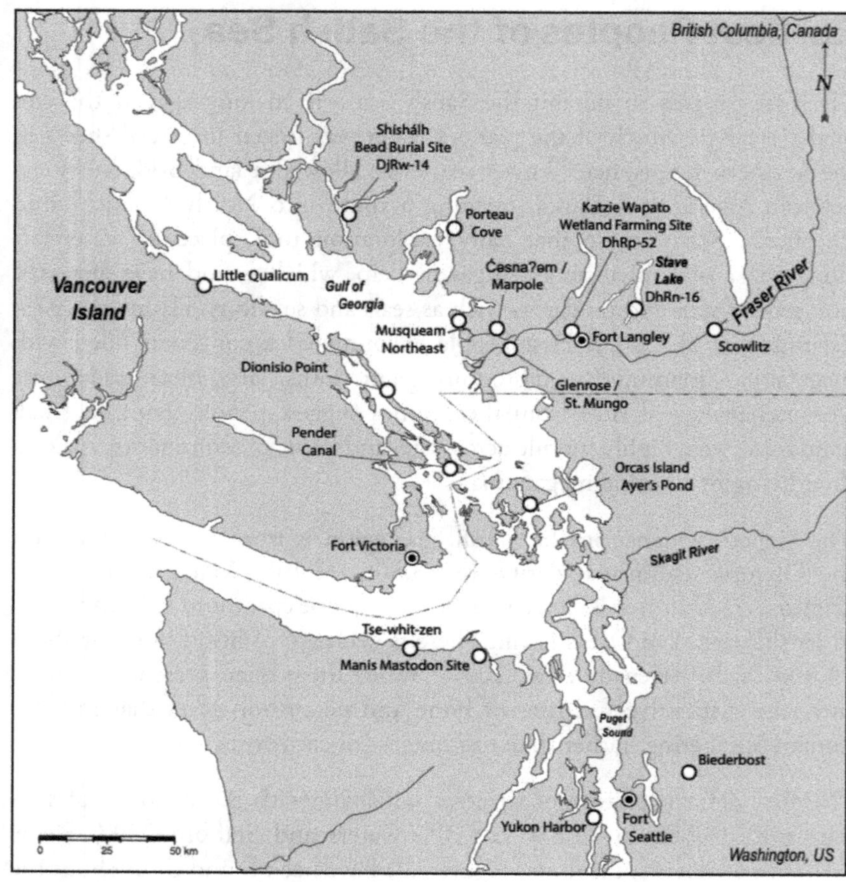

Figure 7.1 Central Salish Sea, with archaeological sites mentioned in the text

Source: Bill Angelbeck.

Toolkits reveal that these early groups hunted smaller mammals and birds, and used tools for processing plant foods (Boldurian and Cotter 1999: 114). Clovis peoples also exhibited cultural and ritual aspects, applying red ochre paint to tools and using it during burial rituals. Another early culture, known as Protowestern Tradition (13,200–9,500 cal BP) was also present in the Northwest, exhibiting a toolkit involving large stemmed and lanceolate points (McLaren et al. 2018).

Figure 7.2 Spearheads from the Paleoamerican period

Notes: Clovis spearhead from Yukon Harbor, site 45–KP–139; Western lanceolate from Stave Lake, site DnRh-46; Western Stemmed from Stave Lake, site DhRn-16.

Sources: Croes et al. (2008); McLaren (2017).

The end of this early period is marked by a shift to hunting other mammal species, primarily because of the decline of megafauna at the end of the Pleistocene. Indeed, as some archaeologists argue (e.g. Martin 1984; Surovell et al. 2005), human predation contributed to the rapid decline of most megafauna populations, which were dwindling due to changes in climate with the Holocene era. It is certain that older sites will be found throughout the region.

Indeed, recent studies indicate that people have inhabited North America 21,000 to 23,000 years ago, with footprints found in New Mexico, and the common passage would have been along the shores through the Northwest (Bennett et al. 2021). This would be in accordance with the predictions of Indigenous archaeologists like Paulette Steeves (2021) who maintains that such finds could extend much further back in time, given the time depth indicated in oral histories.

Early Holocene Lifeways

With the onset of the Holocene, peoples in the region gradually shifted their lifeways in adapting to changing environments. Peoples remained highly mobile, continuing much of the year in small bands that traversed the region as part of seasonal rounds throughout their territories. Prominent tools begin to take on regionally distinctive qualities. In the Salish Sea, there are a host of names used by archaeologists to characterise the lifeway of the period from about 10,000 years ago to about 4,500 years ago, including: the Archaic period (Ames and Maschner 1999); Old Cordilleran Culture (Matson and Coupland 1995); Olcott Phase (Chatters et al. 2011) or the Pebble Tool Tradition (Carlson 1996a, 1996b). Common tools included large willow-leaf-shaped spearheads, knives and large cobble choppers. Evidence suggests that the specific focus of subsistence and economic efforts varied throughout the Salish Sea.

Porteau Cove in Howe Sound contains one of the earliest sites in the region, located within a rock shelter about 85 m above the shoreline. The site dates to 9,707 to 9,544 cal BP (Reimer 2012). Another of the early sites in the region is at Bear Cove, at the north end of Vancouver Island, which was initially occupied about 8,500 years ago (Carlson 1979, 2003). The assemblage of animal bones revealed a marine focus on fishing and the hunting of sea mammals.

The Glenrose site (Figure 7.1) also contains early evidence of habitation at a coastal site, about 8,500 to 5,500 years ago (Matson 1976). Evidence indicates that the inhabitants were mobile but within a broad territory, with travel annually throughout the seasons. They built light shelters, and typically operated as small, egalitarian groups that drew upon a wide range of resources. In the Fraser River area, the bones of hunted land mammals outnumbered those from the sea (Matson and Coupland 1995: 70–4).

The faunal remains at Glenrose reveal that the inhabitants occupied the site during late spring and early summer as a generalised hunting and fishing camp, with much of the remains coming from land mammals such as elk and deer. Salmon, eulachon and bay mussel shells were also fished and collected. The use of plants during these early periods becomes more difficult with increasing antiquity and therefore has not been studied as well, as with animal use.

Evidence at numerous sites indicates that peoples engaged in long-distance trade networks at this time. This is trackable with obsidian, a volcanic tool stone highly prized for its sharpness and shapeability. Its composition allows for identification as to its original montane source. Trade from places as far as Mount Edziza or Anahim Lake in British Columbia (BC) or Three Sisters, Oregon, is evident from sites that age back nearly 10,000 years in the region (Carlson 2013; Springer et al. 2018).

Peoples of the earliest periods are sometimes contrasted as being simple hunter-gatherers, as opposed to the complex hunter-gatherers or cultivators of the historic Northwest Coast. As stated by Mackie et al. (2011: 90): 'Under this narrative, the founding archaeological cultures are too easily conceived of as an impoverished default condition: complex hunter-gatherers stripped of their complex trappings.'

Instead, these ancient cultures should be considered as being as complex as any other culture, as befitting their choices within their environment to be highly mobile and lighter in possessions. Indeed, elements of these mobile lifeways continue in later millennia as part of seasonal forays throughout their territories, to which we'll turn next.

By 4,500 years ago, with increasing stabilisation of the shoreline, beaches show evidence of large shell middens—with soils that are black due to organic content from occupation and dense with disposed shells from mussels and clams, providing greater preservation for some organic artefacts in the archaeological record. This is regarded as the beginning of the early Pacific period in many areas.

Cultures of the Pacific Period

The Pacific period encompasses the last few millennia of the Holocene (Ames and Maschner 1999). About 4,500 years ago, Salishan peoples began to maintain larger settlements for their winter homes. While previous cultures were diverse in their economic practices, involving terrestrial and marine resources, some groups began to focus intensively on certain resources during the Pacific period, notably on salmon. This allowed for substantial settlements and elaborate architecture, including the presence of a distinctive art form (Figure 7.3).

Figure 7.3 Human figurine carved from elk antler, Glenrose site

Note: From the Glenrose site, St Mungo Phase, ca. 4,300 to 3,300 BP.

Sources: Holm (1990: 61); Matson (1976).

Most assessments of cultures at this period consider that the size of groups was increasing, given the extent and intensity of occupation at coastal sites. Croes and Hackenberger (1988) offer that storage of surplus goods increased, particularly for smoked salmon and shellfish, which would have been available near coastal villages. Such processes contributed to population growth in the region, allowing for greater efforts towards other endeavours (Ritchie et al. 2016).

In the Sunshine Coast, shíshálh area (site DjRw-14), the burial of four individuals indicates how wealthy certain people became during that time. The grave of one individual contained over 350,000 stone and shell beads, arranged in strands over the body, alongside red ochre. The amount of labour involved to create such goods is extensive (Coupland et al. 2016: 308).[4]

4 Consider, for instance, this in contemporary terms, that if a bead takes an hour to shape and drill from shale or other stone, multiple the hourly wage for that craftsperson by 350,000 beads (likely higher than our current minimum of $15). Whatever the wage determined, it readily reveals wealth in the millions of contemporary dollars in labour—which was buried with them.

These elaborate burials indicate a shift from more egalitarian distributions to instances of great wealth. The burials convey status and exhibit the value of those individuals to the community. Notably, the adjacent infant burial is distinctive for not containing any beads or grave goods. This suggests that the status of those with large numbers of beads had to be earned, not simply inherited.

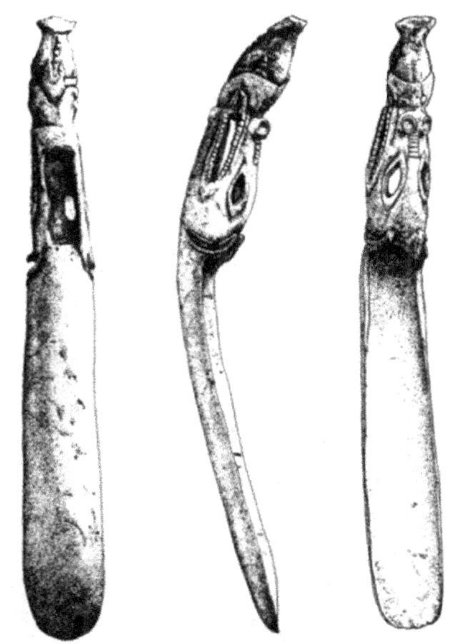

Figure 7.4 Wolf-rockfish spoon carved from elk antler, Pender Canal site

Note: Recovered from the Pender Canal, site DeRt-2, ca. 3,600 BP.

Source: Carlson and Hobler (1993: 46).

At Pender Canal in the Gulf Islands, within an excavated cemetery from 3,500 years ago, graves revealed individuals with labrets, an adornment piercing the lower lip. Since only a few bore such items, archaeologists commonly interpret labrets as demarcating higher status. The diversity of items uncovered also reveals some elements of their funerary ceremonialism.

Carlson and Hobler (1993: 49) inferred that people may have been feasting the dead, given the presence of elaborately carved elk-antler spoons (Figure 7.4). Indeed, these artefacts from over 3,000 years ago exhibit connections with the ethnographically documented and continuing practices of burning food to feed ancestors ritually (McHalsie 2007: 118–21).

The initial evidence of social inequality apparent at this time intensified in later periods, particularly during the Marpole period. This culture began about 2,400 years ago and lasted over a millennium, possibly to 1,100 years ago, ca. AD 900 (Mitchell 1971; Burley 1980; Clark 2013). This period is often regarded as a cultural florescence, given the presence of exquisite art forms. Stone sculpture was highly elaborate, especially with seated human figure bowls (Duff 1975) (Figure 7.5), seemingly used in ceremonies, some with ochre staining. These sculptures embody the distinctive elements of classic Northwest Coast art styles, features that have continued until this day.

Figure 7.5 Marpole-era human figure bowl with rattlesnake spine, from Victoria

Source: Duff (1975: 54–5, 172).

During this period, there are indications of increased regional interaction, involving the exchange of ideas and art forms, as evidenced at Dionisio Point (Grier 2003). This increase in exchange indicates alliance formation between communities, as households intermarried across long distances (Suttles 1987). The archaeological site which defined this culture contains the remains of a large Salishan village situated on the north arm of the Fraser River, known to Musqueam as Cəsna'əm. The increasing storage of salmon, dried berries and other foods during this period undoubtedly contributed to population rise, ceremonialism and the arts.

The shed-roof plank house common among Coast Salish peoples had its origins during this period. House layouts reveal large post and beam architecture. Multiple hearths suggest that numerous family groups composed the household (Coupland et al. 2009). Typical artefacts included unilaterally barbed harpoons for sea mammal hunting, nipple-topped hand mauls for woodworking, and chipped stone points. Bone tools are common as well, with awls and needles for making clothing and bone hooks for fishing.

In one burial mound at Scowlitz dating to 1400 cal BP, grave goods include numerous exotic materials, such as dentalium shell beads from the outer coast, black abalone shell pendants likely from California and copper pendants probably from Alaska. Due to the wide range of items within that mound, it is evident that Marpole peoples participated in a regional exchange network (Blake 2006).

Items of high value, such as carved-stone bowls, appear to have been exchanged among elites.[5] Indeed, the rank of elite individuals appears to have increased. While labrets were generally no longer used, cranial deformation, applied during infancy, became a marker of status. Some archaeologists interpret the practice as markers of a stratified, elite class (Matson and Coupland 1995; Angelbeck and Grier 2012).

Although large villages became established, the inhabitants were still mobile much of the year, with seasonal forays for hunting and fishing within various family camps from spring to fall. During winter, households gathered in plank house villages. Such large villages are not generally characteristic of foragers, whose way of life is simply too mobile to allot time for the building of substantial residences or storage facilities in any one location.

Commonly, for the main wood to construct their buildings, they sought western red cedar. The wood was preferred for its resistance to rain and weathering. Cedar splits readily along the grain—planks of wood could be cleaved off from standing trees. They even steam-bent cedar planks to form watertight boxes for cooking and storage, often carved with elaborate art on the outside. From cedar logs, Salishan carvers formed canoes, house posts and roof beams.

They stripped cedar bark as well, and processed it for weaving baskets, mats and clothing; plus, they worked cedar roots and withes to make ropes, baskets and nets for fishing and other uses. Often, Salishan cedar-gathering practices have left their mark on trees that remain alive and standing even hundreds of years later. These are called culturally modified trees (Stewart 1984a, 1984b; Earnshaw 2019), and these are present throughout the forests of the Northwest.

5 For example, a human seated figure bowl found on Vancouver Island contained images of rattlesnakes (see Figure 7.5), which are only found upriver in the Interior Plateau. On the other hand, a bowl around at Yale, in the Fraser Canyon, exhibits imagery of a seal (Duff 1975).

In rare cases, archaeologists have recovered cedar artefacts, such as baskets, tool handles, nets and boxes. These artefacts are preserved within waterlogged sites, which create anaerobic conditions, preventing organic deterioration. In the Gulf of Georgia, archaeologists have excavated wet sites at the Musqueam Northeast site, in southern Vancouver, and at Little Qualicum site (Bernick and Wigen 1990), on eastern Vancouver Island, as well as in nearby sites in Washington (for example, Ozette and Hoko River). These sites document the heavy use of cedar and other organic items.

Over 90 per cent of artefacts at wet sites were made from wood and plant fibres, while the remaining artefacts (less than 10 per cent) were made predominantly from shell, bone or stone (Croes 2003). A recent collaboration between Ed Carriere, a Suquamish elder, and archaeologist Dale Croes (2017) has revealed the continuity of weaving styles over generations, from 2,000-year-old woven artefacts recovered in Puget Sound.

The Late Pacific Period

The beginning of the Late Pacific period, or the Gulf of Georgia phase (Mitchell 1971; Ames and Maschner 1999), about 1600 cal BP, involves a switch from the Marpole period to different forms of tools, arts and social organisation, as well as new ways interring their dead. Communities expanded their use of the landscape, drawing upon more specific ecological niches along river channels, bays and marshes (Thompson 1978).

Clam gardens have largely dated to this period, although their construction may indeed have been earlier (Lepofsky et al. 2015). Many groups also used fires to alter the forest landscapes, enhancing the growth and productivity of berries and other plants or burning to expand camas prairies (Lepofsky and Lyons 2003); evidence suggests that this practice has origins millennia ago, as archaeologically shown on Valdes Island, dating back 4,500 years (Derr 2012). People also invested intensive efforts in the processing of camas, unearthed in broad roasting pits throughout the region (Lyons and Ritchie 2017).

There is also increasing evidence of the cultivation of plants and environments in recent decades (Deur and Turner 2005), especially with the evidence of wapato farming in Katzie territory at site DhRp-52 (Hoffman et al. 2016). Such studies have contributed to our understanding of how Coast Salish and other Northwest Coast peoples were cultivators of certain resources, combined with substantial storage practices; this is one reason that they blur anthropological boundaries between foragers and agriculturalists.

A new technology was widely adopted at the beginning of the Late period: the bow and arrow. Some economic opportunities likely broadened with this technology (Angelbeck and Cameron 2014; Rorabaugh and Fulkerson 2015), and this likely aided their expansion into further ecological niches. The spear and dart were still present but likely for more limited uses, such as in hunting specific species or for mêlée combat.

Salishans occupied plank house villages throughout the region. Peoples up the Fraser River, such as Stó:lō territory, lived in semisubterranean pit houses, the common winter structure of Interior Plateau peoples (Lepofsky et al. 2009). This sharing of architectural styles indicates the degree of interaction and relationship between Interior and Coast Salish peoples. In shifting to different seasonal camps throughout the year, people participated within a variety of social structures, some egalitarian (like small family gathering camps) and some hierarchical (as with prominent fishing camps during peak runs) (Angelbeck 2017).

The practice of cranial deformation of infants continued, but in the Late period, expanded to encompass the majority of individuals in the society (Angelbeck and Grier 2012). Yet, it was still a marker of high status. In this way, cranial deformation is a way to archaeologically track the development of the 'inverted pear' shape of their social structure, as famously phrased by Wayne Suttles (1987) to indicate how the Coast Salish had more noble individuals than commoners and slaves.

Suttles suggested the 'inverted pear' structure was the opposite to that of a pyramid, common in hierarchical and capitalist societies. Wealth was broadened for the bulk of society through potlatching distributions. About 600 to 800 years ago, the population appears to have increased substantially, which surely contributed to these shifts in social dynamics throughout the region (Ritchie et al. 2016).

The changes in social organisation are also reflected in the care invested in cemeteries. The manner of burial shifted from in-ground burials within shell middens to above-ground burial cairns and mounds. Some comprise very large cemeteries including hundreds of cairns that form expansive mortuary landscapes, such as those found along the coast of southern Vancouver Island (Mathews 2014).

Art also reflects distinctive changes in the Late period. Some noted a shift from the elaborate curvilinear art of Marpole to more geometric and naturalistic forms in the Late period (Mitchell 1971: 49; Thom 1998). Even so, a broad analysis of art in the Coast Salish world recognises strong continuities that maintain aesthetic and artistic threads across the Salish region throughout late millennia (Holm 1990).

This Late period is also marked by the proliferation of fortifications, indicating that warfare occurred (Moss and Erlandson 1992; Angelbeck 2016). On scores of remote and naturally protected coastal spits, bluff promontories or rocky headlands, Salishans further enhanced such natural defences with the construction of trenches, embankments and palisades. Walled forts appear related to the adoption of the bow and arrow about this time in the shift towards projectile combat.

Warfare also contributed to changes in social organisation, such that some survivors could be taken as slaves, although historically the number of slaves in the Coast Salish area was comparatively less than northern groups, where raiding was a more prominent aspect of their lifeway, especially during the early colonial period (Donald 1997).

After contact with Europeans, communities suffered greatly with repeated waves of smallpox epidemics, some of which reached the region as early as 1782, prior to the arrival of the Spanish into the area in 1790 (Harris 1994). Many bands were reduced substantially, with in some cases a loss of over 80 per cent of the community members. Some communities found it necessary to amalgamate their villages to accommodate such losses (Carlson 2010).

It appears that warfare increased with the increased disruptions to Coast Salish lifeways due to the fur trade and increased access to firearms, combined with the loss of populations due to disease. The Kwakwaka'wakw Lekwiltok (Ligwiłda'x̱w) took advantage of such events, raiding villages throughout the Salish Sea for several decades after European contact, often capturing slaves (Carlson 2010: 215–6).

These new developments contributed to communities shifting their village locations to more naturally protected areas, such as along river sloughs rather than open on the seashore. In the decades after contact, with the onset of the fur trade, Coast Salish peoples adopted new materials and technologies, such as iron knives, axes, kettles, mirrors and glass, as well as firearms.

Some communities also moved closer to fur trade forts to take advantage of their position as intermediaries, even marrying those in the forts, fitting their long tradition of marrying distant others for increased access to resources or security. For these marriages, women engaged in prominent roles as intermediaries between settler and Indigenous communities (Wellman 2017). In such ways, historical developments are associated with and manifested in changing patterns in the archaeological record.

Contemporary Context of Archaeological Heritage

The presence of archaeological sites in the Salish Sea is important, as is the heritage of all peoples. To descendant First Nation communities, however, there is often an obligation to their ancestors, shown through their care and concern for sacred sites and cemeteries. For instance, among the Hul'qumi'num, they are 'socially obliged to undertake the stewardship of their ancestral family remains so that they may maintain the appropriate reciprocal relations between the living and the spirit world' (McLay et al. 2004: 156).

The presence of archaeological sites and culturally valued places are chronotopes (*sensu* Basso 1996) in the landscape wherein the legacy of ancestors past perpetuates into the present. Archaeological sites are present manifestations of how Indigenous ancestors occupied and related to their territories in the past. In protecting or maintaining these sites, such actions embody an active engagement between people with landscapes and ancestors, which further enhances the tradition for the future.

Oral histories about such sites often provide a 'poetics of place' that helps Salishans to anchor their identities, providing a tangible connection between people and territory (Elsey 2013). For such reasons, Salishan peoples can exhibit a stronger connection to their territories when compared to settler populations, who are peoples defined by colonial mobility.[6] It is incumbent

6 The dislocation of colonisers in this region is apparent in the naming practices of settlers, which often are not about the distinctiveness of places here, but references to the colonial homeland; for example, Victoria, New Westminster, Surrey. On the other hand, those places named after the Indigenous groups or individuals provide a linguistic quality that provides local character and distinguishes the region from other areas of North America; such as, Nanaimo, Chilliwack, Squamish, Duwamish, Seattle, etc.

upon settler peoples to respect the heritage in the landscape, if they are serious about reconciliation and making progress towards the goals outlined in the Truth and Reconciliation Commission of Canada (2015).

This extends to how extractive industries—logging and mining principally—operate throughout the Northwest, withdrawing resources from the territories of Indigenous peoples, in many cases damaging or destroying their heritage. While archaeological surveys and investigations do document the presence of sites in areas slated for development, such studies can often read as simply remnant documentary glimpses of what was destroyed to permit resource extraction or building construction.

In such matters regarding handling the excavation of ancient sites, especially when necessary to unearthing of the remains of Salishan ancestors, it is important to recognise the primacy of descendant peoples' views. Archaeologists, whether of settler or Indigenous origin, can aid in addressing such situations as part of the application of their craft (Shanks and McGuire 1996). In turn, Indigenous elders and knowledgeable community members provide proper protocols consistent with how heritage matters ideally should be handled (Angelbeck and Grier 2014).

Such approaches are considered variously as collaborative, community-oriented, or Indigenous archaeologies (e.g. Martindale and Lyons 2014), and these contribute to better understandings of the archaeological record overall. Indeed, some case studies indicate that Indigenous participation in the archaeological investigations can even be a healthy or therapeutic practice that facilitates connections with the material artefacts, or the 'belongings' or 'heirlooms' of their ancestors (Schaepe et al. 2017). It is clear that Indigenous peoples are increasingly asserting their rights towards their heritage in the context of contemporary development and resource extraction, pursuing changes in how heritage is protected and how archaeology is conducted (Klassen et al. 2009).

The evidence of Salishan presence in the landscape is especially important for most Indigenous groups, since many in BC are engaged in ongoing treaty negotiations with the provincial and federal governments over land claims and usage. Extractive industries often actively erase the Indigenous heritage from the landscape, removing the evidence of their longevity of presence.

If we are to be committed to decolonisation, all of us should be conscious and consider how we handle issues about Indigenous heritage, ancestral sites and especially cemeteries, so that these are not acts of erasure, but ones of mutual respect between settler and Indigenous peoples as Canadians living within and around the Salish Sea.

Conclusion

The ancestors of Coast Salish peoples have inhabited their territory for over 14,000 years (Table 7.1). Cultures have varied throughout this time, beginning long ago with the earliest inhabitants, the nomadic hunters of Pleistocene megafauna. Subsequent generations began to restrict their seasonal forays, increasingly demarcating and settling into what became their traditional territories.

Figure 7.6 *Sulsultin* (spindle whorl), depicting two-headed snake from Chemainus

Table 7.1 Archaeological chronology of the Salish Sea

Years ago	Northwest Coast chronology	Salish Sea chronology	Some historical developments or details
<250	POST-CONTACT	Post-contact	Contact at 1790 CE (Spanish); 1792 CE (British); period of warfare ensues after contact; firearms introduced.
500	LATE PACIFIC	Sí:yá:m	Rise of powerful chiefs on the mainland Fraser River area.
1,000		Late period	Shift to above-ground mortuary structures.
1,500	MIDDLE PACIFIC	Late Marpole (Garrison)	Period of warfare; bow and arrow introduced; defensive sites and large burial features constructed.
2,000		Middle Marpole (Beach Grove)	Permanent winter plank house villages and markers of hereditary status appear.
2,500		Early Marpole – Old Musqueam	Villages concentrated at Fraser Delta.
3,000	EARLY PACIFIC	Locarno Beach	Increased coastal settlement and subsistence focus; some sites with shallow basins as remnants of shelters. Higher status individuals appear to wear labrets.
3,500			
4,000		Charles (St Mungo/ Mayne)	Burial of prominent individual with 380,000 beads in Shíshálh area.
4,500			Intensifying use of coastal sites with denser shell middens appearing.
5,000	ARCHAIC	Old Cordilleran /Olcott/ Pebble Tool Tradition	Highly nomadic culture with light shelters, given short duration of occupations at seasonal camps. Chopper tools were prominent, along with bi-pointed or leaf-shaped spearheads.
5,500			
6,000			
6,500			
10,500	PALEO-AMERICAN	Protowestern/ Clovis	Protowestern stemmed points found in the Stave Lake area.
12,000			Several Clovis spearheads found throughout region, especially Puget Sound.
12,500			Bison hunted and processed on Orcas Island.
13,500		Pre-Clovis	Hunted mastodon on Olympic Peninsula.
14,000			

Source: Author's tabulation.

With the onset of the Pacific period, local groups established villages throughout the area, and created forms of arts, practices and ceremonies that reveal the varied symbols expressing their cultural identities (Figure 7.6). These changed over the generations while maintaining solid threads of continuity throughout, linking descendants with ancestors, generation after generation.

This has largely been an archaeological account of Coast Salish history. Such knowledge can and should be expanded with Salishan oral histories of these origins (Arnett, this volume). Indeed, Rudy Reimer/Yumks (2012: 45–7) has presented an archaeological chronology in combination with Skwxwú7mesh oral historical conceptions of the past, as follows.

The *Sxwexwiyam* ('Mythical Time') is correlated with the late Pleistocene and early Holocene from 15,000 to 7,000 years ago. In *Xaay Xays,* 'Age of Transformation', mythical figures like Xe'els transformed the world and 'set things right', equated with Holocene stabilisation of environments and sea levels. *Syets* ('Recent Time') relates to events over the last few millennia.

Reimer's integration of oral histories with archaeological chronologies reveals commonalities, showcasing the utility of Indigenous histories for archaeological understandings. Much can be gained by collaborating and integrating knowledges and approaches toward such understandings of the past and protecting such heritage.

This archaeological overview, like the oral histories, reveals immense changes over long periods of time in both environment and culture. Yet, continuities of culture are evident throughout these changes. Peoples still seasonally access resources throughout their territories as done since the earliest inhabitants. Spiritual practices evident in Clovis period, as with the use of red ochre, continue to appear in burials 4,000 years ago and later, and are integrated in contemporary rituals.

Burnings for the ancestors, as at Pender Canal cemetery over 3,500 years ago, are still a common practice in reburial rituals. Artworks reveal connecting threads of style to the earliest zoomorphic and anthropomorphic carvings found in the archaeological record. In this way, we can witness the how traditions of the past interweave with present forms of Salishan identity and expression.

References

Ames, K.M. and H.D.G. Maschner, 1999. *Peoples of the Northwest Coast: Their Archaeology and Prehistory*. London: Thames and Hudson.

Angelbeck, B., 2016. 'The Balance of Autonomy and Alliance in Anarchic Societies: The Organization of Defences in the Coast Salish Past.' *World Archaeology* 48(1): 51–69. doi.org/10.1080/00438243.2015.1131620

——, 2017. 'Applying Modes of Production Analysis to Non-State, or Anarchic, Societies: Shifting from Historical Epochs to Seasonal Microscale.' In R.M. Rosenwig and J.J. Cunningham (eds), *Modes of Production and Archaeology*. Gainesville, Florida: University Press of Florida.

Angelbeck, B. and I. Cameron, 2014. 'The Faustian Bargain of Technological Change: Evaluating the Socioeconomic Effects of the Bow and Arrow Transition in the Coast Salish Past.' *Journal of Anthropological Archaeology* 36: 93–109. doi.org/10.1016/j.jaa.2014.08.003

Angelbeck, B. and C. Grier, 2012. 'Anarchism and the Archaeology of Anarchic Societies: Resistance to Centralization in the Coast Salish Region of the Pacific Northwest Coast.' *Current Anthropology* 53(5): 547–87. doi.org/10.1086/667621

——, 2014. 'From Paradigms to Practices: Pursuing Horizontal and Long-Term Relationships with Indigenous Peoples for Archaeological Heritage Management.' *Canadian Journal of Archaeology* 38(2): 519–40.

Basso, K., 1996. *Wisdom Sits in Places: Landscape and Language Among the Western Apache*. Albuquerque: University of New Mexico Press.

Bennett, M.R., et al., 2021. 'Evidence of Humans in North America During the Last Glacial Maximum.' *Science* 373: 1528–31. doi.org/10.1126/science.abg7586

Bernick, K. and R.J. Wigen, 1990. 'Seasonality of the Little Qualicum River West Site.' *Northwest Anthropological Research Notes* 24(2): 153–59.

Blake, M., 2006. 'Fraser Valley Trade and Prestige as Seen From Scowlitz.' In W.C. Prentiss and I. Kuijt (eds), *Complex Hunter-Gatherers: Evolution and Organization of Prehistoric Communities on the Plateau of Northwestern North America*. Salt Lake City, Utah: University of Utah Press.

Boldurian, A.T. and J.L. Cotter, 1999. *Clovis Revisited: New Perspectives on Paleoindian Adaptations from Blackwater Draw, New Mexico*. Philadelphia: University of Pennsylvania Press. doi.org/10.9783/9781934536728

Burley, D., 1980. *Marpole: Anthropological Reconstructions of a Prehistoric Northwest Coast Culture Type*. Burnaby: Archaeology Press.

Carlson, C., 1979. 'The Early Component at Bear Cove.' *Canadian Journal of Archaeology* 3: 177–94.

——, 2003. 'The Bear Cove Fauna and the Subsistence History of Northwest Coast Maritime Culture.' In R.L. Carlson (ed.), *Archaeology of Coastal British Columbia: Essays in Honour of Professor Philip M. Hobler*. Burnaby: Archaeology Press.

Carlson, K.T., 2010. *The Power of Place, the Problem of Time: Aboriginal Identity and Historical Consciousness in the Cauldron of Colonialism*. Toronto: University of Toronto Press.

Carlson, R.L., 1996a. 'Introduction to Early Human Occupation in British Columbia.' In R.L. Carlson and L.D.Bona (eds), *Early Human Occupation in British Columbia*. Vancouver: UBC Press.

——, 1996b. 'The First British Columbians.' In J. Hugh and J.M. Johnston (eds), *The Pacific Province: A History of British Columbia*. Vancouver: Douglas and McIntyre.

——, 2001. *A Sto:lo-Coast Salish Historical Atlas*. Seattle: University of Washington Press.

——, 2013. 'Trade and Exchange in Prehistoric British Columbia.' In T.G. Baugh and J.E. Ericson (eds), *Prehistoric Exchange Systems in North America*. New York: Plenum Press.

Carlson, R.L, and P.M. Hobler, 1993. 'The Pender Canal Excavations and the Development of Coast Salish Culture.' *BC Studies* (99): 25–52.

Carriere, E. and D. Croes, 2017. *Re-Awakening Ancient Salish Sea Basketry: Fifty Years of Basketry Studies in Culture and Science*. Richland, Washington: Northwest Anthropology (*Journal of Northwest Anthropology* Memoir 15).

Chatters, J.C., J.B. Cooper, P.D. LeTourneau and L.C. Rooke, 2011. *Understanding Olcott: Data Recovery at Sites 45SN28 and 45SN303, Snohomish County, Washington*. Everett, Washington: AMEC Earth & Environmental, Inc.

Clark, T.N., 2013. *Rewriting Marpole: The Path to Cultural Complexity in the Gulf of Georgia*. Ottawa: University of Ottawa Press.

Coupland, G., D. Bilton, T. Clark, J.S. Cybulski, G. Frederick, A. Holland, B. Letham and G. Williams, 2016. 'A Wealth of Beads: Evidence for Material Wealth-Based Inequality in the Salish Sea Region, 4000–3500 Cal B.P.' *American Antiquity* 81(2): 294–315. doi.org/10.7183/0002-7316.81.2.294

Coupland, G., T. Clark and A. Palmer, 2009. 'Hierarchy, Communalism and the Spatial Order of Northwest Coast Houses: A Comparative Study.' *American Antiquity* 74: 77–106. doi.org/10.1017/S000273160004751X

Croes, D., 2003. 'Northwest Coast Wet-Site Artifacts: A Key to Understanding Resource Procurement, Storage, Management, and Exchange.' In R.G. Matson, G. Coupland and Q. Mackie (eds), *Emerging from the Mist: Studies in Northwest Coast Culture History*. Vancouver: University of British Columbia Press.

Croes, D.R. and S. Hackenberger, 1988. 'Hoko River Archaeological Complex: Modeling Prehistoric Northwest Coast Economic Evolution.' *Research in Economic Anthropology*, 3, 19–85.

Croes, D.R., S. Williams, L. Ross, M. Collard, C. Dennler and B. Vargo, 2008. *The Projectile Point Sequences in the Puget Sound Region*. Burnaby: Archaeology Press.

Derr, K.M., 2012. *Intensifying with Fire: Floral Resource Use and Landscape Management by the Precontact Coast Salish of Southwestern British Columbia*. Pullman: Washington State University, Department of Anthropology.

Deur, D. and N.J. Turner (eds), 2005. *Keeping it Living: Traditions of Plant Use and Cultivation on the Northwest Coast of North America*. Seattle: University of Washington Press.

Donald, L., 1997. *Aboriginal Slavery on the Northwest Coast of North America*. Berkeley: University of California Press. doi.org/10.1525/9780520918115

Duff, W., 1975. *Images, Stone, B.C.: Thirty Centuries of Northwest Coast Indian Sculpture*. Seattle: University of Washington Press.

Earnshaw, J.K., 2019. 'Cultural Forests in Cross Section: Clear-Cuts Reveal 1,100 Years of Bark Harvesting on Vancouver Island, British Columbia.' *American Antiquity* 29: 1–15. doi.org/10.1017/aaq.2019.29

Elsey, C., 2013. *The Poetics of Land and Identity among British Columbia Indigenous Peoples*. Halifax: Fernwood Publishing.

Erlandson, J.M., M.H. Graham, B.J. Bourque, D. Corbett, J.A. Estes and R.S. Steneck, 2007. 'The Kelp Highway Hypothesis: Marine Ecology, the Coastal Migration Theory, and the Peopling of the Americas.' *Journal of Island and Coastal Archaeology* 2(2): 161–74. doi.org/10.1080/15564890701628612

Fiedel, S.J., 2014. 'The Clovis Era Radiocarbon Plateau.' In A.M. Smallwood and T.A. Jennings (eds), *Clovis: On the Edge of a New Understanding*. College Station, Texas: Texas A&M University Press.

Fladmark, K.R., 1979. 'Routes: Alternative Migration Corridors for Early Man in North America.' *American Antiquity* 44: 55–69. doi.org/10.2307/279189

Frison, G.C., 1987. 'Prehistoric, Plains-Mountain, Large-Mammal, Communal Hunting Strategies.' In M.H. Nitecki and D.V. Nitecki (eds), *The Evolution of Human Hunting*. New York: Springer. doi.org/10.1007/978-1-4684-8833-3_6

Grier, C., 2003. 'Dimensions of Regional Interaction in the Prehistoric Gulf of Georgia.' In R.G. Matson, G. Coupland and Q. Mackie (eds), *Emerging from the Mist: Studies in Northwest Coast Culture History*. Vancouver: UBC Press.

Harris, C., 1994. 'Voices of Disaster: Smallpox around the Strait of Georgia in 1782.' *Ethnohistory* 41(4): 591–626. doi.org/10.2307/482767

Hoffmann, T., N. Lyons, D. Miller, A. Diaz, A. Homan, S. Huddlestan and R. Leon, 2016. 'Engineered Feature Used to Enhance Gardening at a 3800-year-old site on the Pacific Northwest Coast.' *Science Advances* 2(12): e1601282. doi.org/10.1126/sciadv.1601282

Holm, M., 1990. Prehistoric Northwest Coast Art: A Stylistic Analysis of the Archaeological Record. Vancouver: University of British Columbia (PhD thesis).

Kenady, S.M., M.C. Wilson, R.F. Schalk and R.R. Mierendorf, 2007. 'Late Pleistocene Butchered *Bison antiquus* from Ayer Pond, Orcas Island, Pacific Northwest: Age Confirmation and Taphonomy.' *Quaternary International* 233(2): 130–41. doi.org/10.1016/j.quaint.2010.04.013

Klassen, M.A., R. Budhwa and R. Reimer, 2009. 'First Nations, Forestry, and the Transformation of Archaeological Practice in British Columbia, Canada.' *Heritage Management* 2(2):199–238. doi.org/10.1179/hma.2009.2.2.199

Lepofsky, D. and N. Lyons, 2003. 'Modeling Ancient Plant Use on the Northwest Coast: Towards an Understanding of Mobility and Sedentism.' *Journal of Archaeological Science* 30: 1357–71. doi.org/10.1016/S0305-4403(03)00024-4

Lepofsky, D., D.M. Schaepe, A.P. Graesch, M. Lenert, P. Ormerod, K.T. Carlson, J.E. Arnold, et al., 2009. 'Exploring Sto:lo-Coast Salish Interaction and Identity In Ancient Houses And Settlements In The Fraser Valley, British Columbia.' *American Antiquity* 74(4): 595–626. doi.org/10.1017/S0002731600048988

Lepofsky, D., N.F. Smith, N. Cardinal, J. Harper, M.C. Morris, E. White, R. Bouchard, D. Kennedy, A.K. Salomon, M. Puckett and K. Rowell, 2015. 'Ancient Shellfish Mariculture on the Northwest Coast of North America.' *American Antiquity* 80(2):236–59. doi.org/10.7183/0002-7316.80.2.236

Lyons, N. and M. Ritchie, 2017. 'The Archaeology of Camas Production and Exchange on the Northwest Coast: With Evidence from a Sts'ailes (Chehalis) Village on the Harrison River, British Columbia.' *Journal of Ethnobiology* 37(2): 346–67. doi.org/10.2993/0278-0771-37.2.346

Mackie, Q., D. Fedje, D. McLaren, N. Smith and I. McKechnie, 2011. 'Early Environments and Archaeology of Coastal British Columbia.' *Trekking the Shore*: 51–103. doi.org/10.1007/978-1-4419-8219-3_3

Martin, P.S., 1984. 'Prehistoric Overkill: The Global Model.' In P.S. Martin and R.G. Klein (eds), *Quaternary Extinctions*. Tucson: University of Arizona Press.

Martindale, A. and N. Lyons, 2014. 'Introduction: Community-Oriented Archaeology.' *Canadian Journal of Archaeology* 38(2): 425–33.

Mathews, D., 2014. Funerary Ritual, Ancestral Presence, and the Rocky Point Ways of Death. University of Victoria (PhD thesis).

Matson, R.G., 1976. *The Glenrose Cannery Site*. Ottawa: National Museum of Man (Mercury Series 52). doi.org/10.2307/j.ctv17454

Matson, R.G. and G. Coupland, 1995. *The Prehistory of the Northwest Coast*. New York: Academic Press.

McHalsie, A., 2007. 'We Have to Take Care of Everything That Belongs to Us.' In B.G. Miller (ed.), *Be of Good Mind: Essays on the Coast Salish*. Vancouver: University of British Columbia Press.

McLaren, D., 2017. 'The Occupational History of the Stave Watershed.' In M.K. Rousseau (ed.), *Archaeology of the Lower Fraser River Basin*. Burnaby: Archaeology Press.

McLaren, D., D. Fedje, A. Dyck, Q. Mackie, A. Gauvreau and J. Cohen, 2018. 'Terminal Pleistocene Epoch Human Footprints from the Pacific Coast of Canada.' *PLoS ONE* 13(3): e0193522–30. doi.org/10.1371/journal.pone.0193522

McLaren, D., D. Fedje, Q. Mackie, L.G. Davis, J. Erlandson, A. Gauvreau and C. Vogelaar, 2020. 'Late Pleistocene Archaeological Discovery Models on the Pacific Coast of North America.' *PaleoAmerica* 6(1): 43–63. doi.org/10.1080/20555563.2019.1670512

McLay, E., K. Bannister, L. Joe, B. Thom and G. Nicholas, 2004. '*'A'lhut tu tet Sulhween*—Respecting the Ancestors: Understanding Hul'qumi'num Heritage Laws and Concerns for the Protection of Archaeological Heritage.' In C. Bell and V. Napoleon (eds), *First Nations Cultural Heritage and Law: Case Studies, Voices, and Perspectives*. Vancouver: UBC Press.

Miller, B.G., B. Angelbeck, M. Malone, R. Mierendorf and J. Perrier, 2020. *The Upper Skagit Indian Tribe Historical Atlas*. Sedro-Woolley, Washington: Upper Skagit Indian Tribe.

Mitchell, D.H., 1971. 'Archaeology of the Gulf of Georgia Area: A Natural Region and its Culture Types.' *Syesis* 4: 1–228.

Moss, M., 2011. *Northwest Coast: Archaeology as Deep History*. Washington, DC: Society for American Archaeology.

Moss, M. and J. Erlandson, 1992. 'Forts, Refuge Rocks, and Defensive Sites: The Antiquity of Warfare along the North Pacific Coast of North America.' *Arctic Anthropology* 29: 73–90.

Reimer, R., 2012. The Mountains and Rocks are Forever: Lithics and Landscapes of Skwxwú7Mesh Uxwumixw. Hamilton, Ontario: McMaster University (PhD thesis).

Ritchie, M., D. Lepofsky, S. Formosa, M. Porcic and K. Edinborough, 2016. 'Beyond Culture History: Coast Salish Settlement Patterning and Demography in the Fraser Valley, BC.' *Journal of Anthropological Archaeology* 43: 140–54. doi.org/10.1016/j.jaa.2016.06.002

Rorabaugh, A. and T. Fulkerson, 2015. 'Timing of the Introduction of Arrow Technologies in the Salish Sea, Northwest North America.' *Lithic Technology* 40: 21–39. doi.org/10.1179/2051618514Y.0000000009

Schaepe, D.M., B. Angelbeck, D. Snook and J.R. Welch, 2017. 'Archaeology as Therapy: Connecting Belongings, Knowledge, Time, Place, and Well-Being.' *Current Anthropology* 58(4): 502–33. doi.org/10.1086/692985

Shanks, M. and R.H. McGuire, 1996. 'The Craft of Archaeology.' *American Antiquity* 61(1): 75–88. doi.org/10.1017/S0002731600050046

Springer, C., D. Lepofsky and M. Blake, 2018. 'Obsidian in the Salish Sea: An Archaeological Examination of Ancestral Coast Salish Social Networks in SW British Columbia and NW Washington State.' *Journal of Anthropological Archaeology* 51: 45–66. doi.org/10.1016/j.jaa.2018.04.002

Steeves, P., 2021. *The Indigenous Paleolithic of the Western Hemisphere*. Lincoln: University of Nebraska Press. doi.org/10.2307/j.ctv1s5nzn7

Stewart, H., 1984a. *Cedar: Tree of Life to the Northwest Coast Indians*. Vancouver: Douglas & McIntyre.

——, 1984b. 'Culturally Modified Trees.' *The Midden* 16(5): 7–9.

Surovell, T., N. Waguespack and P.J. Brantingham, 2005. 'Global Archaeological Evidence for Proboscidean Overkill.' *Proceedings of the National Academy of Sciences* 102(17): 6231–6. doi.org/10.1073/pnas.0501947102

Suttles, W., 1987. 'Affinal Ties, Subsistence, and Prestige Among the Coast Salish.' In W. Suttles (ed.), *Coast Salish Essays*. Seattle: University of Washington Press.

Thom, B., 1998. 'The Marpole-Late Transition in the Gulf of Georgia Region.' *The Midden* 30: 3–7.

Thompson, G., 1978. *Prehistoric Settlement Changes in the Southern Northwest Coast: A Functional Approach*. Seattle: University of Washington (Reports in Archaeology 5).

Truth and Reconciliation Commission of Canada, 2015. *Honouring the Truth, Reconciling for the Future: Summary of the Final Report of the Truth and Reconciliation Commission of Canada*. Ottawa: Truth and Reconciliation Commission of Canada.

Waters, M.R. and T.W. Stafford, 2007. 'Redefining the Age of Clovis: Implications for the Peopling of the Americas.' *Science* 315(5815): 1122–26. doi.org/10.1126/science.1137166

Wellman, C., 2017. *Peace Weavers: Uniting the Salish Coast through Cross-Cultural Marriages*. Pullman, Washington: Washington State University Press.

8

Ligwiłda'x̱w Expansion into Northern Coast Salish Lands in the Nineteenth Century

Marie Mauzé

This chapter is based on research I undertook 40 years ago, when I started fieldwork in the Ligwiłda'x̱w[1] community (part of the Kwakwa̱ka̱'wakw community) of Cape Mudge, on Quadra Island, British Columbia. Since then, I have kept an ongoing relationship with several Ligwiłda'x̱w families I visit on a regular basis. About six months before my arrival in Cape Mudge in January 1980, a great event had taken place in the village with the opening of the Kwagiulth Museum (now the Nuyumbalees Cultural Centre) in June 1979. The museum was built to house part of the Potlatch Collection confiscated by the Canadian Government in 1922 following the illegal organisation of a potlatch, an institution banned since 1884.[2] In the context of the repatriation of confiscated ceremonial regalia by the then National Museum in Ottawa and the re-appropriation of their cultural heritage in the 1980s and 1990s, elders and Native consultants' interests led me to undertake research on the history of the Ligwiłda'x̱w. Several of the elders and consultants were aware that their group had been kept on the margins of ethnographic attention when compared to other Kwakwa̱ka̱'wakw communities.

1 Also spelled Liqwiltokw' (in reference to the people) or Liq'wala (in reference to the language).

2 Because of internal conflicts within the Kwakwaka'wakw community, the Potlatch Collection was divided between two local museums, the Kwagiulth Museum and the U'Mista Cultural Centre. Located in Alert Bay, the latter opened in November 1980.

This chapter examines the expansion of the Ligwiłda'x̱w into Coast Salish territory in the first half of the nineteenth century. It analyses the phases of this expansion as well as the causes that likely led to such geographic and social changes. It relies on a variety of sources, including published historical and ethnographic literature as well as unpublished archival material, and data collected from Wiweka'yi and Wiwekam elders in the early 1980s.

The text does not only focus on internecine wars between the Ligwiłda'x̱w and the Coast Salish people but also discusses how potential enemies made peace through marriage alliances and exchanged ceremonial privileges considered immaterial property (Lévi-Strauss 1982: 174) or intangible property (Thom and Bain 2004). The transfer of privileges, which was central in the social and ritual life of both tribal groups, was validated by the distribution of material wealth.

The first observations of the Indigenous peoples and their locations in the northern part of the Salish Sea come from the journals of the British and Spanish explorers who arrived in the area in the last decade of the eighteenth century. Accounts by traders and early settlers provide additional information on the identity and location of Native groups between the 1820s and the 1860s. Ethnographic and linguistic data based on Indigenous voices recorded between the late 1880s and the early 1980s by authors Franz Boas, Edward Curtis, Homer Barnett, Wilson Duff and Herbert Taylor, Joy Inglis, Dorothy Kennedy and Randy Bouchard, and me provide a clearer picture of population movements in the nineteenth century. The cross-referencing of sources, while being a challenging methodological task because of the varied categories of available data, provide an overview of the movements of the Ligwiłda'x̱w, and their relations with their Coast Salish neighbours, although for much of the period the Indigenous groups involved are not clearly localised and identified.

What follows is adapted from Chapters 2 and 3 of my book *Les Fils de Wakai: Une histoire des Lekwiltoq* (1992), which is a revised version of my PhD thesis entitled 'Enjeux et jeux du prestige : des Kwagul méridionaux aux Lekwiltoq (côte nord-ouest du Pacifique)' (1984). Dorothy Kennedy and Randy Bouchard (2015) produced a well-documented report a few years ago on the delineation of the northern Coast Salish–Ligwiłda'x̱w territory in the nineteenth century. Research on the history of the Salish people relying both on Native and non-Native accounts has contributed to a better understanding of the complex relations between the Kwakwa̱ka̱'wakw and northern Coast Salish people, in which the Ligwiłda'x̱w came to play a key role. Publications in the field of colonial history provide complementary insights into the process of Lig̱wiłda'x̱w expansion into northern Coast Salish territory. They include various topics such as pre-contact spreading of smallpox in the Puget Sound and Strait of Georgia areas and the ensuing demographic collapse (Harris 1994; Boyd 1999). The history of aboriginal and non-aboriginal relations in the nineteenth century (Lutz 1999, 2008; Storey 2016), and the detailed study of the Battle of Maple Bay involving allied Salish groups against the Lig̱wiłda'x̱w (Angelbeck and McLay 2011), bring to light additional information.

Wakashan Roots

The territorial expansion of the Lig̱wiłda'x̱w is a phenomenon unmatched in scope and speed when compared to other population movements of the Pacific Northwest after first contact. A cross-check of archaeological, linguistic, ethnographic and historical data (McMillan 2003: 247, 258) shows this movement to be part of a wider context of migrations that began at least 2,000 years ago in the area currently occupied by Salish and Wakashan peoples.

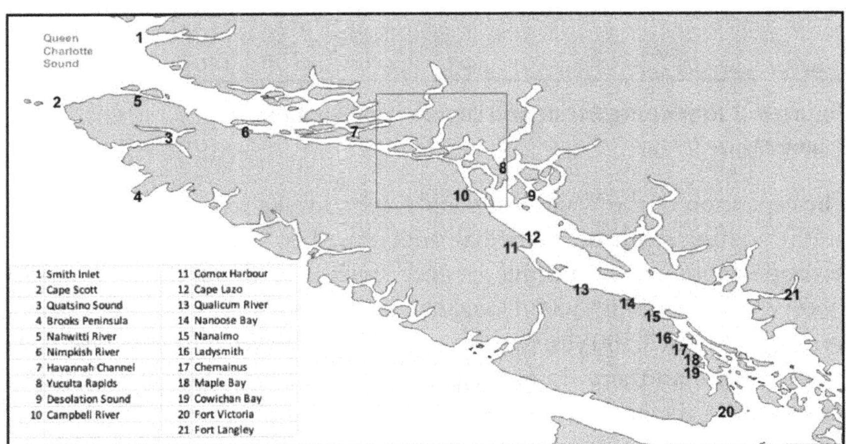

Figure 8.1 Vancouver Island and vicinity

Note: Inset shown in Figure 8.2.

Source: Moshe Rapaport.

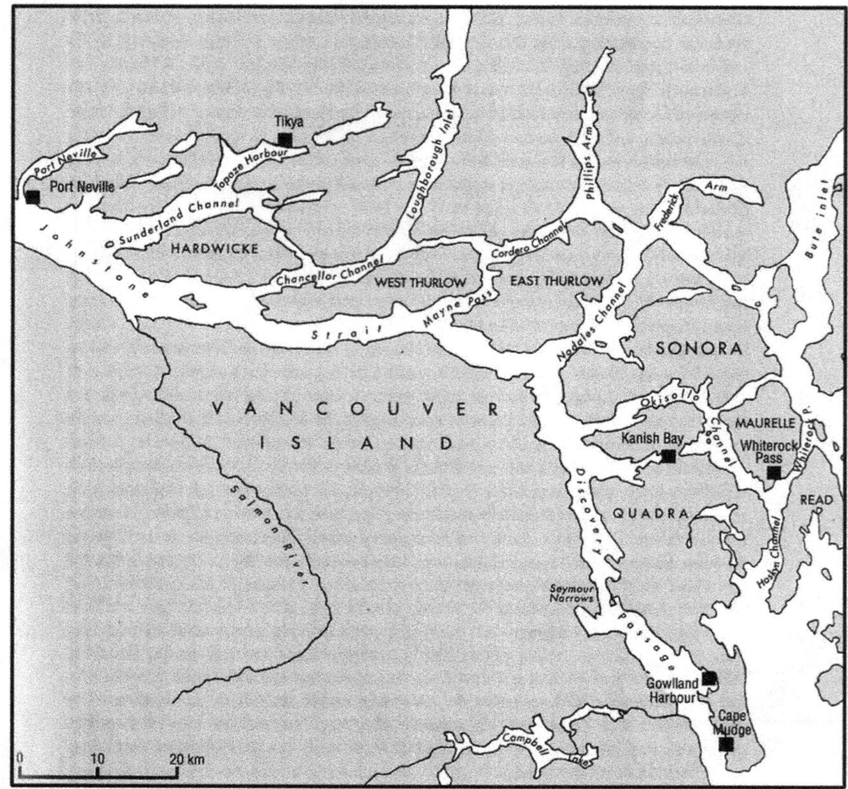

Figure 8.2 Johnstone Strait and Discovery Passage
Source: Mauzé (1992).

The expansion of the Wakashan peoples (Kwakwa̱ka̱’wakw and Nuu-chah-nulth) was the result of intertribal wars caused by the acquisition of new territories holding rich fishing grounds (salmon and eulachon) (Swadesh 1948: 84–6; Ferguson 1984; Donald and Mitchell 1994: 117). The ancient territory (around 2,500 years ago) occupied by Kwak’wala speakers was likely limited to the northern tip of Vancouver Island, from Brooks Peninsula to the mouth of the Nahwitti River.

Later these groups moved toward Queen Charlotte Sound and the continental coastline, displacing the Tsimshian north and the Salish south, with the exception of the Nuxalk (Mitchell 1990: 357). Around 1830, the Kwakwa̱ka̱’wakw territory covered the northwest coast of Vancouver Island as far as Quatsino Sound, extending east to Smith Inlet and south to Havannah Channel (Donald and Mitchell 1994: 113–4; Galois 2012: 237–44).

Archaeological work undertaken in the northeastern region of Vancouver Island suggests that the mid-eighteenth century 'was a period of conflict and displacement, perhaps dating the period of Salish withdrawal, or [marking] a period when village sites were changed' (Kennedy and Bouchard 2015: 39).

The development of the maritime fur trade and the establishment of trading posts in the first decades of the nineteenth century resulted in the territorial reorganisation of the inhabitants of this vast region. It also led to an intensification of war expeditions, undertaken with the sole purpose of capturing slaves to sell for profit.

The slave trade allowed for the acquisition of various types of goods (furs, guns, blankets, etc.) (Donald 1997: 149–54).

Lig̱wiłda'x̱w Expansion

In the late eighteenth century (from 1770–1800), the Lig̱wiłda'x̱w were the southernmost Kwakwa̱ka̱'wakw group. They may have already made inroads into the territory of Comox Coast Salish speakers. Oral traditions associate the Lig̱wiłda'x̱w people and their ancestor Wakai with several locations such as the Cape Scott region at the farthest tip of Vancouver Island, two villages at the mouth of the Nimpkish River, and Tekya (Topaze Harbour) in the region of Sunderland Channel. Tekya (Tikya) is a site of great significance for the Lig̱wiłda'x̱w. It is the place where they settled after the great flood, and which they made their winter village for several decades (Assu with Inglis 1989: 20; Mauzé 1992: 147–8).

Observations by early British and Spanish explorers on language, behaviour and possession of firearms suggest that the Lig̱wiłda'x̱w had abandoned their Nimpkish River village by 1792 and had already settled in Tekya, which they made their main village (Mauzé 1992: 39–49; Kennedy and Bouchard 2015: 40–52).[3] The Lig̱wiłda'x̱w advanced from Topaze Harbour, encroaching on the Coast Salish territory of the Island K'ómox and the Homalco toward the south (Johnstone Strait and Discovery Passage) and east (Loughborough Inlet and Phillips Arm) (Mauzé 1992: 57–61; Kennedy and Bouchard 2015: 52–82). In the course of this movement the Lig̱wiłda'x̱w—whose core group was the Wiwaka'yi—underwent transformations either through internal splits or by the absorption of exogenous groups, both Coast Salish and

3 See Newcombe (1923); Lamb (1984); Kendrick (1991); Jane (1971).

Kwakwa̱ka̱'wakw.[4] The new historical context prompted the Ligwiłda'x̱w to resort to warlike actions to reach economic goals, while at the same time it fostered solidarity among the subgroups. Such developments are evident throughout the nineteenth century in the social history of the Ligwiłda'x̱w (Mauzé 1992: 79–99, 151–90; Galois 2012: 233–76).

Waging War on the Coast Salish

While they were advancing into the southern Johnstone Strait area, from Salmon River (Xwésam)[5] to Cape Mudge, the Ligwiłda'x̱w conducted long-distance raids in the Georgia Strait and Puget Sound, killing and capturing victims. During the first decades of the nineteenth century, they continually harried neighbouring populations, demonstrating their superiority by use of firearms and gradually forcing them back to the limits of their territory. In clearing an area of its inhabitants, they were able to move into their territories, occupying new sites. Around 1850, they were in a position to control both sides of Johnstone Strait and Discovery Passage.

Ligwiłda'x̱w raiders had been waging war in the southern part of the Salish Sea before the Hudson's Bay Company established a land-based trading post on the Fraser River in 1827. The creation of Fort Langley provided new opportunities to enrich themselves by taking slaves from a set of groups gathering around the trading post. These groups included the Skagit, the Kwantlem, the Tlalam (Klallam), the Songhees, the Semaino (Stz'uminus), the Cowichan, the Squamish and the Noosack (Maclachlan 1993: 59, 65, 111, 152, 202–5; Mauzé 1992: 71–4).[6] The Ligwiłda'x̱w who were considered at the time as 'the largest slave-holding tribe among the Kwakwa̱ka̱'wakw, with the largest number coming from Georgia Strait and the Fraser region' bartered their slaves with the Kwaguł and Na̱'witi (Nahwitti), who were part of the northern slave trade network (Donald 1997: 229, 142).

4 One segment of the Kwixa, previously part of the larger Kwaguł group, merged with the Ligwiłda'x̱w and settled at Port Neville, then later in the region of Philips Arm (Mauzé 1992: 89–92; Galois 2012: 250–8). According to the first surveys carried out by the Hudson's Bay Company, the Ligwiłda'x̱w tribe included six subgroups around 1830–40: the Wiweka'yi, the Wiwakam, the Komenox, the Kwixa, the Tlaluis and the Xaxamatsis. The Tlaluis were said to have split off from the Kwixa (Mauzé 1992: 82, 89–92; Galois 2012: 236) They may also be of Salish origin (Kennedy and Bouchard 2015: 73, 81).

5 As transcribed by Kennedy and Bouchard (1983: 167).

6 See following sources: Maclachlan (1993); McKelvie (1947: 34, 50); Waite (1977: 17–18); Maud (1978, vol. 2: 49–50); Curtis (1913: 33–5); Curtis (1915: 106–8); Amoss (1978: 10); Tate (n.d.: 9–13); Boas (1889: 325).

While biased and not free of negative comments regarding the Lig̱wiłda'x̱w, the *Fort Langley Journals* (1827–1830) (Maclachlan 1993) provide a great number of entries referring to attacks by the 'ferocious' Yewcultas (Lig̱wiłda'x̱w). The Coast Salish people, with few or no firearms, proved easy prey for the Lig̱wiłda'x̱w, and served to some extent as a stockpile for captives of war, mostly women and children.[7] In 1829, the situation seemed so critical that chief trader Archibald McDonald was in favour of selling guns and ammunition to local Natives 'to promote the ruin of that detestable tribe', opining that 'the Complete (sic) annihilation of this truly barbarous banditti would be no loss to the human race' (Maclachlan 1993: 111–2). Protecting the Natives who came to exchange goods at the trading post or giving them the means to defend themselves enabled the company to make a profit.

Coast Salish warriors of different groups retaliated against the frequent attacks by the Lig̱wiłda'x̱w and defeated them several times. The most stinging defeat for the northern raiders occurred in Maple Bay in Cowichan territory sometime around 1840–1850. It involved a large alliance between tribes from the Puget Sound and the Gulf of Georgia (Cowichan and Nanaimo) including hundreds of warriors who had obtained arms and ammunition from Fort Langley (Maclachlan 1993: 228). They engaged in a coordinated canoe battle and succeeded in destroying the Lig̱wiłda'x̱w fleet and killing its warriors. While hardly mentioned by the Lig̱wiłda'x̱w, the famous battle is vividly remembered in Coast Salish oral accounts to this day. The latter attack was supposedly the last great battle on intertribal wars, but did not end the cycle of intertribal conflict (Angelbeck and McLay 2011).[8]

The building of Fort Victoria in 1843 changed the situation and encouraged the Lig̱wiłda'x̱w to use new raiding tactics. Instead of bringing war elsewhere the Lig̱wiłda'x̱w intercepted groups navigating from the Alaska Panhandle along the Inside Passage through Johnstone Strait and Georgia Strait on their way to visit, trade and work in Victoria or further south in Puget Sound (Lutz 2008: 167–71; Storey 2016: 51).

7 In addition to the issue of limited access to weapons, the weakness of the Salish populations resulted from a smallpox epidemic in 1782 in the region around Georgia Strait and Puget Sound (Harris 1994; Galois 2012: 235).

8 For a detailed analysis based on Coast Salish accounts of the Maple Bay Battle, and the building of an alliance of Coast Salish groups, formed under specific historical circumstances, see Angelbeck and McLay (2011). The Maple Battle is mentioned in Boas (1889); Curtis (1913: 33–5; 1915: 108–110); and Tate (n.d.: 9).

In the 1840s, the Ligwiłda'x̱w had taken over several K'ómox villages located at the extreme northern end of the Salish Sea, along Discovery Passage, and were thus able to take control of the sea route used by the northern tribes—the Haida, the Stikine (Tlingit), the Nuxalk and the Tsimshian—to reach Victoria. Thereafter, the Ligwiłda'x̱w, instead of engaging in regular trade economy in Fort Victoria, carried out raids on any boats that came within their range to seize manufactured goods and take slaves (Mauzé 1992: 74–6).[9]

Conquest of Discovery Passage

There is a gap of over 30 years between the observations of the British and Spanish explorers in 1792 and those of the Hudson's Bay Company traders in the 1820s. At the end of the 1820s, the *Fort Langley Journals* (1827–1830) refer to the 'Yukletaws', 'Yewkultas' or 'Yucultas' (Maclachlan 1993) as northern tribes with no further location details. By the end of the 1820s, the Youcattas (Ligwiłda'x̱w) already controlled the region around Seymour Narrows (?u?ta̱wi).[10]

The journal of William F Tolmie (Tolmie 1963) placed the Hachaamdsis (Xaxamatsis)[11] in Johnstone Strait, and three other Ligwiłda'x̱w subgroups (Wiwaka'yi, Wiwakam, Tlaluis) in Desolation Sound, which at the time comprised Loughborough Inlet, Bute Inlet and Cordero Channel, in the northern part of the Gulf of Georgia.

James Douglas (1840), future chief factor of the Hudson's Bay Company in Fort Victoria, located the Neekultas (Ligwiłda'x̱w) on Vancouver Island, across from (West) Thurlow Island and in Loughborough Inlet. According to Douglas, the linguistic boundary at that time between Kwakwa̱ka̱'wakw and Salish speakers passed through Port Chatham across from Nodales Channel, at the junction between Johnstone Strait and the Discovery Passage (Mauzé 1992: 58).

9 See Barrett-Lennard (1862: 43–4); Drucker (1953); Curtis (1915: 113–4); Douglas (1853–59); and Mayne (1862: 245–6).

10 Xwémkwu in Comox language (Kennedy and Bouchard 1983: 167).

11 The real identity of the Xaxamatsis is still an enigma. It is likely that the Xaxamatsis included members of a Salish village located up the Salmon River and members of the Ligwiłda'x̱w group (see Mauzé 1992: 92–6; Kennedy and Bouchard 2015: 67).

The conquest of Discovery Passage likely accelerated with the establishment of a trading post at Fort Victoria (1843), affording the Lig̱wiłda'x̱w a position from which to exert control over the vessels using this sea route to bring Indigenous groups from the north to Fort Victoria. The other route, through Cordero Channel, along the continental coastline, was also controlled by Lig̱wiłda'x̱w raiders: there are reports of taxing travellers from their village on Big Bay on Stuart Island (Yuculta Rapids); they were quick to kill any who refused to comply with their demands (Duff 1960). The Lig̱wiłda'x̱w secured a strategically advantageous position for themselves, allowing them to control both passageways from north to south.

Attacks designed to drive the K'ómox from their villages took place after a series of raids and counter-raids involving the Lig̱wiłda'x̱w and Coast Salish groups, in which the role played by the K'ómox was ambivalent (Galois 2012: 268). According to Native historians James Smith and Billy Assu, raids were conducted out of several Lig̱wiłda'x̱w bases. They also reported that the Lig̱wiłda'x̱w drove the K'ómox out of Gowlland Harbour (Gwigwak'ulis) in Discovery Passage from their bases at Green Point (Loughborough Inlet) and Yuculta Rapids (Duff 1960). Smith and Assu mentioned that Lig̱wiłda'x̱w had established a village at Whiterock Pass (tatapa'ulis),[12] between Maurelle and Read Island (Duff 1960; Drucker 1953). Seen as a crucial step in their conquest of the passage, the Lig̱wiłda'x̱w (Wiwakam) would occupy the village at Whiterock (Dawson 1887: 75).

Other K'ómox sites were taken over along Discovery Passage. Kanish Bay ('Qanis) on the western coast of Quadra Island became a winter village where the different Lig̱wiłda'x̱w groups gathered for potlatches and ceremonies (Duff 1960; Drucker 1953). There are reports of a raid carried out by the Lig̱wiłda'x̱w around 1835, where the K'ómox Eeksam of Campbell River ('łamataxw)[13] were all but exterminated while gathered to fish on the banks of the river; the survivors fled and joined the Pentlatch at Comox Harbour (Meade 1980: 20).

The Lig̱wiłda'x̱w attack on Gowlland Harbour, where the K'ómox had taken refuge, was a decisive moment in the fight to take control of Discovery Passage (Curtis 1915: 110–2). The inhabitants of Point Mudge eventually

12 T'át'pu7us in Comox Coast Salish (Kennedy and Bouchard 2015: 156). Transcribed in the Kwak'wala language as t'et'epe7úyas by Kennedy and Bouchard (ibid.). This was a firmer Homalco settlement.

13 Tl'ám̓atexw in Comox language (Kennedy and Bouchard 1983: 167).

resettled with the Pentlatch. The last K'ómox villages (Campbell River, Gowlland Harbour and Point Mudge) came to be abandoned as a result of the attacks and raids carried out by Lig̲wiłda'x̱w warriors.

According to testimonies collected by Hudson's Bay Company's employees, the definitive occupation of Discovery Passage, with the settlement of the Wiweka'yi at Cape Mudge, did not occur until the 1840s. In his 1839 survey of the populations around the Gulf of Georgia, James Murray Yale of Fort Langley mentioned the Tsiloths (Catloch or K'ómox) residing 'on the east coast of Vancouver Island, at Point Mudge' (Yale 1838–39). In May 1840, Douglas (1840) noted in his journal the presence of three K'ómox villages off Point Mudge, one of which was surrounded by a palisade. In September 1841, George Simpson, travelling aboard the *Beaver*, commanded by Captain WH McNeill, made the same observation (Simpson 1847: 186), without giving exact details of their location. Some years later, in December 1847, J Thorne, captain of the *Beaver*, made a stopover in Discovery Passage and traded there with the Lig̲wiłda'x̱w in the area around Seymour Narrows (Mauzé 1992: 60).

In February 1852, the crew of the *Beaver* stopped at Duncan Bay, north of Campbell River to cut firewood. The crew members warned the Natives of their presence by firing a few shots, and were met by several boats of the Uculty (Lig̲wiłda'x̱w) tribe (Ship Records n.d.). The following year, in January 1853, at Point Mudge, Douglas reported that around 100 people had spent the entire day alongside their vessel. He also recorded that the K'ómox resided at Point Hope (Cape Lazo), near Comox Harbour, suggesting that the K'ómox had left Point Mudge to settle further south (Douglas 1853–59).

According to information collected from fur traders, the Lig̲wiłda'x̱w occupied the area around Discovery Passage between 1842 and 1847; possibly around 1845 according to Taylor and Duff (1956: 63); or between 1845 and 1852–53. In 1853, the Wiweka'yi had established a village at Cape Mudge, a mile from Point Mudge. According to Chief Harry Assu, the Wiweka'yi decided to settle at Cape Mudge rather than Campbell River due to the abundance of blueback, young coho salmon, offshore from the first of these villages (Mauzé 1992: 62).

In the 1850s a Lig̲wiłda'x̱w subgroup had control over a salmon-fishing site on the Qualicum River after its inhabitants were wiped out by disease or wars against the Opichesaht. This Lig̲wiłda'x̱w camp was itself almost destroyed by fire and its occupants killed during an attack by the Haida in 1855, who were travelling in the area, in retaliation for a raid the latter had suffered near Cape Mudge (Walkem 1914: 42–3).

The reputation of the Lig̱wiłda'x̱w as a renowned group of warriors came to an unhappy and bloody end in 1860, when the gunboat *Forward* was dispatched to Cape Mudge to seek restitution of stolen goods on behalf of traders who had been attacked in Georgia Strait. When the Lig̱wiłda'x̱w refused to comply, the ship responded by opening fire, their cannons blowing the boats moored along the beach to pieces and destroying the defensive wall that surrounded the village (Mayne 1862: 245–6).

Two years later, the Wiwaka'yi suffered heavy losses after contracting smallpox following an ambush on a Haida group who had returned from Victoria with the disease (*Daily Colonist*, 1 July 1862; Boyd 1999: 188–9). Northern tribes had in fact been coming to Victoria and spending several months there since the gold rush of 1858. In April 1862, a sailor disembarked in Victoria carrying smallpox. Instead of taking quarantine measures, the colonial authorities evicted the northern Native people from the city, causing the epidemic to spread the entire length of British Columbia and killing several thousands of Indigenous people (Boyd 1999; Van Rijn 2006; Storey 2016).

Further South Radiation

Long before establishing a village at Cape Mudge, some of the Lig̱wiłda'x̱w who still lived in Tekya and others in Port Neville seem to have claimed fishing rights on Comox Harbour. Narratives about a conflict between the K'ómox and the Lig̱wiłda'x̱w reported that the latter had moved to Comox Harbor to catch herring (Curtis 1915: 108). They continued to camp in Comox for fishing after the Wiwaka'yi had moved to Cape Mudge and the K'ómox south to join the remnants of the Pentlatch.

Regardless of whether they actually planned to settle here on a longer-term basis or not, the colonial authorities ensured this did not happen, concerned as they were to protect their colonists, who in 1862 had begun to arrive and establish permanent settlements in the Comox Valley. Rumors regarding Indigenous peoples' hostile behaviour were well spread among colonists who also were afraid of tensions ready to burst out between aboriginal groups (Lutz 2008: 168–9.)

There were around 70 colonists in the region at the time, who employed Native peoples as a labour force, while remaining wary of the thefts that they had committed (including stealing potatoes from the fields). Disputes sometimes required the intervention of the colonial authorities (Hayman 1989: 123–4), especially in this very case, when acrimonious relations between Lig̱wiłda'x̱w and the K'ómox were felt, according to colonial discourse, as nuisances for the peace of the settlers. In 1865, by order of the Governor of Vancouver Island, Arthur Kennedy, three gunboats were dispatched to Comox to defend the colonists there against the Natives, who were deemed dangerous. The commander was also asked by the K'ómox chief to dispose of 50 men in the event that the Lig̱wiłda'x̱w would attack his people.

Whether or not the situation was really threatening, Admiral Denman, who led the expedition, was charged with driving the Lig̱wiłda'x̱w back toward Cape Mudge and prohibit them from ever returning. They agreed to be escorted back to Cape Mudge, though by Denman's recommendation they were permitted to return to Comox Harbour to fish, on the condition that they would not seek to settle there permanently. In March 1866, Robert Brown, who led the Vancouver Island Exploring Expedition, visited Cape Mudge, 'now a very miserable collection of 10 or 12 huts' to see that all the inhabitants had left; some of them had traveled to Comox Harbour to assert their long-standing fishing rights (Brown 1866: 19; see Mauzé 1992: 62–4).

Between the 1850s and 1860s, the Lig̱wiłda'x̱w held a territory comprising winter villages and summer sites spread across a great distance between Port Neville and Tekya (Topaze Harbour) to the north and Qualicum and even a bit further to the south in Georgia Strait. The Wiwakam occupied several sites in Loughborough Inlet and relocated in Campbell River at the turn of the twentieth century.

Around the same time, two related Kwakwa̱ka̱'wakw subgroups, the Kwexa (Kwiakah) of Phillips Arm, and the Walatsama of Salmon River, relocated to the Comox Harbour area (Mauzé 1992: 88, 90, 96). According to Chief Harry Assu, the K'ómox and Lig̱wiłda'x̱w groups shared sites on Hornby Island and Denman, and the Island Halkomelen from Cowichan, Ladysmith, Chemainus and Nanaimo camped near the village of Cape Mudge during the summer months and fished salmon (Assu with Inglis 1989: 12, 14; Mauzé 1992: 194; Thom 2005: 362–5).

Although some of these were fishing sites only occupied on a seasonal basis, in the mid-nineteenth century the Ligwiłda'x̱w controlled the northern part of the Salish Sea including Johnstone Strait, Discovery Passage and a part of Georgia Strait. By 1873 the Ligwiłda'x̱w had expanded as far south as Nanoose Bay where they had a village and spent several months a year showing that access to resources fuelled hostilities between neighbouring groups (Kennedy and Bouchard 2015: 115).

Making Peace through Marriage and Ritual

Relations between the Coast Salish and the Ligwiłda'x̱w were marked during the first half of the nineteenth century by hostile attacks fuelled by slave raiding. The Gulf of Georgia and Puget Sound Indigenous groups suffered during several decades of the Ligwiłda'x̱w raids and conquest of new territories with access to rich fishing grounds. The relations between the Ligwiłda'x̱w and Island K'ómox fluctuated, ranging from outright hostility to military collaboration, participation in Ligwiłda'x̱w potlatches, and marriage alliances.[14]

Oral traditions indicate that marriages and potlatches were organised to make peace, to ease access to resources, as well as to transfer chiefly names and prerogatives. Intermarriages between K'ómox and Wiwka'yi families, and between Halkomelen and Wiwaka'yi families, continued through the first half of the twentieth century (Assu with Inglis 1989: 13–14).

Among the Kwakwa̱ka̱'wakw, where social statuses were governed by an internal hierarchical system of ranks, marriages were a means of obtaining status through the transfer of names and dancing privileges. Chiefs were responsible for family property including names, songs, dances and narratives about how the ancestor of his group obtained the right to use a dance in ceremonies. For the K'ómox, marriage alliances with the Ligwiłda'x̱w were considered a source of prestige.

In 'Family Stories', Boas's main collaborator, George Hunt, recounted the story of a young K'ómox man who by marrying a Wiwaka'yi woman received a name and a seat in one of the *numaym* of the Wiwaka'yi as well

14 An early case is reported in the Fort Langley Journals of a marriage between a Ligwiłda'x̱w woman and a Kwantlen chief before the company arrived which was followed by a period of peace between the two tribes (Maclachlan 1993: 203).

as the right to the Sisiuł (double-headed serpent) dance in the Winalagilis ritual of the winter ceremonies (Boas and Hunt 1921: 951–4). Potlatches and winter ceremonies were also organised in Comox by Lig̱wiłda'x̱w high-ranking chiefs.

In 1876, a Wiwak a'yi ai chief from Cape Mudge named Johnny Chiceete (Chickite) was initiated in the Hawinalał dance, also part of the Winalagilis ritual. At the end of the potlatch, Chiceete burnt a large war canoe (McKechnie 1972: 29–37).[15] Conversely, privileges belonging to the K'ómox were transferred through marriage to the Lig̱wiłda'x̱w. Around the turn of the 1900s, a noble Wiwaka'yi family from Cape Mudge earned the privilege of using the Sxwayxwey mask through marriage to a K'ómox woman.[16]

In 1941, the Walatsama of Salmon River merged with the K'ómox, and the reserve that had been established for the Walatsama in 1886 became a de facto Comox reserve as well (Kennedy and Bouchard 1983: 167; Mauzé 1992: 96). Following the amalgamation of the two First Nations bands, the K'ómox were placed under the administration of the colonial Kwawkewlth Agency. Since then, the Comox reserve has been considered the southernmost Lig̱wiłda'x̱w settlement (Assu with Inglis 1989: 11). Today the K'ómoks First Nation is a member of the Kwakiutl District Council.[17]

Kwak'wala eventually became the dominant language in Comox (Kennedy and Bouchard 1983: 17, 23) and has remained the spoken language in ceremonies and potlatches. Today, however, 'the K'ómoks First Nation and its members are undertaking the huge task of language revitalization', and collaborating with 'sister nations Homalco, Tla'amin, and Klahoose, who share the same traditional language' (K'ómoks First Nation n.d.).

The K'ómox now consider themselves as being part of two main cultures: Northern Coast Salish as well as Kwakwa̱ka̱'wakw (K'ómoks First Nation n.d.). The artist Andy Everson of K'ómox and Kwakwa̱ka̱'wakw (K̲wik̲wa̱sut'inu̱x̱) descent, who lives on the Comox reserve, considers that the Comox area 'was a point of intersection for two large cultural groups:

15 A traditional demonstration of chiefly wealth.

16 The initiator of this marriage between John Dick from Cape Mudge and Maggie Frank from Comox was the Wiwaka'yi ai chief Sewish (Mauzé 1992:105). The Sxwayxwey mask was included in the Potlatch Collection confiscated from the Kwakwa̱ka̱'wakw people in 1922 following the illegal organisation. Along with other ceremonial regalia, the mask was returned in 1979 to the Kwagiulth Museum in Cape Mudge. See also Kennedy and Bouchard (1983: 53).

17 See www2.gov.bc.ca/gov/content/environment/natural-resource-stewardship/consulting-with-first-nations/first-nations-negotiations/first-nations-a-z-listing/k-moks-first-nation-comox-indian-band.

the Salish and the Kwakwa̱ka̱'wakw' (Everson 2021: 369). His own work, as well the monumental carvings and paintings in the Comox Big House, testify to the strong connections and identification to Kwakwa̱ka̱'wakw culture.

Conclusion

The southward migration of the Ligwiłda'x̱w into the Salish Sea has a long history. We do not know all the causes at the origin of this expansion movement, but we do understand the complexity of their relations with their Coast Salish neighbours, which were sometimes marked by open hostilities, sometimes by occasional alliances, according to circumstances. This movement was at first part of a more general matter of access to the rich waters of the Salish Sea. With the establishment of the Hudson's Bay Company's land-trade posts it was further motivated by the slave trade, procuring goods of European origin, including weapons, which signalled their superiority in war matters.

Attempting to break up the hegemony of the Ligwiłda'x̱w, and to allay the anxieties of the settlers who started to appropriate Indigenous lands in the Comox Valley, the Royal Navy put a stop to further displacement of the Coast Salish population. Around the 1860s, fishing sites occupied seasonally were no longer the object of major conflicts, a trend amplified when marriage alliances reinforced kinships ties.

What began as warfare for control of fisheries and obtaining material wealth resulted in the exchange of intangible wealth between the Ligwiłda'x̱w and the K'ómox.

References

Amoss, P., 1978. *Coast Salish Spirit Dancing. The Survival of an Ancestral Religion.* Seattle: University of Washington Press.

Angelbeck, B. and E. McLay, 2011. 'The Battle of Maple Bay: The Dynamics of Coast Salish Political Organization through Oral Histories.' *Ethnohistory* 58(3): 359–92. doi.org/10.1215/00141801-1263821

Assu, H. with J. Inglis, 1989. *Assu of Cape Mudge. Recollections of a Coastal Indian Chief.* Vancouver: UBC Press.

Barrett-Lennard, C.E., 1862. *Travels in British Columbia with a Narrative of a Yacht Voyage around Vancouver Island.* London: Hurst and Blackett.

Boas, F., 1889. 'Notes on the Snamaimuq.' *American Anthropologist* A2: 321–8. doi.org/10.1525/aa.1889.2.4.02a00050

Boas, F. and G. Hunt, 1921. 'Ethnology of the Kwakiutl.' In *Thirty-Fifth Annual Report of the Bureau of American Ethnology.* Washington, DC: Smithsonian Institution.

Boyd, R., 1999. *The Coming of The Spirit of Pestilence: Introduced Infectious Diseases and Population Decline Among Northwest Coast Indians, 1774–1874.* Vancouver: UBC Press.

Curtis, E., 1913. *The North American Indian: Salishan Tribes of the Coast*, vol. 9. Nordwood: Plimpton Press.

——, 1915. *The North American Indian: The Kwakiutl*, vol. 10. Nordwood: Plimpton Press.

Dawson, G., 1887. 'Notes and Observations on the Kwakiool People in the Northern Part of Vancouver Island and Adjacent Coast […].' *Proceedings and Transactions of the Royal Society of Canada for the Year 1897* 5(2): 63–98.

Donald, L., 1997. *Aboriginal Slavery on the Northwest Coast of North America.* Berkeley: University of California Press. doi.org/10.1525/9780520918115

Donald, L. and D.H. Mitchell, 1994. 'Nature and Culture on the Northwest Coast of North America: The Case of Wakashan Salmon Resources'. In E.S. Burch Jr and L.J. Ellana (eds), *Key Issues in Hunter-Gatherer Research.* Oxford: Berg.

Everson, A., 2021. 'Afterword. Between this World and That.' In A. Glass (ed.), *Writing the Hamatsa: Ethnography, Colonialism, and the Cannibal Dance.* Vancouver: UBC Press.

Ferguson, B., 1984. 'A Reexamination of the Causes of Northwest Coast Warfare.' In B. Ferguson (ed.), *Warfare, Culture and Environment*, London: Academic Press.

Galois, R., 2012. *Kwakwa̱ka̱'wakw Settlements, 1775–1920: A Geographical Analysis and Gazetteer.* Vancouver: UBC Press.

Gough, B., 1984. *Gunboat Frontier: British Maritime Authority and the Northwest Coast Indians.* Vancouver: UBC Press.

Harris, C., 1994. 'Voices of Disaster: Smallpox around the Strait of Georgia in 1782.' *Ethnohistory* 41(4): 591–626. doi.org/10.2307/482767

Hayman, J. (ed.), 1989. *Robert Brown and the Vancouver Exploring Expedition*. Vancouver: University of British Columbia.

Jane, C. (transl.), 1971. *A Spanish Voyage to Vancouver & the Northwest Coast being the Narrative to the Voyage Made in the Year 1792 by the Schooners 'Sutil' and 'Mexicana' to explore the Strait de Fuca*. New York: Da Capo Press.

Kendrick, J. (transl.), 1991. *The Last Spanish Exploration on the Northwest Coast of America*. Spokane: The Arthur H. Clark Company.

Kennedy, D. and R. Bouchard, 1983. *Sliammon Life, Sliammon Lands*. Vancouver: Talons Books.

——, 2015. Homalco Rights and Title Interests: Thurlow, Sonora and Maurelle Islands. Report prepared by Dorothy Kennedy and Randy Bouchard for the Homalco First Nation, Campbell River, British Columbia, 14 October 2014 (Attested 19 March 2015). Original held by Bouchard & Kennedy Research Consultants, Mill Bay, BC. On file as an Expert Report with the Supreme Court of British Columbia, Vancouver Registry, No. S – 150306, Cape Mudge Indian Band v. British Columbia (Ministry of Aboriginal Relations and Reconciliation) and Homalco Indian Band.

K'ómoks First Nation, n.d. 'Language Revitalization Efforts.' K'ómoks First Nation site. Viewed 16 June 2023 at: komoks.ca/cultures/.

Lamb, W.K., (ed.). 1984. *The Voyage of George Vancouver, 1791–1795*, vol. 2. London: Haklyut Society.

Lévi-Strauss, C., 1982. *The Way of the Masks*. Vancouver: Douglas & McIntyre Ltd.

Lutz, J., 1999. 'Inventing an Indian War: Canadian Indians and American Settlers in the Pacific West, 1854–1864.' *Journal of the West* 38(3): 7–13.

——, 2008. *Makúk: A New History of Aboriginal-White Relations*. Vancouver: UBC Press.

Maclachlan, M., 1993. *The Fort Langley Journals, 1827–1830*. Vancouver: UBC Press.

Maud, R., 1978. *The Local Contributions of Charles Hill-Tout. The Salish People: The Squamish and the Lillooet*, vol. 2; *The Sechelt and the South-Eastern Tribes of Vancouver Island*, vol. 4. Vancouver: Talonbooks.

Mauzé, M., 1992. *Les Fils de Wakai: Une Histoire des Indiens Lekwiltoq*. Paris: Editions Recherche sur les Civilisations.

Mayne, R.C., 1862. *Four Years in British Columbia and Vancouver Island*. London: John Murray.

McKechnie, R.E., 1972. *Strong Medicine*. Vancouver: J.J. Douglas Ltd.

McKelvie, B.A., 1947. *Fort Langley. Outpost of an Empire*. Montreal: Vancouver Daily Province.

McMillan, A.D., 2003. 'Reviewing the Wakashan Migration.' In R.G. Matson, G. Coupland and Q. Mackie (eds), *Emerging from the Mist: Studies in Northwest Coast Culture and History*. Vancouver: UBC Press.

Meade, E., 1980. *The Biography of Dr. Samuel Campbell R.N. Surgeon and Surveyor including the Naming and Early History of Campbell River*. Vancouver: Edward F. Meade.

Mitchell, D., 1990. 'Prehistory of the Coasts of Southern British Columbia and Northern Washington.' In W.C. Sturtevant (ed.), *Handbook of North American Indians*, vol. 7. Washington: Smithsonian Institution Scholarly Press.

Newcombe, C.F. (ed.), 1923. *Menzies' Journal of Vancouver Voyage: April to October 1792*. Victoria: King's Printer (Archives of British Columbia Memoir 5).

Simpson, G., 1847. *Narrative of a Journey Around the World during the Years 1841 and 1842*, 2 vols. London: H. Colburn.

Storey, K., 2016. *Settler Anxiety at the Outposts of Empire: Colonial Relations, Humanitarian Discourses and the Imperial Press*. Vancouver: UBC Press.

Swadesh, M., 1948. 'Motivations in Nootka Warfare.' *Southwestern Journal of Anthropology* 4: 76–93. doi.org/10.1086/soutjanth.4.1.3628474

Taylor, H.C. and W. Duff, 1956. 'A Post-Contact Southward Movement of the Kwakiutl.' *Research Studies of the State College of Washington* 24(1): 56–66.

Thom, B. and D. Bain, 2004. *Aboriginal Intangible Property in Canada: An Ethnographic Review*. Ottawa: Industry Canada.

Tolmie, W.F., 1963. *The Journals of William Fraser Tolmie, Physician and Fur-Trader*. Vancouver: Mitchell Press.

Van Rijn, K., 2006. '"Lo! The Poor Indian!" Colonial Responses to the 1862–63 Smallpox Epidemic in British Columbia and Vancouver Island.' *Canadian Bulletin of Medical History* 23(2): 541–60. doi.org/10.3138/cbmh.23.2.541

Waite, D.E., 1977. *The Fort Langley Story*. Vancouver: Don Waite Publishing.

Walkem, W., 1914. *Stories of Early British Columbia*. Vancouver: News Adviser.

Manuscripts

Brown R., 1866. 'The Land of the Hydahs. A Spring Journey due to North. The Voyage of the Goldstream to Queen Charlotte Islands and with a Reconnaissance of the Coast of British Columbia and the Eastern Coast of Vancouver Island.' Victoria, BC: Provincial Archives (Add. Mss 794, vol. 4/4).

Daily Colonist, 1862. Victoria: Provincial Archives (1 July).

Douglas, J., 1840. 'Diary of a Trip to the Northwest Coast.' Victoria: Provincial Archives (Transcript April 22–October 2, 1840, A/B/40).

——, 1853–59. 'Miscellaneous Notes B/20.' Victoria: Provincial Archives.

Drucker, P., 1953. 'Drucker's Field Notebooks'. Washington, DC: Smithsonian Institution Archives (ms 4516:2, vol. 6); BC Archives (Mss 870).

Duff, W., ca. 1960. 'Southern Kwakiutl', Victoria: Provincial Archives (Unpublished Field Notes).

Mauzé, M., 1984. Enjeux et Jeux du Prestige : des Kwagul Méridionaux aux Lekwiltoq (Côte Nord-ouest du Pacifique). Paris: L'École des hautes études en sciences sociales (EHESS) (Thèse pour le doctorat de 3ème cycle).

Ship Records, n.d. '*Beaver*, Sloop Converted Into a Brig. Hudson's Bay Company Archives.' Manitoba: Provincial Archives (C1/207, 208).

Tate (Family), n.d. 'Indian Legend of the Nanaimo Chief.' Victoria: Provincial Archives (Add Mss 303, vol. 1.).

Thom, B., 2005. Coast Salish Senses of Place: Dwelling, Meaning, Power, Property and Territory in the Coast Salish World. Montreal: McGill University (PhD thesis).

Yale, J.M., 1838–39. 'Census of Indian Population.' Manitoba: Provincial Archives (Hudson's Bay Company Archives, B.223).

9

At the Edge of the Salish Sea: The Nuu-chah-nulth and their Relatives in Broader Context

Alan D. McMillan

Vancouver Island, the largest island on the Pacific coast of North America, forms the western side of the Salish Sea. Its substantial landmass encompasses considerable geographic, cultural and linguistic diversity. This chapter focuses on the people of its western coast, the Nuu-chah-nulth and Ditidaht, along with their Makah relatives across Juan de Fuca Strait to the south. Networks of kin ties and trade, however, linked the people of the west coast with those across the island and beyond, requiring a perspective that extends to the entire island and adjacent waterways.

This chapter discusses some distinct west coast cultural features, along with broader interactions with the peoples of the Salish Sea. Archaeological research documents cultural patterns that were deeply rooted in place, persisting over millennia. Despite this overall stability, the long history of the Nuu-chah-nulth and their relatives also featured wars, territorial shifts and the introduction of new goods and ideas through contact with their neighbours. Such dynamic cultural processes continued into recent and modern times.

Vancouver Island's Peoples: The Nuu-chah-nulth and their Neighbours

The Nuu-chah-nulth language belongs to the southern branch of the Wakashan language family. Linguists distinguish three closely related languages in this division, although the people consider themselves a single cultural unit, often referred to as the 'West Coast people' or 'Westcoasters' (McMillan 1999; Coté 2010; Arima and Hoover 2011). Nuu-chah-nulth, the largest of the three, extends as a string of dialects from Cape Cook, northwest of Kyuquot Sound, to southeast of Barkley Sound (Figure 9.1).

The Nuu-chah-nulth were never a single political unit, consisting instead of numerous politically independent villages scattered along this rugged coastline. Even today, following a long series of amalgamations, 14 nations are represented by the Nuu-chah-nulth Tribal Council.

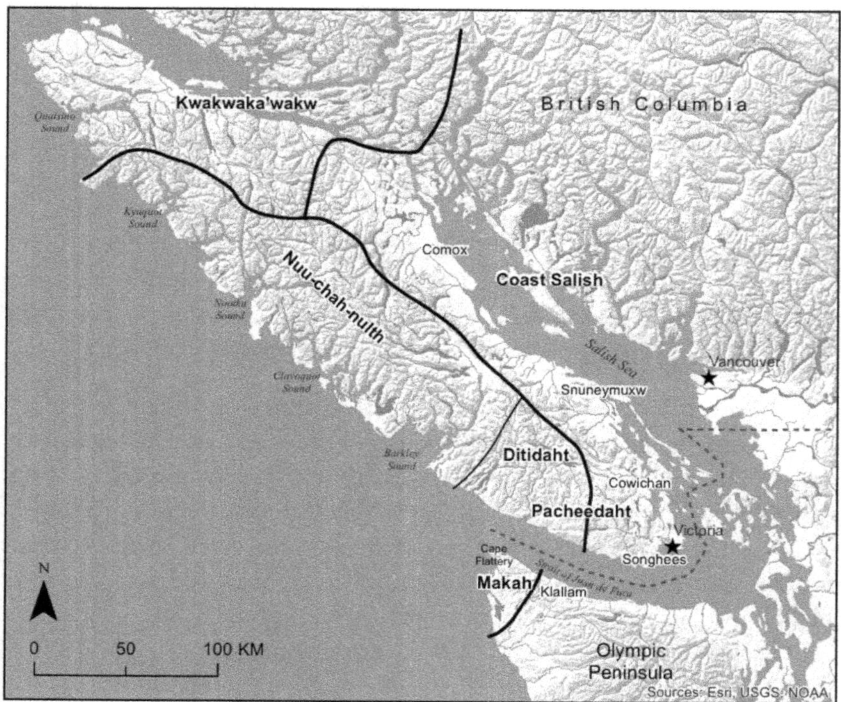

Figure 9.1 Major linguistic and political groups mentioned in the text

Note: Kwakwa̱ka̱'wakw territory is shown as prior to their southward expansion into the northern Salish Sea early in the nineteenth century.

Source: Alan D. McMillan.

To the southeast, along Nitinat Lake and the adjacent outer coast, are the Ditidaht. Today the Ditidaht are members of the Nuu-chah-nulth Tribal Council, although the southernmost Ditidaht-speakers, the Pacheedaht around Port Renfrew, are politically autonomous. Across the entrance to Juan de Fuca Strait, on the Olympic Peninsula, are the third group, the Makah. The Makah homeland around Cape Flattery juts into the open Pacific, providing ready access to maritime resources and a strategic location for trade.

The Ditidaht, Pacheedaht and Makah controlled the entrance to Juan de Fuca Strait, and thus into the Salish Sea. Prior to the nineteenth-century imposition of an international border separating these people, large canoes frequently traversed this waterway, bringing goods to trade or guests to a feast. Numerous ties of kinship, commerce and shared ceremonial life linked the people on both sides of the strait.

To the north are the Northern Wakashan languages, particularly Kwakwala, spoken on northern Vancouver Island and the adjacent islands and mainland. Speakers of this language, collectively known as the Kwakwa̱ka̱'wakw, border the Nuu-chah-nulth on northwestern Vancouver Island around Quatsino Sound and occupy the northeastern portion of the island and the adjacent islands in Queen Charlotte Strait (Figure 9.1). A nineteenth-century southern expansion of several Kwakwa̱ka̱'wakw groups brought them into the northern Salish Sea, where they displaced some Salish peoples and merged with others (Kennedy and Bouchard 1990; Mauzé, this volume).

In the heart of the Salish Sea, with territories covering most of the east coast of Vancouver Island, the Gulf Islands, Puget Sound and the mainland extending up major rivers such as the Fraser, are the Coast Salish people, speaking various languages in the coastal branch of the widespread Salishan family. Individual Salishan languages of eastern Vancouver Island include K'ómox in the north; Halkomelem (although the island dialect of this language is generally rendered as Hul'q'umi'num'), spoken by groups such as the Cowichan and Snuneymuxw (Nanaimo); and the Straits Salish, including the Songhees around Victoria and the Klallam across Juan de Fuca Strait.

Numerous social, economic, ceremonial and kinship ties link the Indigenous peoples of Vancouver Island and environs, despite the differences in language and the division into numerous independent political units. Many people spoke two or more languages, as a result of intermarriage and to facilitate trade. In a recent study of the Makah, Reid (2015) employed the concept of 'borderlands' to place the Makah at the centre of a social world that in the late eighteenth and early nineteenth centuries extended along the coast from northern Vancouver Island to the Columbia River and east into the Salish Sea.

Borderlands, in this use, refer to 'spaces shared and contested by distinct peoples' that are linked through networks of social, political and economic relationships (Reid 2015: 14). The Nuu-chah-nulth world can also be examined through this framework. A later section of this chapter ('Contact, Trade and Kin Ties') provides examples of individual and group mobility, networks of economic exchange, widespread kin ties, bilingualism and cultural hybridity. As social constructs, borders and networks of contact were historically contingent, changing over time.

Table 9.1 presents a timeline of significant stages and events in Nuu-chah-nulth culture history, as discussed in this chapter.

Table 9.1 Timeline of Nuu-chah-nulth culture history

5,000 to 3,000 years ago: Earliest dates from Nuu-chah-nulth-area sites. Those in Barkley Sound are associated with raised landforms (often directly behind later villages) reflecting occupation when sea levels were several metres higher than today. Reliance on fishing and marine mammal hunting (including whaling) characterises even this earliest known period.
4,000 to 2,000 years ago: Distinctive stone artefacts from the Shoemaker Bay site, at the top of Alberni Inlet, indicate close ties with cultures of the Salish Sea and possibly the presence of Salish people.
3,000 years ago: Basketry in waterlogged deposits at the Hoko River site suggests long-term residence of the Makah on the Olympic Peninsula.
2,000 to 250 years ago: Late Period sites in Barkley Sound, Nootka Sound and Hesquiat Harbour, associated with near-modern sea levels, document Nuu-chah-nulth life prior to European contact. Large shell middens demonstrate the importance of shellfish in the diet. Fishing, particularly for herring, rockfish and salmon, was a major part of west coast life. A variety of marine mammals, including seals, sea lions, porpoises and whales, were hunted. Late in this period, several houses at Ozette, a village on the outer Olympic Peninsula, were sealed under wet clay when the hillside behind gave way. The waterlogged conditions led to excellent preservation of wooden and bark implements, offering a more complete view of life prior to contact with Europeans.
Late 18th century: First recorded contact with European ships, beginning with the Spanish under Pérez in 1774 and the British under Cook in 1778. Beginning of an intensive, although short-lived, maritime fur trade fueled by the European demand for sea otter pelts.
1840s to 1860s: Increased Euro-Canadian colonisation and settlement. The establishment of Fort Victoria on southern Vancouver Island in 1843 particularly affected the Makah and the southern Nuu-chah-nulth groups.
1871: When British Columbia joined Canada, responsibility for 'Indians and land reserved for Indians' shifted to the federal government. The *Indian Act* was passed in 1876.

1880s: Small pockets of land were designated as 'Indian reserves' throughout Nuu-chah-nulth territory.
Late 19th century: A commercial fur sea hunt, involving many Nuu-chah-nulth, Ditidaht and Makah men, provided lucrative new economic opportunities. Seasonal work in the fish canneries and the hop fields drew many into the Salish Sea area.
1958: Formation of the West Coast Allied Tribes, which later became the Nuu-chah-nulth Tribal Council, to represent the common interests of the related peoples along Vancouver Island's west coast. Fourteen Nuu-chah-nulth nations make up the Tribal Council today.
2011: The Maa-nulth Agreement, between five Nuu-chah-nulth nations and the governments of Canada and British Columbia, came into effect. The five Nuu-chah-nulth nations are now self-governing and in control of their lands, freed from the restrictions of the *Indian Act*.

Source: Alan D. McMillan's tabulation.

The Ancient Past

The west coast of Vancouver Island has received considerably less archaeological attention than more populated areas such as the Salish Sea. Large portions of Nuu-chah-nulth territory lack significant excavated data with radiocarbon dates. Nootka Sound and nearby Hesquiat Harbour, along with Makah territory on the Olympic Peninsula, are areas of more extensive research. More recent work, including both site inventory and excavation at a number of former village sites, has been focused on Barkley Sound.

Rising sea levels have made traces of the earliest occupation difficult to recover. The oldest radiocarbon-dated deposits extend only to about 5,000 to 2,500 years ago. Several such sites have been investigated in Barkley Sound, in each case on a raised landform adjacent to a major Late period village, corresponding to a time when relative sea levels stood several metres higher than today. These raised elevation sites show significant differences in artefact forms from the adjacent Late period sites (McMillan 1998, 2003).

Several distinctive artefact types considered defining traits of the Locarno Beach stage in the Salish Sea (Mitchell 1990; Angelbeck, this volume) have counterparts in contemporaneous deposits in Barkley Sound (McMillan 1998; McMillan and St Claire 2005). Shoemaker Bay, at the end of the long Alberni Inlet that extends from Barkley Sound far into the centre of the island, seems to have been closely linked with cultural developments in the Salish Sea throughout its 4,000-year occupation (McMillan and St Claire 1982). Broad cultural similarities spanning Vancouver Island at this time may indicate earlier population distributions or extensive cultural contact.

As the southernmost in a long chain of peoples speaking Wakashan languages, the ancestors of the Makah must have at some point in their history arrived from the north. Archaeological evidence, however, suggests a lengthy tenure in their present homeland. The Hoko River site, along Juan de Fuca Strait at the eastern edge of traditional Makah territory, offers some insight on this issue.

The stone artefacts excavated from the site's 'dry' portion closely resemble those from contemporaneous Locarno Beach sites in the Salish Sea. The 'wet' portion of the site, however, has preserved basketry artefacts that may provide more sensitive indicators of ethnicity. Through detailed comparisons with basketry from other Makah-area and Salish Sea sites, Croes (1989, 1995) has convincingly argued that the Hoko River basketry reflects Wakashan, rather than Salishan, manufacture, placing the ancestors of the Makah on the Olympic Peninsula by 3,000 years ago.

Late period sites in Barkley Sound, associated with modern sea levels, date to within the last two millennia. Many became major historic villages, with names and histories known through oral traditions. Several clearly show the outlines where large traditional plank houses once stood, with the large size and central placement of some dwellings thought to indicate status distinctions (Mackie and Williamson 2003; McMillan and St Claire 2012).

Oral histories tell of great whaling chiefs who lived in these large houses and directed the economic and ceremonial activities of the community. The excavated materials from these sites, along with Yuquot in Nootka Sound and a cluster of sites in Hesquiat Harbour, can be placed in a complex termed the 'West Coast culture type' (Mitchell 1990), considered to be the archaeological expression of Nuu-chah-nulth culture prior to contact with Europeans. Small bone points, interpreted as components of a wide variety of fishing implements, dominate these assemblages.

A much more complete view of this Late period, however, comes from Ozette, the southernmost of the large Wakashan villages, on the outer coast of the Olympic Peninsula. Not long before contact with Europeans, the hillside above a portion of the village gave way, destroying several traditional plank houses but preserving the architectural elements and the house contents under a thick layer of wet clay (Kirk 2015). While only discarded objects of bone, shell and stone tell of life at most village sites, the Ozette houses contain almost the complete material possessions of the occupants at the time the disaster struck.

The small bone points that are so common at most west coast sites were found at Ozette as parts of intact fishing tackle, complete with their wooden shanks and bark wrapping. The excellent preservation of organic materials allows unparalleled insights into the lives of these west coast people, including their art and beliefs. Images of whales and thunderbirds adorn many household items, reflecting the cultural importance placed on whaling (Daugherty and Friedman 1983; Kirk 2015; McMillan 2019).

West Coast culture, as revealed archaeologically, was heavily based on the sea. Fish played a dominant role in the diet. Excavated Barkley Sound sites reveal a wide range of fish species taken, from small herring and anchovy to giant bluefin tuna (McMillan and St Claire 2005, 2012; McKechnie 2012). Herring and rockfish were particularly important, with salmon playing an increased role in the most recent period. Marine mammals, including various species of seals, sea lions, porpoises and whales, also featured heavily. A successful whale hunt, with the whalers particularly targeting humpbacks and greys, provided an economic windfall (Drucker 1951; Arima and Hoover 2011; McMillan 2015).

Fur seals were also commonly taken, as evidenced by their numerous skeletal elements at almost all Nuu-chah-nulth archaeological sites (McMillan 1999: 140; Crockford et al. 2002; McMillan and St Claire 2005, 2012). As the northward migration of the fur seal herds took them close to the shore as they passed Cape Flattery, their remains are particularly common in Makah-area sites (Kirk 2015: 63–4). The deep deposits of discarded shells that make up most former village sites also attest to the vital dietary role of shellfish, particularly the large outer coast California mussel.

Many early historic accounts express admiration for the large, sturdy, seaworthy canoes, constructed from the huge west coast cedar trees, that allowed ocean travel (Figure 9.2). West coast hunters ventured considerable distances from land in pursuit of marine mammals. An oral tradition recorded early in the twentieth century tells of a whaler and his crew being so far out to sea that the tops of the snow-capped mountains were all that was visible of the land (Sapir et al. 2009: 67–8). Similarly, large halibut could be taken from banks far offshore. Nuu-chah-nulth fishing parties set off at dusk from their outer coast villages in Barkley Sound and paddled all night to reach the halibut banks by dawn (Sapir and Swadesh 1955: 41; Sproat 1987: 151). Halibut and whale bones in west coast archaeological sites indicate that such excursions were a long-established practice.

Figure 9.2 'At Nootka', a staged photograph by Edward S. Curtis, 1915

Note: This canoe, of a size used for fishing or sealing, shows the typical Nuu-chah-nulth form. Larger canoes were used for whaling and for transporting people and goods along the coast.

Source: Library and Archives Canada C-0020852.

This emphasis on open-ocean fisheries and marine mammal hunting distinguishes Nuu-chah-nulth adaptations from those in the Salish Sea, which focused more on terrestrial, riverine and estuarine resources. In a study that compared mammalian faunal remains from a large sample of archaeological sites around southern Vancouver Island and adjacent areas, McKechnie and Wigen (2015) demonstrated that assemblages from west coast sites are dominated by marine mammal remains, with terrestrial mammals relatively uncommon, whereas the reverse holds true in assemblages from the Salish Sea. This study excluded whales from the samples considered, making the actual differences even more pronounced. Although environmental differences explain much of this long-standing distinction, cultural traditions also have to be considered.

For the Nuu-chah-nulth and their relatives, whaling was not only a major economic activity but played a central role in chiefly prestige, power and identity (Drucker 1951; Arima and Hoover 2011; McMillan 2015). Oral traditions relate accounts of famed whaling ancestors (e.g. Sapir et al. 2004), chiefly names often refer to whaling or supernatural beings associated with whaling and the art is dominated by images related to whaling (Coté 2010; McMillan 2015, 2019). Although these traits strongly distinguish the peoples of the west coast from those of the Salish Sea, long-standing contact and interactions led to some transfer of traits. The Klallam, Salish neighbours of the Makah along Juan de Fuca Strait, acquired from the Makah not only the practice of whaling and its technology, but also associated ritual activities (Gunther 1927: 204).

Contact, Trade and Kin Ties

The large Nuu-chah-nulth, Ditidaht and Makah canoes allowed travel and transport of goods over great distances along the coast. Trade, social events such as reciprocal feasting and intergroup marriages, and occasional raids or more prolonged warfare all led to contacts with distant groups. In the early historic documents, the Makah (generally referred to as the 'Classet' or 'Klazzart' from the Nuu-chah-nulth word '*Tl'aa7as7atẖ*', or 'people of the outside coast') appear as far north as Nootka Sound (McMillan 1999: 6; Reid 2015: 24). The three powerful chiefs of the early fur trade period—Maquinna at Nootka Sound, Wickaninish at Clayoquot Sound and Tatoosh at Cape Flattery—were all linked through ties of marriage, thus facilitating alliances and trade (Marshall 1993: 213). Early historic accounts frequently mention the arrival of canoes bearing people and goods from one of these chiefly territories to another.

Ethnographic and historic accounts describe extensive trade involving a variety of foodstuffs and raw materials. The Nuu-chah-nulth of Barkley Sound traded dried halibut, herring and cedar bark baskets to the people of southern Vancouver Island in exchange for camas bulbs and swamp rushes for making mats (Sproat 1987: 58). The Makah travelled to Barkley Sound to obtain large cedar planks for house construction, particularly valuing those prepared with painted designs, transporting them across the open sea on their large canoes (Banfield 1858a). The Makah took whale blubber and dried halibut to the Nuu-chah-nulth, receiving in exchange dried salmon, cedar bark, dentalium shells, canoes and slaves, many of which they then traded south along the coast to the Columbia River (Swan 1870: 31).

Jewitt's early nineteenth-century account at Nootka Sound notes the Makah ('Kla-iz-zarts') bringing 'great quantities' of whale oil, whale sinew, *yama* (salal berry) cakes, camas roots, elk hides and sea otter pelts to trade (Jewitt 1967: 78–9). The Makah also made trading expeditions into the Salish Sea as far as Puget Sound, trading their dried halibut for spring salmon and clams (Reid 2015: 37). In 1792, Tatoosh impressed a Spanish expedition with his detailed geographical knowledge of the length of Juan de Fuca Strait (Reid 2015: 37).

Some accounts of cultural contacts describe armed conflict, from simple raiding to longer-term warfare, often involving groups at some distance. A well-documented example is the lengthy series of battles between the Barkley Sound Nuu-chah-nulth and the Klallam Salish (Arima et al. 1991: 208–13, 222–30; Sapir et al. 2009: 291–4). In these hostilities, which spanned several decades, the destruction of Nuu-chah-nulth villages by the Klallam was followed by a devastating attack by a combined Nuu-chah-nulth and Ditidaht force on the Klallam home territory (McMillan and St Claire 2012: 17–8).

Wars also involved the allied Ditidaht and Pacheedaht against such Vancouver Island Salish as the Saanich (Arima et al. 1991: 309–11). The Makah also waged occasional bloody skirmishes with their Klallam neighbours (Swan 1870: 50–1), as well as their Ditidaht relatives (Arima et al. 1991: 300–4). These hostilities caused only brief interruptions in Makah trading practices, which could be maintained through use of intermediaries until peaceful relations could be restored (Swan 1870: 30).

Extensive trade patterns were clearly well established when Europeans arrived in the late eighteenth century and were very likely of long duration. However, the perishable trade items mentioned in the ethnographic accounts leave little trace in the archaeological record, making it difficult to discern the antiquity of these practices. Most archaeological studies of trade rely on the presence of lithic tools made from distinctive non-local materials, although these are rare in west coast archaeological contexts.

Obsidian, for example, which does not occur naturally on Vancouver Island, can be accurately sourced through its trace elements. Small numbers of obsidian artefacts from the early components at Ts'ishaa in Barkley Sound and Shoemaker Bay at the head of Alberni Inlet were traced to sources in central Oregon (McMillan and St Claire 1982: 70, 2005: 86–87). Such long-distance movement of material for tools suggests down-the-line trading rather than any direct contact. It does, however, show that the people of western Vancouver Island were part of a broad trade network that existed by at least 2,500 years ago.

Dentalium shells, valued for their decorative uses, are another example of materials found archaeologically that indicate trade networks. The northern Nuu-chah-nulth, particularly the Kyuquot and Ehattesaht, had important dentalium beds in their offshore waters and developed ingenious devices to procure the shells from considerable depths (Drucker 1951: 111–3). Other groups had to rely primarily on trade to obtain these valued items. As the people of the Salish Sea lacked direct access to dentalium, any occurrence in archaeological sites indicates trade with the west coast.

Large numbers of dentalium shells placed with human burials in midden sites are considered a characteristic feature of the Marpole stage, dated to about 2,000 years ago, in the Salish Sea (Mitchell 1990: 345; Angelbeck, this volume). At one Gulf Islands site, over 3,800 dentalium shells were associated with Marpole-period burials (Burley 1989: 22, 24). The abundance of these distinctive shells in such contexts clearly demonstrates the importance of trade connections spanning Vancouver Island at that time. Dentalium continued to be a wealth or prestige item among Salish groups well into the historic period (Suttles and Lane 1990: 493).

At Ozette, the excellent preservation of basketry in waterlogged deposits allows glimpses into contacts and trade patterns shortly before European arrival. Discoveries of basket types judged to be 'foreign' in a Makah context led Croes (2003) to posit widespread trade relations. Particularly noteworthy were a number of distinctive coiled baskets, a type characteristic of the Salish Sea area, that he interprets as trade items.

In addition to lengthy canoe voyages, major trade and travel corridors across Vancouver Island linked the west coast with the Salish Sea. In the north, important overland trails led from the heads of Kyuquot and Nootka Sounds (Figure 9.3). Both trails began at villages named 'Tahsis', meaning 'doorway' (Drucker 1951: 228). After a trek across the mountains, travellers from Nootka Sound reached Woss and Nimpkish Lakes, then descended the Nimpkish River to its mouth, reaching the main village of the 'Namgis ('Nimpkish') people, one of the Kwakwa̱ka̱'wakw groups of northeastern Vancouver Island. The Spanish of the 1791 Malaspina expedition were told that Nuu-chah-nulth trading parties walked the trail from Tahsis 'carrying on their shoulders their small canoes', which they used to traverse the two large lakes (Cutter 1991: 95).

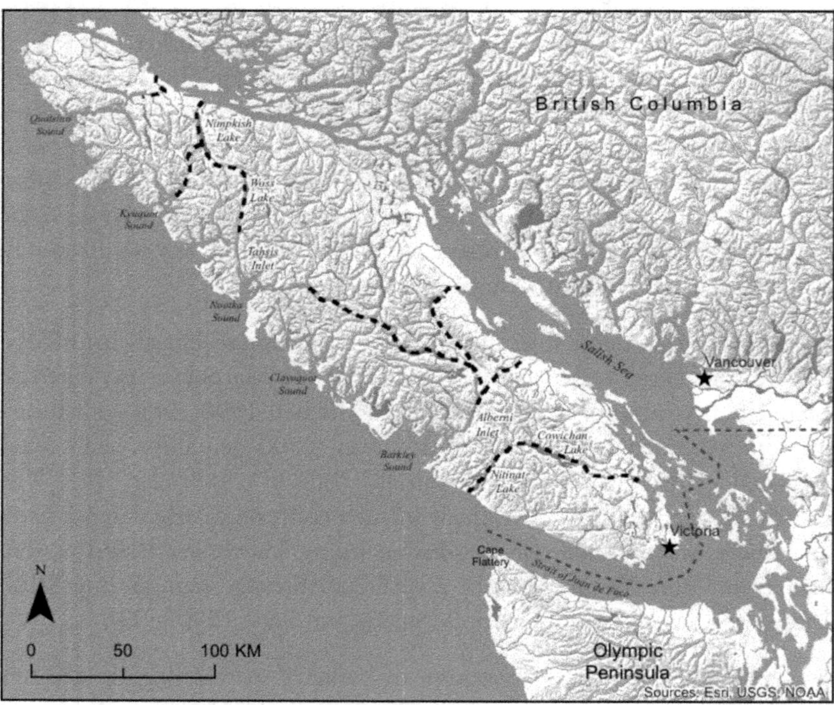

Figure 9.3 Major Indigenous travel corridors across Vancouver Island
Source: Alan D. McMillan.

A major segment of the trail was controlled by the Ninelkaynuk, who had a village far up the Nimpkish River prior to their mid-nineteenth-century population decline and amalgamation with the 'Namgis (Galois 1994: 319, 366). Some individuals in that group spoke both Kwakwala and Nuu-chah-nulth as a result of trade ties and strategic intermarriages (Galois 1994: 319). Moziño (1970: 63), at Nootka Sound in 1792, noted the importance of such kinship in maintaining 'the long-standing commercial relations' across the island.

The term 'grease trail' commonly applied to this route reflects an important trade commodity from the Kwakwa̱ka̱'wakw to the Nuu-chah-nulth. Oil rendered from the eulachon, a small oily smelt that was not available on western Vancouver Island, was highly valued as a condiment for dried foods. An ethnographic account tells of a Nuu-chah-nulth chief who sent a party of young men across the trail to buy eulachan oil as a high-prestige commodity to serve at a feast (Drucker 1951: 375). Other foodstuffs and luxury items featured prominently in this cross-island trade.

Manby (1992: 255), at Nootka Sound with Vancouver in 1793, witnessed the arrival of canoes led by Tatoosh bringing dentalium to trade, which Maquinna then sent up the Tahsis trail to the Nimpkish River. Jewitt (1996: 80) noted in an 1804 journal entry that traders arriving via the trail brought salmon and dried clams to Nootka Sound. 'Namgis researcher Diane Jacobsen lists her people's trade items to the west coast as eulachan oil, fish roe, dried clams, mica, wolf and sea otter skins, yellow cedar bark robes, and cedar bark headdresses and neck rings, in exchange for such west coast goods as whale meat, halibut and dentalium shells (Stafford and Jacobsen 2005: Appendix 2). She notes the importance of strategic marriages and the transfer of such valued intangible possessions as names, songs and dances in strengthening trade ties.

As few of these items would preserve in the archaeological record, we have limited glimpses into the early use of this trade route. However, an archaeological impact assessment along a portion of the trail led to the discovery of several sites, including a rock shelter near Woss Lake with evidence of repeated short-term occupation (Stafford and Jacobsen 2005; Stafford and Christensen 2007). As the rock shelter is located far inland and directly along the trail, it provides the best evidence for ancient use of this important travel corridor. Excavation revealed charcoal and calcined bone, presumably from a hearth, along with simple chipped stone tools and flakes, from a level dated to over 6,000 years ago (Stafford and Christensen 2007).

A small piece of obsidian from higher levels is estimated by the excavators to date to over 2,000 years ago (Stafford and Christensen 2007: 20). Although it could not be traced to a specific source, the absence of known obsidian flows on Vancouver Island indicates that this material had been transported over a considerable distance (Stafford and Christensen 2007: 36). Additional types of stone tools from the upper levels demonstrate that people continued to use this location as a camping stop along the trail into quite late times. Doubtlessly a wide variety of perishable goods, including desirable foodstuffs as documented ethnographically, passed in both directions along this route.

After contact with Europeans in the late eighteenth century, this long-established travel corridor took on new importance as the Nuu-chah-nulth used it to distribute items of European manufacture in exchange for furs valued by the new arrivals. At Nootka Sound in 1792, the American traders aboard the ship *Columbia* were told of this overland trail and that Maquinna himself frequently travelled it (Howay 1990: 265). The Spanish with Malaspina learned that Maquinna and his people took copper and other European goods over the Tahsis trail to trade to the Kwakw̱a̱ka̱'wakw for sea otter pelts (Cutter 1991: 95).

Menzies, who arrived with Captain Vancouver at the mouth of the Nimpkish River in 1792, observed that the Kwakw̱a̱ka̱'wakw were well supplied with muskets, which he determined had been obtained through trade with Maquinna, the 'grand agent' of commerce (Newcombe 1923: 80). In fact, he noted that the villagers at the mouth of the Nimpkish River at this time attributed 'most of the Articles of European Manufactory in their possession' to trade with the people of Nootka Sound 'by some inland communication' (Newcombe 1923: 88).

Although the Tahsis trail is particularly well documented, other vital travel routes crossed Vancouver Island. From the head of Muchalat Inlet, which extends westward from Nootka Sound, a trail ran along the centre of the island to the Alberni Valley and upper Alberni Inlet (Drucker 1951: Map 1). Hunters from the two areas occasionally met inland along this trail (Sapir et al. 2009: 361, note 53). From Alberni, it was a short journey over a low mountain pass to the Salish Sea, providing relatively easy access for the

Nuu-chah-nulth groups around Barkley Sound. One main route, partly by land and partly by water, led to Horne Lake and down the Qualicum River to the coast; another branch led northward to the K'ómox (Hayman 1989: 30–1). Drucker (1951) noted that the Nuu-chah-nulth of upper Alberni Inlet had a number of cultural traits that distinguished them from the outer coast Nuu-chah-nulth, attributing these distinctions to extensive contact and kin ties with their Coast Salish neighbours. One Alberni group may have descended from formerly Salish peoples who adopted Nuu-chah-nulth language and culture after their partial displacement by arrivals from Barkley Sound (Boas 1891: 584; McMillan and St Claire 1982: 12–4; Arima et al. 1991: 76–8;).

An oral narrative tells of traders from Qualicum arriving in the Alberni Valley by the overland trail with blankets of deerskin, racoon and mountain goat wool to trade. In exchange, they received robes of yellow cedar bark and marriage partners to strengthen ties between the two groups (Sapir et al. 2009: 106–8). Cross-island contact did not always involve peaceful trade and cultural borrowing; warriors from the K'ómox, Qualicum, Snuneymuxw and other groups raided across the mountains, killing many people at Alberni (Arima et al. 1991: 311–2; Hayman 1989: 129).

To the south, a major travel corridor across the island involved poling canoes up the Cowichan River, skirting the series of falls near its head to reach Cowichan Lake, among the mountains in the centre of the island (Figure 9.4; Rozen 1985: 216; Hayman 1989). After paddling their canoes to the lake's western end, travellers took a short overland trail to Nitinat River, then descended that river and Nitinat Lake to arrive at the Ditidaht village of Whyac on the outer coast (Figure 9.3; Hayman 1989). This was a long-established and frequently travelled corridor that became a major route for nineteenth-century Euro-Canadian exploration (Pemberton 1860; Hayman 1989). Both the Cowichan Salish and the Ditidaht hunted inland around Cowichan Lake, with contact leading to intermarriage and kin ties that linked the east and west coasts. In 1858, WE Banfield wrote of an 'Indian trail' with 'considerable intertraffic' between Cowichan and the west coast via a large interior lake, along which the Cowichan brought potatoes to exchange for halibut and whale oil (Banfield 1858b).

Figure 9.4 Two men pole a traditional Nuu-chah-nulth-style canoe, through the rapids, upper Cowichan River

Note: This was part of a major travel corridor across Vancouver Island, from Cowichan Bay on the Strait of Georgia to the mouth of the Nitinat River on the west coast.

Source: Alberni Valley Museum PN03469.

Explorer Robert Brown, who traversed the island from Cowichan Bay to Whyac in 1864, referred to the resident population at Cowichan Lake as the 'Masolomo' (Hayman 1989), his rendering of the Hul'qumi'num word *muthe'lumuhw*, applied to the people of the west coast (Rozen 1985: 306). Brown noted that these people came 'from Nitinat', indicating that they would have been Ditidaht-speakers (Hayman 1989: 58). Brown's guide on this expedition was Kakalatza, the chief of Somenos village on the lower Cowichan River, who frequently travelled this route to the west coast to visit his son at Whyac.

The extent of kin ties across the island is clear in Brown's comment on his guide: 'His wife is a Masolomo, & most of his tribe are half Somenos, half Masolomo' (Hayman 1989: 58). In this Late period, some Ditidaht people travelled down the Cowichan River to Somenos Village to plant potatoes (Hayman 1989: 58). Despite traditions of earlier hostilities, the Ditidaht and Cowichan established a network of kin ties that allowed safe travel and access to resources across Vancouver Island.

Contact between different peoples led to exchanges of goods and ideas far beyond the merely material and technological. Songs, chiefly names and ceremonial prerogatives were highly valued properties that could be passed from one group to another, particularly through intermarriage. In this manner, each group could acquire the major rituals of their new allies and incorporate them into their own ceremonial practices. As Rozen (1985: 220) noted:

> Today some Nitinat [Ditidaht] families have the right to use sxwaixwe masks through intermarriage with the Cowichan and some Cowichan families have the right to use the Nootka Wolf Dance.

Similarly, the important Salish sxwaixwe mask and dance was transferred north to the Kwakwa̱ka̱'wakw through intermarriage with the K'ómox (Boas and Hunt 1921: 892–4; Mauzé, this volume), while the K'ómox acquired many elements of Kwakwa̱ka̱'wakw ceremonial life (Boas and Hunt 1921: 951–3; Kennedy and Bouchard 1990: 448–9).

Economic Shifts: European Arrival to Today

Long-standing trade relationships took on new importance in the maritime fur trade of the late eighteenth century. Powerful chiefs sought to control access to the newcomers' ships, forcing other groups to trade through them, and to dominate the surrounding areas to control the availability of furs desired by the outsiders. Maquinna consolidated his economic and political power through his trade monopoly at Nootka Sound, dominating the trade of the more northerly Nuu-chah-nulth. His control of the Tahsis trail across Vancouver Island added to his trade wealth and power.

At Clayoquot Sound, Wickaninish expanded his territory and power through a series of wars, forcing other chiefs to surrender their furs for him to trade. As far south as Barkley Sound, the British and American traders found few furs available as they had been surrendered to Wickaninish (Howay 1990: 79; Gibson 1992: 115). From his base at Cape Flattery, Tatoosh dominated the Pacheedaht across the strait, enlisting their warriors in his military excursions and controlling their trade with outsiders (Reid 2015: 53, 70). The American sailors on a trading ship in Juan de Fuca Strait in 1789 complained that the Pacheedaht and others along the strait had no furs as these had already been gathered by Tatoosh (Howay 1990: 71). With both sides of western Juan de Fuca Strait under Tatoosh's control, the Makah dominated the early maritime fur trade into the Salish Sea.

In the early nineteenth century, the Hudson's Bay Company (HBC) emerged as the dominant trading force. The Makah used their strategic position at Cape Flattery to intercept the HBC ships heading into Juan de Fuca Strait, trading fish, furs and whale oil for a wide range of the company's goods. The Makah intensified whaling, rendering great quantities of whale oil to trade to the HBC, which shipped it to London to be distilled for lighting homes and other buildings (Reid 2015: 98). Small private American trading vessels also stopped to purchase whale oil, as well as furs and fish, from the Makah. The establishment of Fort Victoria on southern Vancouver Island in 1843 bound the Makah even more closely to the HBC.

Large numbers of Makah canoes laden with goods to trade at the fort traversed the strait into the central Salish Sea. As well as bringing whale oil, Makah, Ditidaht and Nuu-chah-nulth trading groups arrived with salmon and potatoes in demand by the officers and labourers at the fort (Reid 2015: 100). In 1852, six years after the international border had been established, the Indian agent for the Makah lamented that they could rarely be found in American territory, instead trading mainly at Vancouver Island (Starling 1852:172).

During the second half of the nineteenth century, the people of the west coast sought new economic opportunities. By the 1850s, the expanding lumber industry required lubricants for the sawmills, a need the Westcoasters met by catching large quantities of dogfish and basking sharks and rendering their livers into oil that could be sold to the coastal trading ships (Swan 1870: 29; Drucker 1951: 12; Arima and Dewhirst 1990: 409). Slightly later, the demand for fur seal skins led commercial sealing schooners to take on board Nuu-chah-nulth, Ditidaht and Makah men with their canoes and sealing gear (Drucker 1951: 13; Arima and Dewhirst 1990: 409; Reid 2015: 178).

The schooners set out on lengthy voyages, seeking the fur seal herds as far away as California and the Bering Sea. This lucrative hunt, in which the Westcoasters were highly adept, allowed some Nuu-chah-nulth, Ditidaht and Makah men to purchase their own schooners (Arima and Dewhirst 1990: 409; Marshall 1993: 300–1; Reid 2015: 179–82). This enterprise expanded the world-view of the sealers, as the voyages often took them to places as distant as Japan and San Francisco.

Around the same time and into the twentieth century, many people began to leave their villages on the west coast for new seasonal economic opportunities around the Salish Sea. Entire families relocated to the fish canneries on the Fraser River, men fishing for the companies while women worked at cleaning and canning the fish. Others worked at picking hops in the fields around Puget Sound (Arima and Dewhirst 1990: 409; Drucker 1951: 13; Reid 2015: 184). Most returned to their home villages for the winter. Such movements and activities brought the Nuu-chah-nulth, Ditidaht and Makah into contact with a wide range of other Indigenous and non-Indigenous peoples.

Today, the people of the west coast, like other First Nations in British Columbia, seek to assert some level of control over their traditional territories, to share in the economic benefits derived from those lands, and to revitalise many of their cultural practices. Five Nuu-chah-nulth nations, comprising the Maa-nulth treaty group, have finalised a modern treaty with the governments of British Columbia and Canada. That agreement, which came into effect in 2011, established five self-governing nations with jurisdiction over their own lands. Several other Nuu-chah-nulth nations are in the process of negotiating treaties or have launched legal action regarding Aboriginal title to their traditional lands.

A pressing challenge for Nuu-chah-nulth governments is to stimulate new economic developments to support programs for adequate housing, educational facilities and social and health services, as well as to provide employment for their members. Declines in the fishing and logging industries that supported so many in the recent past have made other employment options particularly important. Despite some successful ventures, the isolated nature of many west coast communities has restricted opportunities and led to substantial off-reserve movement.

Large numbers of Nuu-chah-nulth, Ditidaht and Makah people now reside around the Salish Sea, in such urban centres as Campbell River, Nanaimo, Victoria, Vancouver and Seattle, while still maintaining ties with their home communities. These modern movements echo earlier connections across Vancouver Island that linked the people of the west coast to the Salish Sea.

References

Arima, E. and J. Dewhirst, 1990. 'Nootkans of Vancouver Island.' In W. Suttles (ed.), *Handbook of North American Indians*, vol. 7: *Northwest Coast*. Washington, DC: Smithsonian Institution.

Arima, E. and A. Hoover, 2011. *The Whaling People of the West Coast of Vancouver Island and Cape Flattery*. Victoria: Royal BC Museum.

Arima, E.Y., D. St Claire, L. Clamhouse, J. Edgar, C. Jones and J. Thomas, 1991. *Between Ports Alberni and Renfrew: Notes on West Coast Peoples*. Gatineau: Canadian Museum of Civilization (Canadian Ethnology Service Mercury Paper 121). doi.org/10.2307/j.ctt22zmdnn

Banfield, W.E., 1858a. 'Vancouver Island: Its Topography, Characteristics, etc. (No. VI, Ohiat and Netinett Sounds).' *Victoria Gazette*, 3 September. Victoria: Public Archives of British Columbia.

——, 1858b. 'Vancouver Island: Its Topography, Characteristics, &c. (IV: Distribution of a Whale by the Netinetts, and Some Accounts of the Cowichans).' *Victoria Gazette*, 28 August. Victoria: Public Archives of British Columbia.

Boas, F., 1891. 'The Nootka.' In *Second General Report on the Indians of British Columbia. Report of the 60th Meeting of the British Association for the Advancement of Science for 1890*. London.

Boas, F. and G. Hunt, 1921. 'Ethnology of the Kwakiutl.' In *Thirty-Fifth Annual Report of the Bureau of American Ethnology*. Washington, DC: Smithsonian Institution.

Burley, D.V., 1989. *Senewélets: Culture History of the Nanaimo Coast Salish and the False Narrows Midden*. Victoria: Royal BC Museum.

Coté, C., 2010. *Spirits of Our Whaling Ancestors: Revitalizing Makah and Nuu-chah-nulth Traditions*. Seattle: University of Washington Press.

Crockford, S.J., S.G. Frederick and R.J. Wigen, 2002. 'The Cape Flattery Fur Seal: An Extinct Species of *Callorhinus* in the Eastern North Pacific?' *Canadian Journal of Archaeology* 26(2): 152–74.

Croes, D.R., 1989. 'Prehistoric Ethnicity on the Northwest Coast of North America: An Evaluation of Style in Basketry and Lithics.' *Journal of Anthropological Archaeology* 8: 101–30. doi.org/10.1016/0278-4165(89)90021-4

——, 1995. *The Hoko River Archaeological Site Complex: The Wet/Dry Site (45CA213), 3,000–1,700 B.C.* Pullman: Washington State University Press.

——, 2003. 'Northwest Coast Wet-Site Artifacts: A Key to Understanding Resource Procurement, Storage, Management, and Exchange.' In R.G. Matson, G. Coupland and Q. Mackie (eds), *Emerging from the Mist: Studies in Northwest Coast Culture History*. Vancouver: UBC Press.

Cutter, D.C., 1991. *Malaspina and Galiano: Spanish Voyages to the Northwest Coast, 1791 and 1792*. Vancouver: Douglas & McIntyre.

Daugherty, R. and J. Friedman, 1983. 'An Introduction to Ozette Art.' In R. L. Carlson (ed.), *Indian Art Traditions of the Northwest Coast*. Burnaby: Archaeology Press, Simon Fraser University.

Drucker, P., 1951. *The Northern and Central Nootkan Tribes*. Washington, DC: Bureau of American Ethnology, Smithsonian Institution (Bulletin 144).

Galois, R., 1994. *Kwakwaka'wakw Settlements, 1775–1920: A Geographical Analysis and Gazetteer*. Vancouver: UBC Press.

Gibson, J.R., 1992. *Otter Skins, Boston Ships, and China Goods: The Maritime Fur Trade of the Northwest Coast, 1785–1841*. Montreal and Kingston: McGill-Queens University Press. doi.org/10.1515/9780773582026

Gunther, E., 1927. *Klallam Ethnography*. Seattle: University of Washington Press.

Hayman, J. (ed.), 1989. *Robert Brown and the Vancouver Island Exploring Expedition*. Vancouver: UBC Press.

Howay, F.W. (ed.), 1990. *Voyages of the 'Columbia' to the Northwest Coast, 1787–1790 and 1790–1793*. Portland: Oregon Historical Society.

Jewitt, J.R., 1967. *Narrative of the Adventures and Sufferings of John R. Jewitt*. Fairfield: Ye Galleon Press.

——, 1996. *A Journal Kept at Nootka Sound*. Fairfield: Ye Galleon Press.

Kennedy, D.I.D. and R.T. Bouchard, 1990. 'Northern Coast Salish.' In W. Suttles (ed.), *Handbook of North American Indians*, vol. 7: *Northwest Coast*. Washington, DC: Smithsonian Institution.

Kirk, R., 2015. *Ozette: Excavating a Makah Whaling Village*. Seattle: University of Washington Press.

Mackie, A.P. and L. Williamson, 2003. 'Nuu-chah-nulth Houses: Structural Remains and Cultural Depressions on Southwest Vancouver Island.' In R.G. Matson, G. Coupland, and Q. Mackie (eds), *Emerging from the Mist: Studies in Northwest Coast Culture History*. Vancouver: UBC Press.

Manby, T., 1992. *Journal of the Voyages of the H.M.S. Discovery and Chatham.* Fairfield: Ye Galleon Press.

Marshall, Y., 1993. A Political History of the Nuu-chah-nulth People: A Case Study of the Mowachaht and Muchalaht Tribes. Simon Fraser University (PhD thesis).

McKechnie, I., 2012. 'Zooarchaeological Analysis of the Indigenous Fishery at the Huu7ii Big House and Back Terrace, Huu-ay-aht Territory, Southwestern Vancouver Island.' Appendix B in A.D. McMillan and D.E. St Claire (eds), *Huu7ii: Household Archaeology at a Nuu-chah-nulth Village Site in Barkley Sound.* Burnaby: Archaeology Press, Simon Fraser University.

McKechnie, I. and R.J. Wigen, 2011. 'Toward a Historical Ecology of Pinniped and Sea Otter Hunting Traditions on the Coast of Southern British Columbia.' In T.J. Braje and T.C. Rick (eds), *Human Impacts on Seals, Sea Lions, and Sea Otters: Integrating Archaeology and Ecology in the Northeast Pacific.* Berkeley: University of California Press. doi.org/10.1525/california/9780520267268.003.0007

McMillan, A.D., 1998. 'Changing Views of Nuu-chah-nulth Culture History: Evidence of Population Replacement in Barkley Sound.' *Canadian Journal of Archaeology* 22(1): 5–18.

——, 1999. *Since the Time of the Transformers: The Ancient Heritage of the Nuu-chah-nulth, Ditidaht, and Makah.* Vancouver: UBC Press.

——, 2003. 'Reviewing the Wakashan Migration Hypothesis.' In R.G. Matson, G. Coupland and Q. Mackie (eds), *Emerging from the Mist: Studies in Northwest Coast Culture History.* Vancouver: UBC Press.

——, 2015. 'Whales and Whalers in Nuu-chah-nulth Archaeology.' *BC Studies* 187: 229–61.

——, 2019. 'Non-Human Whalers in Nuu-chah-nulth Art and Ritual: Reappraising Orca in Archaeological Context.' *Cambridge Archaeological Journal* 29(2): 309–26. doi.org/10.1017/S0959774318000549

McMillan, A.D. and D.E. St Claire, 1982. *Alberni Prehistory: Archaeological and Ethnographic Investigations on Western Vancouver Island.* Penticton: Theytus Books.

——, 2005. *Ts'ishaa: Archaeology and Ethnography of a Nuu-chah-nulth Origin Site in Barkley Sound.* Burnaby: Archaeology Press, Simon Fraser University.

——, 2012. *Huu7ii: Household Archaeology at a Nuu-chah-nulth Village Site in Barkley Sound.* Burnaby: Archaeology Press, Simon Fraser University.

Mitchell, D.H., 1990. 'Prehistory of the Coasts of Southern British Columbia and Northern Washington.' In W. Suttles (ed.), *Handbook of North American Indians,* vol. 7: *Northwest Coast*. Washington, DC: Smithsonian Institution.

Moziño, J.M., 1970. *Noticias de Nutka: An Account of Nootka Sound in 1792* (translated and edited by I. Higbie Wilson). Toronto/Montreal: McClelland and Stewart.

Newcombe, C.F. (ed.), 1923. *Menzies' Journal of Vancouver's Voyage, April to October, 1792*. Victoria: Archives of British Columbia Memoir No. 5.

Pemberton, J.D., 1860. 'From Cowichan Harbour to Nitinat' (letter to Governor James Douglas, 1857). Appendix in *Facts and Figures Relating to Vancouver Island and British Columbia.* London: Longman, Green, Longman, and Roberts.

Reid, J.L., 2015. *The Sea is My Country: The Maritime World of the Makahs, an Indigenous Borderlands People.* New Haven and London: Yale University Press. doi.org/10.12987/yale/9780300209907.001.0001

Rozen, D.L., 1985. Place Names of the Island Halkomelem Indian People. University of British Columbia (MA thesis).

Sapir, E. and M.S. Swadesh, 1955. 'Native Accounts of Nootka Ethnography.' Bloomington: Indiana Research Center in Anthropology, Folklore, and Linguistics. Also published as *International Journal of American Linguistics* 21(4).

Sapir, E., M. Swadesh, H. George, A. Thomas, F. Williams, K. Fraser and J. Thomas, 2009. 'Family Origin Histories.' In E. Arima, T. Klokeid and K. Robinson (eds), *The Whaling Indians: West Coast Legends and Stories* (Part 11: Sapir-Thomas Nootka Texts). Gatineau: Canadian Museum of Civilization (Canadian Ethnology Service Mercury Paper 145).

Sapir, E., M. Swadesh, A. Thomas, J. Thomas and F. Williams, 2004. 'Legendary Hunters.' In E. Arima, T. Klokeid and K. Robinson (eds), *The Whaling Indians: West Coast Legends and Stories: Legendary Hunters* (Part 9: Sapir-Thomas Nootka Texts). Gatineau: Canadian Museum of Civilization (Canadian Ethnology Service Mercury Paper 145).

Sproat, G.M., 1987. *The Nootka: Scenes and Studies of Savage Life* (ed. Charles Lillard). Victoria: Sono Nis Press.

Stafford, J. and T. Christensen, 2007. 'Woss Lake–Tahsis "Grease Trail": Phase II Archaeological Impact Assessment.' Report on file, Permit 2005–246. 'Namgis First Nation, Alert Bay, and British Columbia Archaeology Branch, Victoria.

Stafford, J. and D. Jacobsen, 2005. 'Woss Lake–Tahsis Grease Trail, Archaeological Impact Assessment of Proposed Hiking Trail.' Report on file, Permit 2004–339, 'Namgis First Nation, Alert Bay, and British Columbia Archaeology Branch, Victoria.

Starling, E.A., 1852. 'Report of Indian Agent for District of Puget's Sound.' In *Annual Report of the Commissioner of Indian Affairs for Oregon Territory* 71.

Suttles, W. and B. Lane, 1990. 'Southern Coast Salish.' In W. Suttles (ed.), *Handbook of North American Indians,* vol. 7: *Northwest Coast.* Washington, DC: Smithsonian Institution.

Swan, J.G., 1870. *The Indians of Cape Flattery, at the Entrance to the Strait of Fuca, Washington Territory.* Washington, DC: Smithsonian Institution (Smithsonian Contributions to Knowledge Vol 16). doi.org/10.5962/bhl.title.19016

10

Abundance and Resilience in the Salish Sea

Lissa K. Wadewitz

'There was no way you could starve in this country', Saanich elder Dave Elliot Sr reminisced about his peoples' historic hunting and gathering territories in the 1980s.

> We had so much of everything. It would be impossible to starve. There was so much food, it was everywhere. This is why I say our people were so rich, not to mention the great salmon runs, the deer, the elk and so on. (Poth 1990: 48)

Old Pierre of the Katzie group agreed. 'In earlier times this Fraser River resembled an enormous dish that stored up food for all mankind', he related to an anthropologist in the 1930s (Suttles 1955: 10). The natural abundance of the Salish Sea, its islands and its mainland tributaries have been crucial to how successive human inhabitants have survived in and valued this region. Native peoples used and actively managed this abundance for thousands of years (see Angelbeck, this volume) and this same diversity of resources drew European, Euro-American/Canadian and other newcomers to the area from the late 1700s onward.

The region produced raw materials for distant markets for decades, even as incoming white settlers increasingly displaced Indigenous peoples from their traditional lands and resource procurement sites. Coast Salish communities adapted as best they could but were largely shunted onto reservations (US) and reserves (Canada) by the late 1800s. Natural resource extraction has remained central to the regional economy through the present day, though tourism and high-tech ventures finally diversified it in the late twentieth century.

The result has been booming, vibrant urban centres but an increasingly overtaxed and polluted environment in need of more sustainable stewardship. That recognition, together with crucial legal decisions regarding Native rights and land title has led to heightened awareness and improved cooperation between Native and non-Native entities. Despite hundreds of years of dispossession and oppression, resilient Coast Salish communities on both sides of the border are once again playing crucial roles in managing the abundance of the Salish Sea.

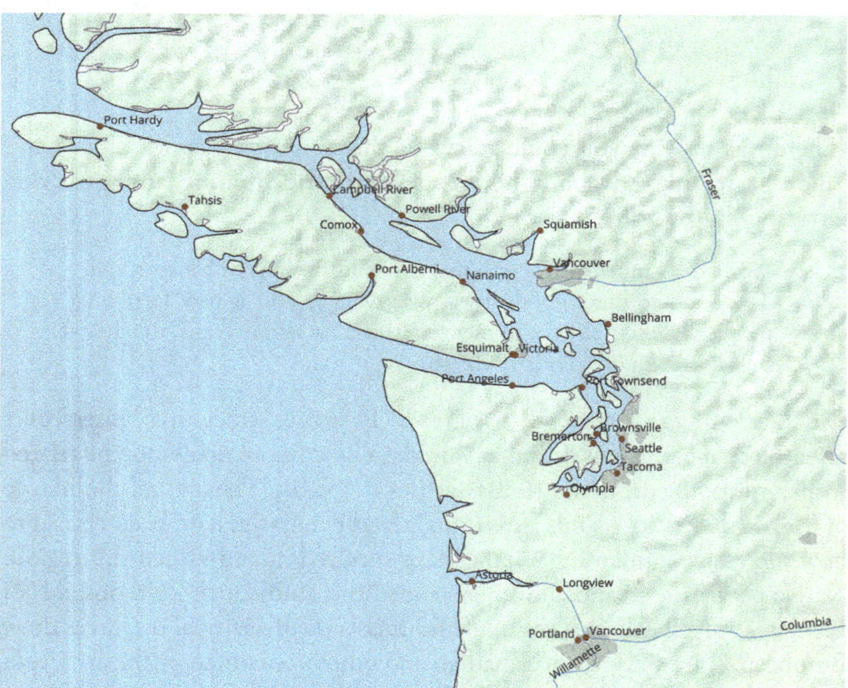

Figure 10.1 Port towns, Salish Sea and surrounding regions
Source: Made with Natural Earth (www.naturalearthdata.com).

Contact and the Fur Trade

More than 200,000 Indigenous people inhabited the Pacific slope[1] in numerous small communities at contact in the late 1700s. They evolved mobile 'seasonal rounds' to take advantage of the rich ecological diversity of the region's coastlines, rivers and adjacent interior lands (Poth 1990: 41–54; Richling 2016: 6–22). Contact with incoming Europeans and Euro-Americans transformed Indigenous lifeways, but regional Native peoples actively negotiated with incoming newcomers to shape their changing world. Despite severe population losses due to the introduction of exotic diseases, Coast Salish peoples successfully adapted to fluctuating circumstances and remained vital players in Native–newcomer relations for many decades.

The voyages of Captain James Cook made Vancouver Island known to European and American fur traders in the late 1770s. After 'discovering' Hawai'i in 1778, Cook landed at Nootka Sound on the west coast of what is now British Columbia's (BC) largest island. This British party spent over a month refitting their vessel and trading with the Nu-chah-nulth people before learning of the tremendous profits to be made selling lush sea otter pelts in China (Cook and King 1784: vol. 2, 220). Cook's crew returned to England with tales of this lucrative trade item as well as crucial navigational instructions. Nootka Sound[2] thus soon became a featured stopover point for hundreds of maritime traders from around the world (Fisher 1996: 125).

Following Cook, several subsequent Spanish and British expeditions increased European knowledge of the region's complicated geography. The Spanish had both peaceful and violent interactions with various groups of Natives throughout the region as they visited different islands and coastlines, mapping and sounding as they went. (Cook 1973: 303–6; McDowell 1998: 48–60; Lutz 2008: 51–3). Sent to both settle the 'Nootka Controversy' that erupted with the Spanish in 1789 and record the still unknown portions of the Northwest Coast, George Vancouver's 1792 expedition performed the most thorough exploration to date by an outside European power (Cook 1973: 129–433). They mapped (and named) what

1 Areas west of the continental divide.
2 One of several deeply indented sounds on the west coast of Vancouver Island.

became Puget Sound and, together with Spaniards Dionisio Alcalá Galiano and Cayetano Valdés y Flores, confirmed that Johnstone Strait[3] was not, in fact, the famed Northwest Passage (Fisher 1992b: 19, 21–47).

Coast Salish peoples initially welcomed the economic opportunities introduced by the maritime fur trade. They became skilled traders who drove hard bargains with their visitors. According to one non-Native trader:

> it appears that the natives are such intelligent traders, that should you be in the least degree lavish, or inattentive in forming bargains, they will so enhance the value of their furs, as not only to exhaust your present stock, but also to injure, if not to ruin, any future adventure. (as cited in Fisher 1992a: 5)

Metal items were initially the most coveted in these years, but soon Native peoples began demanding other items like clothing, blankets, rum, tobacco, molasses and muskets (Fisher 1992a: 5–7).

These interactions also introduced exotic diseases into Coast Salish communities. Indeed, a virulent wave of smallpox struck the Salish Sea in the mid-1770s, prior to any of the regional inhabitants having even laid eyes on a person of European or Euro-American descent. Old Pierre, the Katzie elder mentioned earlier, recounted this moment in his peoples' oral history:

> Then news reached them from the east that a great sickness was travelling over the land, a sickness that no medicine could cure, and no person escape … Then the wind carried the smallpox sickness among them. Some crawled away into the woods to die; many died in their homes. Altogether about three-quarters of the Indians perished. (Suttles 1955: 34)

The signs of destruction were everywhere when Vancouver and his men visited the region in 1792. At the southeastern end of the Strait of Juan de Fuca, Vancouver wrote, 'the scull, limbs, ribs and back bones, or some other vestiges of the human body, were found in many places promiscuously scattered about the beach, in great numbers' (Vancouver 1984, vol 2: 538). Lummi oral traditions confirm this epidemic, noting that survivors relocated to the mainland. Villages on both Orcas and Lopez Islands were similarly decimated and later reestablished on nearby Guemes Island (as cited in Boyd 1999: 31).

3 A narrow water body which connects Queen Charlotte Sound and Queen Charlotte Strait with the Salish Sea.

As sea otter stocks declined due to overhunting, the maritime fur trade transitioned to land in the 1810s and 1820s. The Hudson's Bay Company (HBC) became the central fur trading operation in the far west and soon founded Fort Vancouver north of the Columbia River in 1821, as well as several other forts farther north (Galbraith 1957; Fisher 1992a: 25–6). Many of the forts grew their own food, raised domesticated animals and purchased salmon from the region's Native population, which they salted and exported to Hawai'i. The operations at Vancouver,[4] Langley[5] and Nisqually[6] quickly became the most productive (Rich 1941–44, first series: 205–6, 257–8, 281–2; Bowsfield 1979: 6–7, 12, 24–5, 45, 101–2, 158; Cullen 1979: 26, 34–45; Mackie 1997: 155–243; Cole 2001: 74, 84, 87).

As with the maritime trade, many Native people welcomed the establishment of the HBC forts and dealt with the company on their own terms. Not foreseeing the later flood of European and Euro-American/Canadian settlers, the Coast Salish saw this next phase of encounter as presenting new opportunities for trade and ways to build wealth. Since accumulating trade goods was an important cultural practice, high-status individuals regularly strove to amass enough materials to host a large community gathering (a 'potlatch') at which they gave everything away, thus ensuring a redistribution of wealth and their ongoing high societal rank (Gunther 1927: 306–10; Elmendorf 1960: 328–33, 337–43; Suttles 1987: 23–4).

Wealthy members of the community were often able to purchase slaves, and potentially have more wives, using the labour of both to further enhance their wealth. Access to new trade goods also enabled some lower-status people to host their own potlatches, potentially achieving a higher social rank in the process (Lutz 2008: 76–8, 80–2; Richling 2016: 43–4; Storey 2016: 30). In addition to trade for material objects, both forced and voluntary sexual liaisons between Native women and various newcomers grew increasingly common as the fur trading settlements became more established. Some Native men and women prostituted their female slaves to coastal visitors and resident male fur traders.

Indigenous women also initiated romantic encounters, sometimes to develop ties to the forts and potentially gain higher status. Many intimate relationships turned into lifelong unions, while others were more fleeting.

4 Columbia River.

5 Fraser River.

6 Near Tacoma.

Tellingly, all HBC employees were required to make provision for the Native wives and children they might leave behind. Some of the traders risked societal censure back home and took their Indigenous families with them to Europe or to the east coast of North America (Merk 1931: 101; Van Kirk 1980; Fisher 1992a: 40–2; Lutz 2008: 179–81).

Perhaps the most important consequence of this ongoing influx of newcomers was the continual introduction of new waves of exotic disease into Native communities. Smallpox reappeared on the lower Columbia River in the early 1800s and spread northward to Puget Sound, Vancouver Island, and the Fraser River (Boyd 1999: 39–49). Another fatal disease that available sources simply called the 'mortality' struck the coast in the 1820s or 1830s, while other diseases like malaria, tuberculosis, influenza, dysentery, whooping cough, typhus and typhoid fever all spread through Indigenous communities over the course of the 1800s, with devastating consequences (Boyd 1999: 5; Lutz 2008: 85). Syphilis and gonorrhoea further contributed to declining birth rates among Native women. Historians estimate that these diseases killed off between 80 and 90 per cent of the coastal Native population from the late 1700s to the late 1800s (Boyd 1999: 3, 63–78; Lutz 2008: 89; Igler 2013: 43–71).

Despite the tremendous impacts of disease, Coast Salish peoples adapted and adjusted. Villages regrouped, relocated to new locations and developed different ways to partake in the continually shifting regional economy. They did their utmost to situate themselves in the most advantageous positions possible even as they struggled to comprehend the forces that had been unleashed on their communities (Carlson 2010).

Making Colonial Spaces

Recognising that joint occupation of Oregon country was no longer feasible, the United States and Great Britain finally sat down to negotiate their territorial claims in the mid-1840s. The United States was in an escalating conflict with Mexico, and Britain wanted to avoid any military entanglements over territorial issues. The Oregon Treaty of 1846 extended the international border at the 49th parallel westward to the Pacific Ocean, granting Britain Vancouver Island and the United States the territory between the Columbia River and the new border (Barman 1991: 48–9).

Both countries then took steps to bring their respective charges under stricter control. Despite these efforts, however, the Salish Sea region remained a fluid, distant outpost for several more decades.

After the conclusion of the Oregon Treaty, both Britain and the United States moved forward with their settler-colonial projects by usurping Indigenous lands. Britain granted the HBC proprietorial rights to Vancouver Island in 1849, and HBC Chief Factor James Douglas was soon appointed to facilitate its settlement (Barman 1991: 52–4; Fisher 1992a: 58). As directed by London, Douglas negotiated 14 land sale agreements (often called 'treaties') with tribes near Victoria between 1850 and 1854 (Fisher 1992a: 66–8; Harris 2002: 17–44). Native oral traditions suggest significant confusion surrounded the signing of these documents (Poth 1990: 7–73; Arnett 1999: 30–7).

South of the border, Isaac Stevens, the new governor of Washington Territory, negotiated several treaties with the region's Native peoples between 1854 and 1857. These treaties forced the Natives to relinquish title to over 64 million acres of land and created eight reservations to which they were expected to retreat. In return, the government promised annuity goods, schools and farming instruction in a region largely unsuited to agriculture (Richards 1979; Harmon 1998: 72–102).

On paper, then, government officials on both sides of the international border had succeeded in neatly separating whites from Natives. The reality on the ground, however, was quite different. Once the HBC made Victoria its primary base of operations and more newcomers settled in Seattle, the HBC's polyglot workforce (consisting of men of English, Scottish, French-Canadian, Iroquois, Métis and Hawaiian descent, among others), Coast Salish peoples and new visitors from the far northern coast of what would become British Columbia and southeastern Alaska, all engaged in economic and social exchanges on a regular basis (Mackie 1997: 33–4; Barman 2020).

This latter group, often collectively called 'northern Indians', increasingly made Victoria and Puget Sound an annual destination as part of their seasonal rounds. Between 1853 and the 1880s, two to four thousand Indigenous people annually canoed up to 800 miles southward through Johnstone and Georgia Straits to spend part of the year in Victoria. As a result, emerging Salish Sea communities were centres of mixed-race activity that defied the clear-cut divisions desired by distant colonial government officials (Lutz 2008: 167–71; Reid 2015: 124–63).

The discovery of gold along the Fraser River in 1858 added to Victoria's appeal as upwards of 30,000 eager, non-Native gold seekers poured into the region. Overnight, Victoria was inundated with miners hankering to head to the gold fields. Regional Native peoples capitalised on both the gold rush itself and its attendant demand for labour. In fact, Douglas reported over 4,000 Native peoples visiting Victoria seeking wage labour opportunities in the summer of 1858; this was more than twice the non-Native population.

The influx of gold seekers also prompted the growth of slave raiding and the prostitution of slave women. The Lekwungen reserve in Victoria became the centre of this trade, as people gathered there to buy and sell both sex and slaves (Lutz 2008: 179–81). These events prompted Britain to finally take steps to formalise authority over their claims on the mainland. They created the Crown Colony of British Columbia and extended Douglas's authority over both colonies. The two colonies were combined in 1866 and joined the Canadian Confederation in 1871 (Barman 1991: 61–9, 90–8).

These decades were not without their violent exchanges, particularly as white settlers' intentions to occupy Native lands and extract Native-owned resources became increasingly clear. For instance, dissatisfaction with the Stevens treaties, Euro-American miners' violence against Native women and trespasses on Indigenous lands in Washington led to a major uprising east of the Cascades that spilled west in the Indian War of 1855–56. Responding to white settler anxieties, the US military participated in several violent confrontations with 'northern Indians' in the 1850s as well. The deadly 1856 attack on a Tlingit camp at Port Gamble by the USS *Massachusetts* left over 25 Natives dead and later led to the retaliatory killing of the former collector of customs, Isaac N. Ebey, of Whidbey Island (Wadewitz 2019: 340, 345–8, 359). In BC, tensions surrounding newcomers' interest in Indigenous lands heightened as Douglas attempted to negotiate new land agreements with the Cowichan people on Vancouver Island and relentlessly pursued several Lamalcha villagers from Kuper Island on suspicion of murder in the early 1860s (Arnett 1999: 88–307).

By the 1850s, the vagaries of the Oregon Treaty's border parameters also created problems for the US and Great Britain. The treaty extended the international border westward along the 49th parallel and then south through an undetermined channel, dividing Vancouver Island from the mainland. The problem, however, was that the San Juan Islands created several channels between Vancouver Island and the coast of BC. Initially, the HBC claimed San Juan and Orcas Islands, and Douglas sent men to live there and affirm those property rights (Vouri et al. 2006: 7–8).

In 1859, however, an American miner shot and killed a pig owned by the HBC. Tensions flared and quickly escalated, military forces were called in, and the two countries nearly engaged in the so-called 'Pig War' that year. Fortunately, cooler heads prevailed, and conflict was averted. Still, both sides jointly occupied the island for the next 12 years while negotiations continued. An outside arbitrator was finally called in and the border's current parameters were determined in 1872 (Milton 1869; Cain and Hopkins 1993: 258–68).

Transformations

The colonisation of the Salish Sea region solidified at the turn of the last century. The arrival of transcontinental railroads, technological advances that facilitated the rise of commercial-scale resource extraction and the availability of land all lured increasing numbers of immigrants to the far west. Although Native peoples initially provided vital labour to these fledgling commercial ventures, non-Native migrants soon outnumbered the Indigenous population and vied with them for jobs and space. Ongoing demographic decline among the Coast Salish further fuelled this shift in the workforce. By World War II, the region's extractive economic character—and thus its hinterland status—was firmly established.

One of the most transformative developments for the entire west coast of North America in the late nineteenth century was the completion of transcontinental railroads. In 1879, the Great Northern Railroad extended their line from St Paul, Minnesota to Seattle. The Canadian Pacific Railroad completed their line in 1887, connecting the east to the newly founded city of Vancouver. Both routes facilitated the shipping of goods to market and made travel far faster than ever before. A trip that used to take three to five months now took a mere five to six days—suddenly the Salish Sea and its resources were within orbit of the east in ways that had previously been unthinkable (Barman 1991: 104–12; Schwantes 1993: 38–51; Schwantes 1996: 169–78).

The demographic changes that thus affected BC and Washington in this period were dramatic and cemented the loss of lands and resource access for many of the region's Indigenous communities. Native peoples continued to move through and occupy parts of these burgeoning urban areas, but doing so was becoming increasingly difficult (Thrush 2007: 79–125). Between 1870 and 1890 Seattle's population grew from 1,107 to 42,837, and Tacoma had more than 36,000 inhabitants by 1890. By 1910, Seattle had become the largest city in the Pacific Northwest with a population of more than 237,000 (Schwantes 1996: 233–9). The young city of Vancouver also grew rapidly and soon eclipsed Victoria as the primary destination for new immigrants; by 1911 the population of Vancouver Island and the lower mainland had reached over 260,000 (Barman 1991: 112, 371). In addition to migrants from eastern and midwestern Canada and the US as well as Europe, Chinese and Japanese immigrants were also an important part of the shifting demographics of the late 1800s and early 1900s (Ichioka 1988: 55–6; Roy 1989: 92).

The large numbers of immigrants pouring into BC and Washington State fed the growing labour demands of the Pacific slope's extractive industries. Coal was one of the most important terrestrial resources available. Although the HBC had been extracting coal since the 1840s, coal production on Vancouver Island and around Puget Sound heightened in the latter half of the nineteenth century (Barman 1991: 54–5; Arnett 1999: 47–50; Lutz 2008: 171–3). The Vancouver Coal Mining and Land Company at Nanaimo and Dunsmuir and Sons in Wellington dominated this sector. By the 1890s both companies employed thousands of workers and produced over 900,000 tons of coal per year (Barman 1991: 120–2; Hinde 2003: 14–6). Several coal mining companies operated in Washington Territory as well from the 1860s onward, but by the early 1900s petroleum and natural gas were already eclipsing coal in local markets (Melder 1938: 151–5; Schwantes 1996: 215).

Given the area's heavily forested nature, timber production was also bound to become vital to the regional economy. Initially, the distance to markets and the difficulty of extracting timber from dense forests limited the industry's prospects (Barman 1991: 116–7). By the late 1800s, however, the development of special narrow-gauge logging railroads facilitated access to previously inaccessible stands of trees. The invention of the steam donkey engine and other innovations in sawmilling and tree felling techniques advanced the industry further (Schwantes 1996: 217–8; Rajala 1998: 7–50). Timber companies on the US side of the border were soon able to take advantage of the Great Northern Railroad to sell their lumber products back east. Despite various boom and bust cycles, Washington was the leading lumber-producing state in the entire US for all but one year between 1905 and 1938 (Ficken 1987; Schwantes 1996: 219).

Perhaps one of the most lucrative aquatic extractive industries to evolve in the Salish Sea region was commercial fishing. By the early 1900s there were 46 canneries on the Fraser River, 23 elsewhere in southern BC, and 21 in Washington (Cobb 1921: 152). While the canneries initially depended on Indigenous labour, incoming immigrants largely assumed these jobs after 1900. Time-honoured Native fishing locations were also vulnerable, as competing fishermen occupied prized Coast Salish fishing spots and commercial canners built fish traps on coastal sites. Fish traps were large structures that caught thousands of fish at a time. Such traps were more common in Washington State, and soon hundreds of traps lined the coastlines of nearly all of the Salish Sea Islands. Within a few decades, Coast Salish peoples found it nearly impossible to continue to fish in the spots guaranteed by their treaties and land agreements. Salmon canning and the fresh fish market remained vital extractive businesses in the region for several decades, but signs of depletion from overfishing, the impact of dams and a devastating rockslide on the Fraser River were already evident by the 1910s (Wadewitz 2012: 52–88, 12–143).

Post-World War II Order

World War II transformed the economic bases of Washington and BC and the demands of the wartime economy caused new population influxes. The rise of tourism in the post-war period, and, more recently, a tech boom, have finally helped the region shed its hinterland status. While the Salish Sea region remains reliant on extractive industries to the present day, cities on both sides of the border have become leaders of future-oriented sectors of the global economy. Additional economic, legal, and cultural shifts have also led to a resurgence of interest in and respect for Indigenous cultural practices and rights, contributing to greater recognition of Native power and their role in the region's ongoing development.

World War II had tremendous impacts on manufacturing and employment rates throughout the Salish Sea, causing a tremendous population boom for the west coast. Because the Boeing Airplane Company had large assembly plants in Seattle and Renton, it was well poised to take advantage of military contracts once the US entered the war in 1941. At its peak, Boeing's Washington locations employed nearly 50,000 people and boasted sales of over USD600 million. Aircraft production in BC likewise greatly expanded, and Boeing built a short-lived plant in Vancouver for the duration of the war (Sale 1976: 180–5; Ficken and LeWarne 1988: 131–2; Schwantes 1996: 411). Victoria and Vancouver shipyards employed upwards of 30,000 people at the height of the conflict (Barman 1991: 262). The availability of cheap hydroelectric power prompted the Aluminum Corporation of America and other aluminium producers to build five new plants that directly fed Boeing's aircraft production (Ficken and LeWarne 1988: 131; Stein 2007/08: 3–15). Urban centres experienced significant population growth as people were drawn to coastal cities for wartime jobs (Ficken and LeWarne 1988: 154–6; Barman 1991: 262; Schwantes 1996: 418–9).

After the war's conclusion in 1945, Americans and Canadians worried about a recurrence of the economic depression of the 1930s (Barman 1991: 270; Schwantes 1996: 421–2). However, Boeing's ability to adapt and the onset of the Cold War translated into sustained post-war military spending and high production demands for Seattle-area manufacturers (Kirkendall 1994: 137–49). Still, the state's dependence on natural resource extraction and Boeing was clear, and local businesspeople and government officials pushed to diversify the economy over the next several decades (Ficken and LeWarne 1988: 144–5). BC's economy also continued to boom, as new infrastructure connected remote settlements and resource extraction locations throughout the province's interior (Barman 1991: 270–2).

Logging, fishing and mining remained the primary extractive industries on both sides of the border until environmental concerns and dwindling resources led to policy changes and harvesting restrictions. Increasing prices and a decrease in private timber holdings led Washington's logging industry leaders to pursue reforestation on their lands and deforestation on those of the US Forest Service. Gasoline-powered chainsaws and logging trucks increased the speed of production and access to interior forests, as the largest corporations worked closely with federal managers on sustained yield and reforestation programs that often looked better on paper than on the ground.

The rise of the use of plywood in construction, demand for pulp and the use of new tree species and technologies also helped the timber industry expand and consolidate on both sides of the border in the post-war decades. The large operators played fast and loose with the idea of sustainable yield and regularly overharvested the forests of the provincial and US federal governments. Mining also remained a vital industry in BC, while fishing gradually decreased in importance due to increased regulations, limited licensing and declines in various fish populations (Ficken 1987: 225–7; Barman 1991: 285–9, 330–1; Rajala 1998: 214–21).

By the 1960s and 1970s, these various extractive industries increasingly had to contend with the rise of a vibrant environmental movement and heightened ecological awareness on both sides of the 49th parallel. The passage of the 1973 *Endangered Species Act* in the US had tremendous consequences for the timber industry south of the border. Under this Act, the Fish and Wildlife Service listed the spotted owl as endangered in 1990 and subsequently restricted logging on millions of acres of federal land in order to protect the owl's old growth forest habitat. Increased log exports to Japan, a recession and these limits on logging forced many mills in Washington to close. The listing of several endangered species of fish has further restricted both logging and fishing in Washington (Dietrich 1993). Rampant logging, clear cuts and a provincial decision to allow the largest logging company in BC at the time, MacMillan Bloedel, to cut a significant section of Clayoquot Sound's[7] old growth forest led to an outpouring of concern as thousands of people gathered on the west coast of Vancouver Island in the early 1990s. All summer, activists from around the world arrived to non-violently protest the logging operation. Over 850 people were arrested in these confrontations, making this the second largest act of civil disobedience in Canadian history (Jain 2013; Thomas 2018).

As the twentieth century progressed, tourism became increasingly important to the growing economies of the Salish Sea Islands and adjacent mainland. The rise in automobile ownership, suburbanisation, wartime savings and a vibrant economy after the war all encouraged new levels of tourism. Puget Sound and the San Juan and Gulf Islands drew growing numbers of outdoor enthusiasts, visitors looking to experience Victoria's English flair flocked to Vancouver Island, and the region's burgeoning urban areas lured outsiders with their growing cultural attractions and shopping opportunities (Smith 2012: 67–83; Stewart 2017: 209–51). Pre-COVID, the number of visitors to both Washington State and BC had continued to grow, and state and provincial government officials deemed the tourism sector as vital to their respective economies (Talton 2018).

7 One of several sounds on west coast of Vancouver Island.

Perhaps the biggest economic changes for the Salish Sea and its environs in recent decades has been the rise of the computer, software and other high-tech industries that have proliferated on both sides of the border since the late 1970s. Bill Gates and Paul Allen founded Microsoft in the mid-1970s, relocated their headquarters to Redmond, Washington in 1978, and the brand soon became a household name. Releases of new versions of the Windows operating system in the 1990s and more recent Xbox gaming systems have made Microsoft a software and gaming giant. As of 2017, the company employed 47,000 people in the greater Seattle region and it has announced plans to renovate and expand its Washington campus (Eaton 2011; Meyer 2017). The e-commerce giant, Amazon, also established its headquarters in downtown Seattle in 1994. Now employing over 40,000 workers, Amazon has driven further economic expansion in the city (Easter and Dave 2017; O'Connell 2017).

Similar developments have occurred in BC, especially in Vancouver. MacDonald Dettwiler and Associates Ltd. initiated a high-tech revolution from their small workshop in a Vancouver suburb in 1969 that has since flourished in the city proper. There are currently over 9,000 tech companies in BC that employ 84,000 people and generate more than USD15 billion per year. Microsoft, Sony Pictures Imageworks and Amazon have all recently expanded into new Vancouver locations as well (Littlemore 2018). Economies on both sides of the border thus appear to be diversifying in multiple ways and moving well beyond a dependence on natural resource extraction and airplane construction.

Such growth has brought unintended negative consequences to the region as well. Rapid population increases have led to climbing rents and home prices, not to mention more houselessness; the cost of living in both Seattle and Vancouver is among the highest in North America (Conklin 2015; Brown 2017). The influx of people has also exacerbated environmental challenges, especially in terms of water pollution and declining marine animal populations (Taylor 2016; Rosenberg 2018). Somewhat ironically, the scenery and resources so touted by the region's boosters and appreciated by urban residents appear to be at risk as a result of this recent economic expansion.

In addition to such huge demographic and economic transformations, the Salish Sea has undergone a tremendous cultural shift, as Indigenous peoples have successfully fought for recognition of their land and resource rights in recent decades (see Chapter 13, this volume). This has been especially true in BC since the late 1970s. BC First Nations have long maintained that they still hold legal title to most of the province and the courts have finally agreed. Due to several Canadian legal decisions from the 1970s to the 1990s, the Nisga'a First Nation of northern BC has been able to negotiate a new treaty agreement with the governments of BC and Canada. The revolutionary 2000 agreement recognises Nisga'a land title, facilitates Nisga'a self-governance, and allows for Nisga'a access to, and management of, natural resources. Even as these discussions were under way, the provincial and federal governments, together with representatives of over 50 per cent of BC's First Nations, created the British Columbia Treaty Commission to oversee future mediations in the province. Thus far, seven other First Nations have negotiated new treaties through this process, marking a significant change in Canadian policies (Mills 2008; BC Treaty Commission 2018).

Given that most Washington tribes signed land treaties with the US federal government in the mid-1850s, the legal challenges confronting most Washington tribes have tended to involve regaining access to natural resources. Since salmon have long served as one of the most important subsistence resources for Coast Salish peoples, access to fisheries and the ability to fish at particular times have been contested and contentious since the rise of the commercial fishery in the late 1800s. Native peoples in Washington were systematically pushed off of productive fishing locations, forced to fish only on their reservations for several decades, and regularly arrested for trying to assert their rights to fish (Wadewitz 2012: 52–88). The situation came to a head in the 1960s and 1970s, as Washington Natives staged a series of protests to force the hand of the state.

Judge George Boldt heard a test case, *United States v. Washington*, and ruled in favour of Native treaty fishing rights in 1974. The 'Boldt Decision' recognised the Natives' treaty right to fish and guaranteed the tribes 50 per cent of the total annual catch of salmon. The US Supreme Court upheld the decision in 1979 (Landau 1996: 437–56). Then Governor Booth Gardner and tribal representatives hashed out the 1989 Centennial Accord wherein they committed to cooperating on issues of mutual interest, particularly regarding natural resource management. The accord was affirmed in 1999 and continues to guide state–tribal relations and communications in Washington (Ott 2016; Washington State Governor's Office of Indian Affairs 2018).

Such gains in recognition of Native/First Nation sovereignty and related legal rights have further spilled over into the worlds of archaeology, anthropology and museums. Influenced by changes in the ethics guidelines of international museum organisations like the International Council of Museums (ICOM), US Congress passed the *National Museum of the American Indian Act* in 1989 and the *Native American Graves Protection and Repatriation Act* (NAGPRA) in 1990. Both pieces of legislation require that human remains, funerary materials and sacred and ceremonial objects be returned to culturally affiliated tribes. The passage of these laws, together with ICOM's Code of Ethics, developed in the 1970s and refined and affirmed in 2004, stipulate how museums should treat contemporary Native groups and the artefacts that were often illegally or shamefully acquired in the past.

The museum community is thus currently committed to working with tribes on properly displaying cultural artefacts and conveying their meaning to non-Native audiences. These efforts have transformed museum–tribal relations across the United States, including those of the Burke Museum in Seattle (Peterson 1990; West 2016). As many US tribes have also been able to benefit from the *Indian Gaming Regulatory Act* of 1988 that allows tribes to establish and run casinos, some Washington tribes have used their newfound wealth to build their own cultural centres and educational programs. While not every tribe is reaping casino profits, those that are have more financial resources to apply to their communities, improved abilities to house tribal artefacts themselves and more political influence overall. These legal and economic changes have thus led to significantly increased tribal power and heightened awareness regarding Native culture and the state's conflicted history (Riddle 2008; Wang 2011).

Although Canada has not passed federal legislation along the lines of NAGPRA, similar changes in Native–museum relations and the government's commitment to educating non-Natives about First Nations history and culture have also occurred north of the border. At the prompting of the Assembly of First Nations, the Canadian Museums Association co-sponsored a national conference to explore ways to 'forge a true partnership between museums and First Peoples' in 1988 (Task Force Report on Museums and First Peoples 1992: 12). A task force with regional committees soon followed.

The mandate of the task force was to increase the involvement of First Nations people in the interpretations of their culture and history presented to the public, improve Native peoples' access to museum collections and to repatriate human remains to appropriate Indigenous communities (Assembly of First Nations and Canadian Museums Association 1994). In 2007, the Canadian Government further announced the Indian Residential Schools Settlement Agreement, the largest class action settlement in the country's history. The agreement recognised the legacy of damage and abuse left by Indian residential schools and established a multi-billion-dollar fund to assist former students with recovery. A central part of this process was assigned to the Truth and Reconciliation Commission (TRC); they were to investigate the history of the Indian residential school experience and make recommendations for ways to help survivors, educate the public about the schools and students' experiences there and host national events to share and honour those experiences.

The TRC included 94 calls to action in its 2015 final report. In the spirit of reconciliation, cultural and educational institutions around the country have been actively engaging with First Nations communities to build better relationships and develop effective outreach programs. In BC, the Royal BC Museum, for instance, has several exhibits and educational programs devoted to realising the TRC's recommendations (Frogner 2016). As in Washington State, the recognition of both Native rights and the value of Indigenous peoples and their cultural practices now permeate discussions regarding the history and future of the Salish Sea region.

Conclusion

The Salish Sea Islands and nearby mainland have a long and complicated history, but the relationship between human beings and the natural world has always been at its core. The recent recognition of Coast Salish legal rights to resources and land, as well as their long history of engagement with and management of regional resources is finally being factored into ecological management plans on both sides of the border (for instance, see Pearsall and Schmidt, this volume). Although the dispossession of Native peoples will forever mar this area's history and must be reckoned with in more concrete and tangible ways, perhaps there is hope for the ecological health of the Salish Sea if all hands—Native and non-Native—are willing to work together for restorative purposes.

Timeline

A listing by date of principal events mentioned in this chapter is presented in Table 10.1.

Table 10.1 Post-contact timeline for Salish Sea and vicinity

Mid-1770s	Smallpox epidemic reaches the Pacific Northwest with devastating effects for Indigenous peoples.
1778	Captain James Cook makes contact at Nootka Sound, on the west coast of Vancouver Island.
1789–1795	Nootka Sound Controversy occurs between Great Britain and Spain.
1792	George Vancouver's expedition tours the Puget Sound region.
Early 1800s	Smallpox epidemic again strikes the lower Columbia River, Puget Sound, Vancouver Island and the Fraser River region.
1821	Hudson's Bay Company (HBC) becomes the primary fur trading company in the region and founds Fort Vancouver at the site of present-day Vancouver, Washington.
1846	The Oregon Treaty is signed by Great Britain and the United States. The agreement extends the international border across the 49th parallel to the west coast.
1849	Great Britain grants proprietorial rights to Vancouver Island to the HBC and HBC Chief Factor James Douglas is appointed to oversee the colony.
1850–1854	Douglas negotiates 14 'treaties' with Native peoples near the city of Victoria.
1853–1880s	Two to four thousand 'northern Indians' descend upon Victoria and Puget Sound to trade, work and take occasional retaliatory actions against other Indigenous and Euro-American/Euro-Canadian settlements.
1854–1857	Washington Territorial Governor Isaac Stevens negotiates several treaties with Native peoples living south of the new international border.
1855–1856	An Indian War/uprising breaks out in Washington Territory, with most of the violence occurring east of the Cascade mountains.
1856	The USS *Massachusetts* attacks a Tlingit camp at Port Gamble, killing over 25 Indigenous people. This later leads to a retaliatory killing of a former US collector of customs named Isaac Ebey.
1858	Gold is discovered along the Fraser River in what later becomes the province of British Columbia (BC).
1859	An American miner shoots and kills a pig owned by the HBC, leading both American and British troops to occupy San Juan Island for 12 years in an event known as the 'Pig War'.

Early 1860s	Governor James Douglas attempts to negotiate land deals with the Cowichan First Nation on Vancouver Island and pursues Lemalcha villagers from Kuper Island on suspicion of murder.
1866	Great Britain creates the Crown Colony of British Columbia and combines it with Vancouver Island still under James Douglas's authority.
1871	BC joins the Canadian Confederation.
1872	A German arbitrator determines the current configuration of the international border and ends the so-called 'Pig War' and joint occupation of San Juan Island by American and British troops.
1879	The Great Northern Railroad extends their rail line from St Paul, Minnesota, to Seattle.
1887	The Canadian Pacific Railroad completes a rail line connecting Vancouver BC to the east coast.
1890	Seattle's non-Native population reaches over 42,000 people; Tacoma's settler population climbs to 36,000.
1890s	The Vancouver Coal Mining and Land Company in Nanaimo and the Dunsmuir and Sons company in Wellington grow to employ thousands of workers and to produce over 900,000 tons of coal per year.
Late 1800s	The development of special narrow-gauge logging railroads, the invention of the steam donkey engine and other innovations in sawmilling and tree felling techniques lead to significant growth in the region's logging industry.
Early 1900s	The commercial fishing industry becomes more established, with 46 fish canneries on the Fraser River, 23 elsewhere in southern BC and another 21 facilities in Washington State.
1910	Seattle becomes the largest city in the Pacific Northwest with a population of 237,000.
1911	The non-Native population of the city of Vancouver and Vancouver Island reaches 260,000.
1941	The US enters World War II, leading to a boom in the aircraft, shipyard and aluminium industries. People flock to the region for wartime work opportunities.
1960s and 1970s	Rise of environmental movements in both the US and Canada.
1969	MacDonald Dettwiler and Associates Ltd initiate a high-tech revolution from their small workshop in a Vancouver suburb that has since flourished in the city.
1973	Passage of the *Endangered Species Act* in the US.
1974	US Judge George Boldt issues the 'Boldt Decision' in *United States v. Washington*, recognising Native treaty rights to fish and guarantees the tribes 50 per cent of the total annual catch of salmon.
1978	Bill Gates and Paul Allen found Microsoft and relocate their headquarters to Redmond, Washington.

1979	The US Supreme Court upholds the Boldt decision of 1974.
1988	US Congress passes the *Indian Gaming Regulatory Act* allowing tribes to establish and run casinos.
1988	The Canadian Museums Association co-sponsors a national conference to explore ways to involve First Nations people in museum displays and interpretations, as well as increase Indigenous access to collections and repatriate human remains.
1989	Washington State Governor Booth Gardner and tribal representatives develop a Centennial Accord to cooperate on issues of mutual interest, particularly regarding natural resource management.
1989–1990	US Congress passes the *National Museum of the American Indian Act* and the *Native American Graves Protection and Repatriation Act* (NAGPRA). Both require that human remains, funerary materials and sacred and ceremonial objects be returned to culturally affiliated tribes.
1990	The US Fish and Wildlife Service lists the spotted owl as an endangered species, thus restricting logging on millions of acres of federal land in order to protect the owl's old growth forest habitat.
Early 1990s	Thousands of people from Canada and around the world protest BC's decision to allow MacMillan Bloedel to cut a significant section of Clayoquot Sound's old growth forest. Over 850 people were arrested, making this the second-largest act of civil disobedience in Canadian history.
1994	The e-commerce giant, Amazon, establishes its headquarters in downtown Seattle.
2000	Revolutionary Canadian agreement recognises Nisga'a land title, facilitates Nisga'a self-governance and allows for Nisga'a access to, and management of, natural resources. The provincial and federal governments, together with representatives of over 50 per cent of BC's First Nations, also create the British Columbia Treaty Commission to oversee future mediations in the province.
2007	The Indian Residential Schools Settlement Agreement is announced by the Canadian government. The agreement recognises the legacy of damage and abuse left by Indian residential schools, establishes a multi-billion-dollar fund to assist former students with recovery, and creates the Truth and Reconciliation Commission (TRC) to investigate further.
2015	The TRC issues its final report including 94 calls to action for the government and cultural and educational institutions around the country to build better relations with First Nations communities and educate the public about Indigenous history.
2017	Microsoft employs 47,000 people in the greater Seattle region and announces plans to renovate and expand its Washington campus.
2020–2022	COVID-19 shuts down the border between the US and Canada.

Source: Author's tabulation.

References

Arnett, C., 1999. *The Terror of the Coast: Land Alienation and Colonial War on Vancouver Island and the Gulf Islands, 1849–1863*. Vancouver: Talonbooks.

Assembly of First Nations and Canadian Museums Association, 1994. 'Turning the Page: Forging New Partnerships between Museums and First Peoples.' Ottawa: Task Force Report on Museums and First Peoples (3rd edition).

Barman, J., 1991. *The West Beyond The West: A History of British Columbia*. Toronto: University of Toronto Press.

——, 2020. *On the Cusp of Contact: Gender, Space, and Race in the Colonization of British Columbia*. Madeira Park: Harbour Publishing Co. Ltd.

Bowsfield, H. (ed.), 1979. *Fort Victoria Letters, 1846–51*. Winnipeg: Publications of the Hudson's Bay Record Society (Vol. 32).

Boyd, R., 1999. *The Coming of The Spirit of Pestilence: Introduced Infectious Diseases and Population Decline Among Northwest Coast Indians, 1774–1874*. Vancouver: UBC Press.

British Columbia Treaty Commission, 2018. 'The BC Treaty Commission.' Viewed 31 May 2023 at: bctreaty.ca/

Brown, S., 2017. 'Vancouver Most Expensive City in Canada—But Ranked 107th in the World: Report.' *Vancouver Sun*, 21 June.

Cain, P.J. and A.G. Hopkins, 1993. *British Imperialism: Innovation and Expansion*. London: Longman Group.

Carlson, K.T., 2010. *The Power of Place, the Problem of Time: Aboriginal Identity and Historical Consciousness in the Cauldron of Colonialism*. Toronto: University of Toronto Press.

Cobb, J.N., 1921. 'Pacific Salmon Fisheries: Report of the United States Commissioner of Fisheries for 1921.' Washington, DC: US Government Printing Office (Bureau of Fisheries). doi.org/10.5962/bhl.title.42588

Cole, J.M. (ed.), 2001. *This Blessed Wilderness: Archibald McDonald's Letters from the Columbia, 1822–44*. Vancouver: UBC Press.

Conklin, E.E., 2015. 'The Solution to Seattle's Homeless Problem is Painfully Obvious.' *Seattle Weekly*, 8 September.

Cook, Captain J., FRS and Captain J. King, LLD and FRS, 1784. *A Voyage to the Pacific Ocean: Undertaken by Command of His Majesty for Making Discoveries in the Northern Hemisphere: Performed Under the Direction of Captains Cook, Clerke, and Gore in the Years 1776, 1777, 1778, 1779, 1780, 4 vols*. London: Printed for John Stockdale, Scatchered and Whitaker, John Fielding, and John Hardy. doi.org/10.5962/bhl.title.161773

Cook, W.L., 1973. *Flood Tide of Empire: Spain and the Pacific Northwest, 1543–1819*. New Haven: Yale University Press.

Cullen, M.K., 1979. 'History of Fort Langley, 1827–96.' In J. Brathwaite (ed.), *Canadian Historic Sites: Occasional Papers in Archeology and History.* Quebec: Minister of Supply and Services Canada (No. 20, prepared by the National Historical Parks and Sites Branch, Hull).

Dietrich, W., 1993. *The Final Forest: The Battle for the Last Great Trees of the Pacific Northwest*. New York: Penguin Books.

Easter, M., and P. Dave, 2017. 'Remember When Amazon Sold Only Books?' *Los Angeles Times*, 18 June.

Eaton, N., 2011. 'The History of Microsoft.' *Seattle Post-Intelligencer*, 7 November.

Elmendorf, W.W., 1960. 'The Structure of Twana Culture.' Monographic Supplement No. 2. *Research Studies* 28(3): 1–576.

Ficken, R.E., 1987. *The Forested Land: A History of Lumbering in Western Washington.* Seattle: University of Washington Press.

Ficken, R.E. and C.P. LeWarne, 1988. *Washington: A Centennial History.* Seattle: University of Washington Press.

Fisher, R., 1992a. *Contact and Conflict: Indian–European Relations in British Columbia, 1774–1890*. Vancouver: UBC Press.

——, 1992b. *Vancouver's Voyage: Charting the Northwest Coast, 1791–1795.* Vancouver: Douglas & McIntyre.

——, 1996. 'The Northwest from the Beginning of Trade with Europeans to the 1880s.' In B.G. Trigger and W.E. Washburn (eds), *North America, The Cambridge History of Native Peoples of the Americas*, vol. 1, part 2. Cambridge: Cambridge University Press. doi.org/10.1017/CHOL9780521573931

Frogner, R., 2016. 'Royal B.C. Museum and Archives Official Response Regarding the Truth and Reconciliation Commission's Call to Action.' 24 August.

Galbraith, J.S., 1957. *The Hudson's Bay Company as an Imperial Factor, 1821–1869.* Berkeley: University of California Press. doi.org/10.1525/9780520322714

Gunther, E., 1927. 'Klallam Ethnography.' *University of Washington Publications in Anthropology* 1(5): 171–314.

Harmon, A., 1998. *Indians in the Making: Ethnic Relations and Indian Identities Around Puget Sound.* Berkeley: University of California Press. doi.org/10.1525/9780520926202

Harris, C., 2002. *Making Native Space: Colonialism, Resistance, and Reserves in British Columbia.* Vancouver: UBC Press.

Hinde, J., 2003. *When Coal Was King: Ladysmith and the Coal-Mining Industry on Vancouver Island.* Vancouver: UBC Press.

Ichioka, Y., 1988. *The Issei: The World of First Generation Japanese Immigrants, 1885–1924.* New York: The Free Press.

Igler, D., 2013. *The Great Ocean: Pacific Worlds From Captain Cook to the Gold Rush.* New York: Oxford University Press.

Jain, S., 2013. 'Temperate Protest: Clayoquot Sound, Twenty Years After the "War of the Woods."' *Natural History* 121: 7.

Kirkendall, R.S., 1994. 'The Boeing Company and the Military–Metropolitan Industrial Complex, 1945–1953.' *The Pacific Northwest Quarterly* 85(4): 137–49.

Landau, J.L., 1996. 'Empty Victories: Indian Treaty Fishing Rights in the Pacific Northwest.' In J.R. Wunder (ed.), *Recent Legal Issues for American Indians, 1968 to the Present.* New York: Garland Publishing, Inc.

Littlemore, R., 2018. 'Vancouver's High-Tech Makeover.' *The Globe and Mail,* 12 May.

Lutz, J.S., 2008. *Makúk: A New History of Aboriginal–White Relations.* Vancouver: UBC Press.

Mackie, R.S., 1997. *Trading Beyond the Mountains: The British Fur Trade on The Pacific, 1793–1843.* Vancouver: UBC Press.

McDowell, J., 1998. *José Narváez the Forgotten Explorer: Including his Narrative of a Voyage on the Northwest Coast in 1788.* Spokane: The Arthur H. Clarke Company.

Melder, F.E., 1938. 'History of the Discoveries and Physical Development of the Coal Industry in the State of Washington.' *Pacific Northwest Quarterly* 29(2): 151–65.

Merk, F., 1931. *Fur Trade and Empire: George Simpson's Journal.* Cambridge: Harvard University Press.

Meyer, D., 2017. 'Microsoft is Making Space for 8,000 More Employees at its Headquarters.' *Fortune*, 29 November.

Mills, D. P., 2008. *For Future Generations: Reconciling Gitxsan and Canadian Law.* Saskatoon, SK: Purich Publishing Limited.

Milton, V., 1869. *A History of the San Juan Boundary Question.* London: Cassell, Petter, and Galpin.

O'Connell, J., 2017. 'What Would Happen if Amazon Brought 50,000 Workers to Your City? Ask Seattle.' *The Washington Post*, 19 October.

Ott, J., 2016. 'Governor Booth Gardner and 26 Tribes Sign the Centennial Accord on August 4, 1989.' HistoryLink.org, Essay 20232. Viewed 25 June 2023 at: historylink.org/File/20232

Peterson, J.E. II, 1990. 'Dance of the Dead: A Legal Tango for Control of Native American Skeletal Remains.' *American Indian Law Review* 15: 115–50. doi.org/10.2307/20068668

Poth, J. (ed.), 1990. 'Saltwater People as Told by Dave Elliot, Sr. A Resource Book for the Saanich Native Studies Program' (Rev. ed.). Native Education, School District 63 (Saanich).

Rajala, R.A. 1998. *Clearcutting the Pacific Rain Forest: Production, Science, and Regulation.* Vancouver: UBC Press.

Reid, J.L., 2015. *The Sea is My Country: The Maritime World of the Makahs, an Indigenous Borderlands People.* New Haven: Yale University Press. doi.org/10.12987/yale/9780300209907.001.0001

Rich, E.E. (ed.), 1941–44. *Letters of John McLoughlin: From Fort Vancouver to the Governor and Committee. Publications of the Hudson's Bay Record Society* (3 vols). Toronto: The Champlain Society.

Richards, K., 1979. *Isaac Stevens: Young Man in a Hurry.* Provo, Utah: Brigham Young University Press.

Richling, B., 2016. *The WSÁNEĆ and their Neighbors: Diamond Jenness on the Coast Salish of Vancouver Island, 1935.* Rock Mill's Press.

Riddle, M., 2008. 'The Tulalip Resort Casino Opens on July 20, 1992.' HistoryLink.org, Essay 8842. Viewed 25 June 2023 at: www.historylink.org/File/8842

Rosenberg, M., 2018. 'Orca Death Brings Southern Resident Whale Population to Lowest Level in 34 Years.' *Seattle Times*, 16 June.

Roy, P.E., 1989. *A White Man's Province: British Columbia Politicians and Chinese and Japanese Immigrants, 1858–1914*. Vancouver: UBC Press.

Sale, R., 1976. *Seattle: Past to Present.* Seattle: University of Washington Press.

Schwantes, C.A., 1993. 'Landscapes of Opportunity: Phases of Railroad Promotion of the Pacific Northwest.' *Montana: The Magazine of Western History* 43(2): 38–51.

——, 1996. *The Pacific Northwest: An Interpretive History* (Rev. ed.). Lincoln: University of Nebraska Press.

Smith, D.A., 2012. 'Imagining Victoria: Tourism and the English Image of British Columbia's Capital.' *Pacific Northwest Quarterly* 103(2): 67–83.

Stein, H.H., 2007/08. 'Fighting for Aluminum and for Itself: The Bonneville Power Administration, 1939–1949.' *The Pacific Northwest Quarterly* 99(1): 3–15.

Stewart, H.M., 2017. *Views of the Salish Sea: One Hundred and Fifty Years of Change Around the Strait of Georgia*. Madeira Park: Harbour.

Storey, K., 2016. *Settler Anxiety at the Outposts of Empire: Colonial Relations, Humanitarian Discourses, and the Imperial Press*. Vancouver: UBC Press.

Suttles, W., 1955. *Katzie Ethnographic Notes.* Edited by Wilson Duff. Victoria: BC Provincial Museum, Department of Education (Anthropology in British Columbia Memoir No. 2).

——, 1987. *Coast Salish Essays*. Seattle: University of Washington Press.

Talton, J., 2018. 'A Banner Year for Tourism in Seattle and King County.' *The Seattle Times*, 24 April.

Task Force Report on Museums and First Peoples, 1992. *Museum Anthropology* 16: 12–20. doi.org/10.1525/mua.1992.16.2.12

Taylor III, J.E., 2016. 'Lines that Don't Divide: Telling Tales About Animals, Chemicals, and People in the Salish Sea.' In L. Heasley and D. Macfarlane (eds), *Border Flows: A Century of the Canadian–American Water Relationship*. Calgary: University of Calgary Press.

Thomas, M., 2018. '"So Many People Giving a Damn": War in the Woods Resonates 25 Years Later with New Environmental Battle on the B.C. Coast.' *CBC News,* 12 April.

Thrush, C., 2007. *Native Seattle: Histories from the Crossing-Over Place.* Seattle: University of Washington Press.

Van Kirk, S., 1980. *'Many Tender Ties': Women in Fur Trade Society in Western Canada, 1670–1870*. Winnipeg, Manitoba: Watson and Dwyer Publishing.

Vancouver, G., 1984. *A Voyage of Discovery to the North Pacific Ocean and around the World, 1791–1794* (4 vols). Edited by W Kaye Lamb. London: Hakluyt Society (2nd series, nos. 163–166).

Vouri, M., J. Vouri and the San Juan Historical Society, 2006. *Images of America, San Juan Island*. Charleston: Arcadia Publishing.

Wadewitz, L.K., 2012. *The Nature of Borders: Salmon, Boundaries, and Bandits on The Salish Sea*. Seattle: University of Washington Press.

——, 2019. 'Rethinking the "Indian War": Northern Indians and Intra-Native Politics in the Western Canada–U.S. Borderlands.' *The Western Historical Quarterly* 50(Winter): 339–61. doi.org/10.1093/whq/whz096

Wang, H.L., 2011. 'Casino Revenue Helps Tribes Aid Local Governments.' Morning Edition, *National Public Radio*.

Washington State Governor's Office of Indian Affairs, 2018. 'Centennial Accord.'

West, W.R. Jr, 2016. 'Native America in the Twenty-First Century: Journeys in Cultural Governance and Museum Interpretation.' In B.L. Murphy (ed.), *Museums, Ethics, and Cultural Heritage*. New York: Routledge and ICOM.

Part Three: Society

11

Living on the Salish Sea

Nancy J. Turner and Joan Morris (Sellemah)

Joan Morris, a member of the Songhees First Nation, was born and raised at Tl'ches, the name for the small archipelago gazetteered as Chatham Islands and Discovery Island, at the western edge of the Salish Sea close to the city of Victoria. Her story and her experiences are integral to the Salish Sea and its islands, and closely tied to its history, cultures, languages and ecosystems. Many of the topics covered in this volume tie directly into her personal history; in a sense, she represents an entire population of Salish Sea residents and the lives they have led over the past century and a half.

Joan's family occupied these islands for generations, as one of several Songhees (Straits Salish) settlement locales. During the smallpox epidemic of 1862–63 Tl'ches became a refuge for the Songhees people, enabling them to escape the ravages of this terrible disease. The islands of Tl'ches have been shaped by millennia of careful resource management and subsistence practices, reflected in important resource sites used by Joan and her family: camas (*Camassia* spp.) prairies, tidal marsh root beds, seabird nesting rocks and productive clam beds (Deur and Turner 2005; Turner 2014; Mathews and Turner 2017). The numerous shell middens, burial cairns, house depressions and other archaeological features to be found on the islands remain as evidence of occupation by ancestral Salish (see Angelbeck, this volume).

People of ancient times would have travelled to and from Tl'ches from all around the Salish Sea. Joan's extended family members have lived in many different villages, from Penelekut (Kuper Island) to Pauquachin and Tsartlip on the Saanich Peninsula, and across to the Lummi and Samish sides of the Salish Sea. Joan's older relatives were bilingual or multilingual, speaking not only a dialect of the W̱SÁNEĆ (Straits Salish) language, but also often Halq'eméylem, another Coast Salish language, and sometimes also Chinook Jargon, the trade language containing elements of French, English and several Indigenous languages. Joan often served as a translator for her older family members, because most of them did not speak English at all.

In such a remote location, and with constant travelling on the water in small craft, the lives of Joan and her family could be perilous. Joan still remembers, back in the 1950s, sitting in the bow of a small wooden rowboat as it was getting dark, while her grandmother, Sellemah (after whom Joan is named, and whose name Joan also shared with Nancy), rowed from Tl'ches over to Oak Bay, on the east coast of Victoria. Joan recalls a large ferry—one of the Black Ball vessels—steaming past them. The huge waves from the ferry almost engulfed the little boat, and her grandmother had to fight to keep it from capsizing. Joan, as a child, was oblivious to the danger they were in, and was just enjoying the waves and the excitement they brought.

Joan was cared for by Sellemah and the other 'Old Ones' (*s'áliyluxw*)—her grandfather, her great grandmother Ts'emiykw and other relatives. Joan remembers they would rise early in the morning before the sun was up, chopping wood, lighting the fire and preparing breakfast. Joan recalled (22 January 2013):

> On the island we got up early. It was my responsibility—we had a well in front of the house—to go and get the water. And, it was Grandpa and Uncle Herman, who chopped the kindling to get the stove going, so we could start cooking. Then, once we finished feeding the men, do the dishes, start peeling and getting ready for lunch. Finish that. Clean up, and then we start right into supper. So then we had supper around, maybe five, get the dishes done in a hurry so we could, we were allowed to sit on the floor on a blanket, 'cause that's when the teaching time [was] for sharing of stories. Sxwiy'ám' in our language. Storytelling time.

In the winter, after dishes and chores were done, when Ts'emiykw's sisters from Kuper Island (now Penelekut Island) came to visit, they would talk and exchange stories late into the night, the elders sitting on chairs while the children sat around on the floor, being encouraged to stay quiet and listen. This is how they learned about their history and culture. Discovery Island, adjacent to Chatham, had five Big Houses where people participated in the winter ceremonials. (Many more houses existed over the past millennia.) At this remote location the Songhees were less likely to be raided during the time when potlatches and winter dances were prohibited (Lutz 2008).

Joan's family raised sheep out at Tl'ches for both wool and meat. On the occasion of their precarious encounter in their rowboat with the ferry near Tl'ches, Joan's grandmother Sellemah was bringing a sweater she had just knitted into Victoria to exchange for dishes or food items from some of the city's merchants. Ts'emiykw, Joan's great grandmother, was especially good at bartering. She made yeast bread, and she would go to Chinatown, and barter her bread for the big pans she needed to mix and bake the bread. They were completely self-sufficient for all but a few items like coffee, tea, sugar, flour, canned milk and metal pots and tools.

Joan and Sellemah not only routinely travelled to Oak Bay to exchange Sellemah's knitting for needed goods, but they also frequently rowed over to the Yacht Club at Cadboro Bay, the site of an old Songhees village, and to nearby Ten Mile Point. They harvested seabird eggs (which they boiled or scrambled, like chicken eggs), edible seaweed (*Pyropia* spp., which they sold to Chinese buyers), clams, crabs, sea urchins, chitons, rockfish and salmon. They cultivated and harvested immense quantities of camas bulbs (*Camassia* spp.) from some of the open prairies above the shoreline that they maintained by landscape burning (Turner 1999; Beckwith 2004), and picked salmonberries, trailing blackberries, thimbleberries, salal berries, red huckleberries, Pacific crab apples and other fruits in the summertime, both from Tl'ches and from around Victoria. The men hunted seals, deer and ducks as needed, with great care and respect. They cooked their food over an open campfire or on a wood-burning stove (still a common form of heating in some areas of the Salish Sea). Sometimes, with the camas bulbs and other long-cooking foods, they steamed the food in an underground cooking pit, or earth oven. For the camas, this cooking process allowed the main carbohydrate found in the bulbs (inulin, a sugar complex which is neither very sweet, nor easily digested) to break down into fructose, a sweet and easily digested sugar that was a major source of food energy, at least up until the early 1900s, for all the Salishan peoples of the Salish Sea.

Besides the original food species, Joan's family also grew their own heritage orchard fruits at Tl'ches. Joan recalls that they had two pear trees of different varieties, five kinds of apples and two types of plums, as well as raspberries, boysenberries and loganberries. They also had a vegetable garden at Tl'ches, where they grew onions, carrots, celery, cabbage, turnips, rhubarb and 'lots of potatoes'. There was always barley, too, and split peas for soup. All of these domesticated foods would have been brought in by the colonial officials and traders as early as the 1850s (Turner 2014). Joan also grew up making bannock, roasted over an open fire on a stick or spread out and baked on a flat rock beside the fire. They kept chickens as well as the sheep, and their dear friend 'Grandpa Ned', on Discovery Island, raised rabbits and grew gooseberries, which he exchanged for Joan's family's produce. They all preserved much of their food for winter, cooking it up in big enamel pots, or drying it over the fire. The relatives from Kuper Island, on the east coast of Vancouver Island, and Ucluelet on the west coast, also brought food items to contribute: soapberries (for making a special whipped confection, still enjoyed today), wild strawberries and other valued foods, to trade for wool, mutton or other Tl'ches goods (see McMillan this volume; Wadewitz, this volume).

Figure 11.1 Sellemah Joan Morris with grandmother, Sellemah Elizabeth James

Source: Photo Collection of Sellemah Joan Morris.

In the summertime, Joan travelled with her whole family on the Black Ball ferry from Tl'ches down to the American side of the Salish Sea—Shaw and Vachon islands—and to Tacoma, to pick berries and other farm produce: cucumbers and other vegetables, strawberries, raspberries, logans and blackberries: 'every kind of berry imaginable'. They didn't need passports in those days to cross the border, as they had dual citizenship. The families stayed in little cabins, preparing their own food, while everyone worked in the fields. After the berries were finished for the season, they all travelled to Puyallup and Yakima to pick hops, peaches and later-ripening orchard fruit. They would come back home with buckets of peaches and other fruit to jar for winter.

Joan's experience of living at Tl'ches, travelling by water to different locations, and harvesting the bounty of foods yielded by the ocean and adjacent lands reflects a way of life known to many people all around the Salish Sea (as recounted in several of the chapters of this volume; see also Thom 2005). For example, the Claxton family of Tsawout, at the time when the late Elsie Claxton was a girl, in the 1920s, and later as a wife and mother in the 1930s, used to travel by dugout canoe from East Saanich, north of Victoria, across to Henry Island in the San Juans to undertake reef net fishing for salmon on the American side, after this fishing technology had been banned by the Canadian Government. They would camp with other families for weeks at a time when the sockeye salmon were making their way among the Gulf and San Juan Islands heading for the Fraser River and other, smaller rivers of the Salish Sea, to spawn.

Reef net fishing is unique to the Salish Sea area. The nets, called *sxwala*, are large purse nets made from willow bark (willow is called *sx̱ʷələʔ-íɫch* 'reef net-tree'), strung across two canoes, each with a crew of three. The entrance lines to the reef net are anchored in place with rocks, and dune wildrye grass is tied to the lines to give the appearance to the fish that they are swimming along the bottom. The men in the canoes can see the fish approaching. They wait until sufficient fish have entered the net, then they draw the canoes together and carefully select and lift the fish out, two at a time. The net has a large exit hole built into the far end so that, for every group of fish entering the net, some will always get through. Immediately after the first salmon were caught, they would traditionally hold a First Salmon Ceremony, to honour the salmon and ensure a plentiful catch in the years to come (Claxton and Elliott 1994; Turner and Berkes 2006; Turner and Hebda 2012; Claxton 2015).

The W̱SÁNEĆ also travelled out to different camping and village sites on Salt Spring, Pender, Saturna and adjacent islands, where they hunted, picked berries, harvested cedar bark and tule stems for mats, fished for salmon with reef nets, and harvested edible seaweed which they dried and sold to Chinese buyers who would come around to the camps (Claxton 2015). These seasonal rounds, working in accordance with weather, tides, winds and lifecycles of the fish, clams, berries, medicinal plants and materials were a part of sustainable lifeways, developed over generations and based on rich systems of knowledge, practice and belief (Lantz and Turner 2003). These were applied not only in harvesting and processing these resources but also in sustaining and enhancing them. Tl'ches features in a W̱SÁNEĆ story about the origin of salmon, and the use of a valued spiritual medicine plant, *q'əx̱mín* (*Lomatium nudicaule*, wild celery, or 'Indian consumption plant'):

Origin of Salmon

> Once there were no seals and the people were starving; they lived on elk and whatever other game they could kill. Two brave youths said to each other, 'Let us go and see if we can find any salmon.' They embarked in their canoe and headed out to sea, not caring in what direction they travelled. They journeyed for three and a half months. Then they came to a strange country. When they reached the shore a man came out and welcomed them, saying, 'You have arrived.' 'We have arrived,' the youths answered, though they did not know where they were. They were given food to eat, and after they had eaten their host led them outside the house and said, 'Look around and see what you can see.' They looked around and saw smoke from *q'əx̱mín* [*Lomatium nudicaule*, wild celery, or 'Indian consumption plant'] that the steelhead, sockeye, spring and other varieties of salmon were burning, each for itself, in their houses.
>
> The youths stayed in the place about a month. Their hosts then said to them, 'You must go home tomorrow. Everything is arranged for you. The salmon that you were looking for will muster at your home and start off on their journey. You must follow them.' So the two youths followed the salmon; for three and a half months they travelled, day and night, with the fish. Every night they took *q'əx̱mín* and burned it that the salmon might feed on its smoke and sustain themselves. Finally they reached Discovery Island (*Tl'ches*), where they burned *q'əx̱mín* all along the beach; for their hosts had said to them, 'Burn *q'əx̱mín* along the beach when you reach land, to feed the salmon that travel with you. Then, if you treat the salmon well, you will always have them in abundance.'

> Now that they had plenty of salmon at Discovery Island they let them go to other places – to the Fraser River, Nanaimo, etc. Because their journey took them three and a half months, salmon are now absent on the coast for that period.
>
> The coho said to the other salmon, 'You can go ahead of us, for we have not yet got what we wanted from the lakes.' That is why the coho is always the last of the salmon.
>
> The young men now had salmon, but no good way of catching them. The leaders of the salmon, a real man and woman, taught them how to make *sxwaƚə* (reefnets), and how to use *q'əx̱mín*. They also told the young men how their people should dress when they caught the salmon, and that they should start to use their purse net in July, when the berries were ripe. So today, when the Indians dry their salmon they always burn some *q'əx̱mín* on the fire (or on top of the stove); and they put a little in the fish when they cook it. Also, when they cut up the salmon, before inserting the knife they pray to the salmon, that they may always be plentiful. (Jenness n.d.: 94)

Over the course of Joan's life, tangible changes have occurred at Tl'ches, as to many places around the Salish Sea (discussed in numerous chapters in this volume). She witnessed and experienced the transition from using dugout cedar canoes to wooden rowboats, motorboats and larger fishing boats and other vessels. In 1957, a drinking water shortage (see Moore et al., this volume) forced Joan and the other residents to move away from Tl'ches to the Songhees reserve in Esquimalt. Shortly afterwards her grandmother passed away, and any possibility of her family returning to the islands to live was eliminated. Without residents to care for the fruit trees and gardens, these soon declined, although some of the orchard trees still remain, and the plum trees have formed a small grove, still producing fruit. A Pacific yew tree (*Taxus brevifolia*), which Joan remembered from her childhood, is still standing above the shoreline near Joan's family homestead, perhaps planted there long ago. A major fire, started carelessly by campers on the island in 1962, burned down the family's house, along with some of the trees. Invasive species including Scottish broom and Himalayan blackberry have since spread over wide areas.

Without tending, the camas beds that provided food for hundreds of people in the past have started to deteriorate due to encroaching brush and invasive grasses, although recent removal of some of the invasive species has had promising results in enhancing the growth of camas, chocolate lily, tapertip onion and other native species that had been suppressed (Gomes 2012; Darcy Mathews, University of Victoria, personal communication, 2019; see Barsh and Murphy, this volume). The vast kelp beds Joan remembers from her childhood have diminished in size, and it seems that the seafood species like halibut and sea urchins are not as prevalent either, although recent surveys have shown a diversity of sea life in the surrounding waters (Elena Buscher and Darcy Mathews, University of Victoria, personal communication, 2018). Joan said the seagull eggs are tainted because of pollution. Today, global climate change is threatening to further impact the islands (see Hebda, this volume; Johannessen, this volume), with concerns for erosion of key archaeological sites as well as intertidal and nearshore habitats resulting from projected sea level change. Pollution from sewage (from Victoria's outfalls; with untreated sewage, at least until the facilities were completed in 2021) and from boats, and the looming threat of increased tanker traffic nearby, are all ongoing concerns.

Figure 11.2 Sellemah Joan Morris near plum grove on Tl'ches
Source: Photo by Darcy Mathews.

After being away from Tl'ches for decades, Joan revisited the islands in the early 2000s and since that time has been working with her Nation and a number of ecological restoration specialists, ethnoecologists and archaeologists to document the biocultural attributes of Tl'ches and to undertake ethnoecological restoration and monitoring for impacts from climate change and other factors (Higgs 2003; Senos et al. 2006; Gomes 2012).

The Songhees Nation is one of the Nations included in the Douglas Treaties, but the intention of the treaties and the understandings of the signatories are still under debate (Thom 2005; Lutz 2008, 2020; Turner 2020). A substantial part of Tl'ches is Songhees reserve lands, although there is a Marine Provincial Park on Discovery Island, one of the Tl'ches group, where members of the public can land and camp. There are ongoing concerns from encroachment onto the reserve lands, with people ignoring the *No Trespassing* signs, landing on the beaches, building fires, even when prohibited, and leaving garbage all along the shore. (There was another fire that occurred on Big Chatham Island in the summer of 2018.) This has led to ongoing disputes between the Songhees Nation and the boating recreation community.

Now, however, the Nation itself is starting to run its own tours, hosting boat trips around the waters of their traditional territory, between Oak Bay and Tl'ches, as a first step towards cultural tourism as local economic development (see Wadewitz, this volume). In short, these islands, like all of the others of the Salish Sea, reflect a deep history of occupation and use; they are Indigenous landscapes and Indigenous seascapes (see Lepofsky et al. 2017; Mathews and Turner 2017). This history, and the methods used to sustain and enhance the multiple resources yielded by these ecosystems, are coming to be more widely recognised in both Canada and the United States, as reflected richly in the chapters of this volume.

References

Beckwith, B.R., 2004. The Queen Root of this Clime: Ethnoecological Investigations of Blue Camas (*Camassia quamash, C. leichtlinii; Liliaceae*) Landscapes on Southern Vancouver Island, British Columbia. University of Victoria (PhD thesis).

Claxton, E., Sr (YELḰÁTT̶E) and J. Elliott Sr (STOLȻEȽ), 1994. *Reef Net Technology of the Saltwater People.* Brentwood Bay, BC: Saanich Indian School Board.

Claxton, N. X̱EMŦOLTW̱, 2015. To Fish as Formerly: A Resurgent Journey Back to the Saanich Reef Net Fishery. University of Victoria (PhD thesis).

Deur, D. and N.J. Turner (eds), 2005. *'Keeping it Living': Indigenous Plant Management on the Northwest Coast.* Seattle: University of Washington Press.

Gomes, T., 2012. Restoring Tl'chés: An Ethnoecological Restoration Study in Chatham Islands, British Columbia, Canada. University of Victoria (MA thesis).

Higgs, E.S., 2003. *Nature by Design: People, Natural Process, and Ecological Restoration.* Cambridge: MIT Press. doi.org/10.7551/mitpress/4876.001.0001

Jenness, D., n.d., 'Coast Salish Field Notes.' Ottawa, ON: Canadian Museum of Civilization (Canadian Ethnology Service Archives).

Lantz, T.C. and N.J. Turner, 2003. 'Traditional Phenological Knowledge (TPK) of Aboriginal Peoples in British Columbia.' *Journal of Ethnobiology* 23(2): 263–86.

Lepofsky, D., C.G. Armstrong, S. Greening, J. Jackley, J. Carpenter, B. Guernsey, D. Mathews and N.J. Turner, 2017. 'Historical Ecology of Cultural Keystone Places of the Northwest Coast.' *American Anthropologist* 119(3): 448–63. doi.org/10.1111/aman.12893

Lutz, J.S., 2008. *Makuk: A New History of Aboriginal–White Relations.* Vancouver: UBC Press.

——, 2020. 'Preparing Eden: Indigenous Land Use and European Settlement on Southern Vancouver Island.' In N.J. Turner (ed.), *Plants, People, and Places: The Roles of Ethnobotany and Ethnoecology in Indigenous Peoples' Land Rights in Canada and Beyond.* Montreal: McGill-Queen's University Press. doi.org/10.2307/j.ctv153k6x6.14

Mathews, D.L. and N.J. Turner, 2017. 'Ocean Cultures: Northwest Coast Ecosystems and Indigenous Management Systems.' In P. Levin and M. R. Poe (eds), *Conservation for the Anthropocene Ocean.* London: Academic Press. doi.org/10.1016/B978-0-12-805375-1.00009-X

Senos, R., F. Lake, N.J. Turner and D. Martinez, 2006. 'Traditional Ecological Knowledge and Restoration Practice in the Pacific Northwest.' In D. Apostol (ed.), *Encyclopedia for Restoration of Pacific Northwest Ecosystems.* Washington, DC: Island Press.

Thom, B., 2005. Coast Salish Senses of Place: Dwelling, Meaning, Power, Property and Territory in the Coast Salish World. McGill University (PhD thesis).

Turner, N.J., 1999. 'Time to Burn: Traditional Use of Fire to Enhance Resource Production by Aboriginal Peoples in British Columbia.' In R.T. Boyd (ed.), *Indians, Fire, and the Land in the Pacific Northwest*. Corvallis: Oregon State University Press.

——, 2014. *Ancient Pathways, Ancestral Knowledge: Ethnobotany and Ecological Wisdom of Indigenous Peoples of Northwestern North America*. Montreal: McGill-Queen's University Press.

——, (ed.), 2020. *Plants, People and Places: The Roles of Ethnobotany and Ethnoecology in Indigenous Peoples' Land Rights in Canada and Beyond*. Montreal: McGill-Queen's University Press. doi.org/10.1515/9780228003175

Turner, N.J. and F. Berkes, 2006. 'Coming to Understanding: Developing Conservation Through Incremental Learning in the Pacific Northwest.' *Human Ecology* 34(4): 495–513 (Special issue). doi.org/10.1007/s10745-006-9042-0

Turner, N.J. and RJ. Hebda, 2012. *Saanich Ethnobotany: Culturally Important Plants of the WSÁNEC People*. Victoria: Royal BC Museum Publishing.

12

Population: Past, Present and Future

Moshe Rapaport

Population is not just about numbers. Construed broadly, it is also about origins, settlement, interaction and relations with the environment. Summarising and elaborating on discussions in previous chapters, this chapter reviews the evidence which establishes Indigenous rootedness in the region; the demographic, economic and cultural catastrophes that ensued following contact; demographic profiles of the general and Indigenous populations; population growth projections; the human footprint; and reducing our footprint.

Indigenous Settlement

The Coast Salish are ethnically and linguistically related Indigenous inhabitants of the islands and surrounding coasts in the Strait of Georgia, Puget Sound and Strait of Juan de Fuca, and more distant locations. Archaeological evidence dates human presence in the area to at least 11,000 years ago, but earlier dates are possible, as rising sea levels are likely to have submerged many original settlement sites and upper elevation sites remain to be identified (Lausanne et al. 2019).

Coast Salish have charter stories that recall first ancestors on the land who dropped from the sky or otherwise appeared and established original communities, many of which continue in the same area today. Such accounts, absent in some other Indigenous communities, reflect the highly productive natural environment, settlement in large, permanent winter villages, property notions and long-standing social stratification (Thom 2005).

Winter villages on the islands were more numerous prior to massive depopulation and intertribal warfare in the early nineteenth century. Archaeological data in the form of deep midden deposits with evidence of compact floors and house posts suggest that winter villages have respectable antiquity. Settlements fluctuated over the years according to historical contingencies.

For example, in the 1840s people relocated from Salt Spring, Pender and Mayne Island to the Saanich Peninsula, forming amalgamated villages for protection from northern people attracted to the trading opportunities provided by Hudson's Bay Company (HBC) depots (Suttles 1951). Winter villages in the Southern Gulf Islands had easy access to local resources as well as the great annual sockeye fishery on the Fraser River.

It is important to note that Coast Salish were not the only Indigenous group to have settled in the Salish Sea region. By the time of European contact, the Ligwiłda'x̱w had been migrating from their homebase in Tekya, trading, intermarrying and displacing Coast Salish inhabitants (Chapter 8, this volume). Similarly, the Didiaht were utilising trail routes and kin networks to plant potatoes and trade in Cowichan and other Vancouver Island locations (Chapter 9, this volume).

Explorer records and surviving Indigenous accounts indicate that epidemic disease introduced in the Strait of Georgia resulted in massive depopulation *prior* to the arrival of the first European arrivals. Smallpox is believed to have originated in Mexico in 1779, reaching the Northern Plains in Canada by 1782. From there it diffused to the Columbia River Basin, affecting most if not all the people in the Salish Sea region by 1782–83 (Harris 1994). Measles and other epidemics also took their toll (Boyd 1994).

Smallpox is considered the 'most spectacular' epidemic affecting Indigenous communities (Boyd 1994). For the late 1770s smallpox epidemic, Boyd estimates a mortality of at least 30 per cent, for the Pacific Northwest as a whole. The 1862–63 smallpox epidemic served 'as a final blow to the Native peoples of British Columbia and paved the way for the colonization of their lands by people of European descent' (Boyd 1994).

Based on an 1881 count of 5,452 Coast Salish in the Georgia Strait region, and assuming a population decline of 90 per cent within a century after contact, Harris (1994) estimates a pre-contact population of over 50,000 around the Strait of Georgia and up the Fraser River. While ample evidence confirms the severity of the decline, both authors acknowledge the difficulty of estimating pre-contact population numbers.

As noted by Belshaw (2009), historical population projections could have been compromised by faulty estimates, reliance on elders with imperfect memories, seasonal population movement (in search of eulachon and other species during summer seasons), intertribal raids and warfare, or diseases whose symptoms can also produce scarring, such as syphilis, which was endemic in the New World.

Displacement

The Hudson's Bay Company (HBC) began establishing forts in British Columbia as early as 1805. By 1849 Vancouver Island was proclaimed a British colony, and the HBC placed in charge of immigration and settlement, defended by gunboats. Under direction from the Colonial Office the HBC arranged land sale agreements and established reserves. There was occasional violence when settlers were attacked. In 1863 the village of Hwlumelhtsu on Penelakut island was attacked and burned to the ground by colonial troops and its population scattered (Arnett 1999).

While the so-called treaties in British jurisdiction guaranteed Indigenous people ownership of their villages, 'enclosed fields' and hunting rights and fisheries, American treaties displaced Indigenous people from their ancestral homes, relocating them on substandard lands. This was followed by a number of near genocidal 'wars' (Kluger 2011). Destruction and/or occupation of Native villages became a routine practice any time settlers were killed or ships attacked. Armed resistance by Native populations was weakened by vulnerable coastal locations, tribal warfare and the cumulative effect of epidemic diseases.

An essay by MacAulay (2016) examines displacement following the 1862 smallpox epidemic. Tsimshian, Haida, Tlingit, Heiltsuk and Kwakwa̱ka̱'wakw people were cleared from camps surrounding Victoria because they were seen as susceptible to smallpox and deserving of it because of supposed tendencies to violence, drunkenness and poor hygiene. This in turn resulted in the spread of disease to outlying regions, claiming 20,000 Indigenous lives, opening the Pacific coast for settlement by whites, who would surpass the number of Indigenous people in the 1880s.

It is noteworthy that not a single reservation lies within 20 miles of Seattle, while there are numerous reserves near Vancouver and Victoria. Few reservations/reserves are located on Salish Sea Islands on either side of the border (Penelakut, Galiano, Quadra and Cortes are exceptions). Coast Salish, divided by the 1848 boundary between Canada and the US, have been considered the 'most displaced by the forces of colonization' (Wonders 2010).

The Carlisle Indian Industrial School, founded in 1879 at a former military base, became a model for Indigenous residential schools in the US and Canada. There are, however, some important policy differences affecting residential schools between the two countries. In the US, the 1934 *Indian Reorganization Act* enacted progressive policies reversing assimilationist policies of the Bureau of Indian Affairs. Many boarding schools were closed, while continuing schools supported Indigenous religious freedom and the teaching of Native languages.

In British Columbia the Methodist Women's Missionary Society in Port Simpson set up a residential school for girls as early as 1874. Other churches continued to open schools down the coast, including the Alberni Indian Residential School, Kuper Island Residential School and St Michael's Residential School. The impact of these schools was enormous—children were shipped to these schools from as far away as the Kitlope and Bella Coola and they persisted for three quarters of a century.

Throughout British Columbia, government-funded, church-run residential schools continued to advance assimilationist policies. It was not until the 1950s that policy shifted to the integration of Indigenous students into public schools. Despite this, Indigenous children continued to be removed from their families where child welfare was deemed to be insufficient.

During the 'Sixties Scoop', thousands of Native children were forcibly removed from their families and placed in state-run welfare facilities, often in former boarding schools (Marker 2009).

The Kuper Island School, run by the Catholic Church from 1890–1975, gained notoriety due to kidnapping, sexual abuse, drowning and suicide (see Thom, this volume). Research by Mosby (2014) has revealed unethical experiments at six residential schools including Port Alberni on Vancouver Island between 1942 and 1952 in which nutrition was withheld to study effects on health and teeth of aboriginal children.

In July 2021, 160 unmarked graves were confirmed adjacent to the former Kuper residential school, triggering marches, apologies by the diocese and healing sessions for survivors (CBC News 2021). Diverse theories exist, however, on why these graves remained unmarked and undiscovered, including times of epidemics and high death rates, the use of wooden crosses and cheap headstones, poor record keeping and unwillingness to ship bodies of dead children to their families for cost reasons (Hopper 2021).

The New Settlers

Waves of settlers have each left their distinctive imprints on island landscapes and populations, much as, over a much longer period of time, geologic plates, terranes, glaciation and erosion were instrumental in forming the layers of sandstone and mudstone that overlie the islands today To the Indigenous people who lived, hunted, gathered and fished on Salish Sea Islands, the intricate tie between the health of society and the health of the land and waters upon which they relied was self-evident (Weller 2013: 48–52).

The arrival of Europeans and colonial government changed this fundamental relationship, prioritising individual prescribed ownership of land. Renaming of places by colonial explorers signalled ownership and control by Europeans. The 1876 *Indian Act* racialised Indigenous people from the rest of the country by defining Indian status and creating reserves and residential schools. The potlatch ceremony, in effect the Indigenous institution of self-governance (see Chapters 6 and 7), was banned in 1884, though the practice continued until 1951.

By the late nineteenth century, Indigenous communities and economies had been decimated due to epidemic diseases and government policies, and settler lifestyles revolved around farming, resource extraction and manufacturing. A few of these farms, some of which produced enough to sell to nearby urban centres, are still in production today, but most were divided up in the 1970s for real estate development. Virtually of the islands have been logged, in many areas several times. Iron, copper, gold, coal and limestone deposits provided additional sources of income, as did pulp and paper mills, and brick manufacturing (see Chapter 18).

Non-resident real estate developers began subdividing the islands to meet increasing demand. On Mayne, demand for land caused land prices to double within two years. To keep up with demand, realtors began subdividing large tracts, snapped up by speculators and new residents as soon as they went on the market. Property investors, retirees and vacationers throughout North America and beyond began looking to the Gulf Islands (Weller 2017: 97–98).

By 1974, concern over development led to the passage of the *Islands Trust Act*, aimed to preserve and protect the 'unique amenities and environments' of constituent islands. The Islands Trust includes the Southern Gulf Islands and a portion of the Northern Gulf Islands (see Figure 12.1). The Islands Trust does not include Quadra and Cortes, which are more distant from Vancouver, and less directly threatened by development, or Texada, where continuing mining interests were deemed to be not consistent with the aims of the *Islands Trust Act* (Weller 2013: 62).

Islands Trust and San Juan Islands

The difficulty of obtaining comparable census data for Salish Sea Islands within different jurisdictions is compounded by a tendency in both countries to statistically merge the populations of islands with the adjacent 'mainland' census subdivisions. Two sets of island populations are tabulated separately in recent censuses: Islands Trust (Figure 12.1) and San Juan Islands (Figure 12.2).

Figure 12.1 Islands Trust and vicinity

Note: Contains information licensed under the Open Government Licence—Islands Trust.

Source: Islands Trust (2023a).

Population density is about equal in the Islands Trust and San Juan Islands (even with Islands Trust's larger land area and population), but neither match the comparable mainland regions. The population density of San Juan Islands (in close proximity to population centres) is about eight times higher than British Columbia; whereas the density on the San Juan Islands is lower than in Washington (Table 12.1).

Median age is higher in both island groups, relative to the larger regions they are part of. The amenities in both cases (mild climate, no traffic, lower crime rates) are attractive to retirees and the self-employed. Young people typically leave the islands as soon as they finish school to pursue higher education, work and social opportunities elsewhere. Rentals are scarce, which is why homeownership is relatively high.

Figure 12.2 San Juan Islands and surrounding region

Note: This file is licensed under the Creative Commons Attribution-Share Alike 3.0 Unported licence.

Source: Wikimedia Commons (2021).

Table 12.1 Demographic comparison, Islands Trust and San Juan Islands

	Islands Trust	BC	San Juan Islands	WA
Land area (square km)	798	925,186	450	172,587
People per square km, 2021	39.9	5.4	39.7	45.0
Total population (thousands), 2021	30.5	5000.8	17.8	7705.3
Population growth rate (%), 2016 to 2021	16.4	7.6	6.4	7.3
Median age of population, 2021	58.8	42.8	56.5	38.2
% Homeownership, 2021	84.2	66.8	75.9	64.0

Sources: Derived from OFM (2023a); Islands Trust (2020, 2023b).

Vancouver Island

The size and population of Vancouver Island is on a completely different scale relative to the small islands considered above. Unlike the other islands, Vancouver Island lies on the margins (rather than within) the Salish Sea. It is the sole location in the Salish Archipelago with populous cities, higher education, hospitals and a diversified economy.

Vancouver Island's industries experienced declines due to the COVID-19 pandemic. Forestry was particularly affected by shutdowns and an extended strike by workers. Border and restaurant closures impacted tourism and aquaculture. International enrollments dropped as schools moved to remote learning and travel restrictions (VIEA 2021: 21).

Population of the eight most populous cities and per cent change between 2016 and 2021 are shown in Table 12.2. The metropolitan population of Victoria, Capital of British Columbia, holds the second-place rank among populated cities in British Columbia (Statistics Canada (2022). The city is popular with tourists, retirees and students. Since 2001 the population has grown by about 5 per cent every five years, although growth has slowed slightly in the past few years.

Table 12.2. Population change, Vancouver Island cities

City	Population, 2016	Population, 2021	Percent change, 2016–2021
Greater Victoria	367,770	397,237	+8.0
Nanaimo	104,936	115,459	+6.4
Courtenay	57,950	63,282	+9.2
Duncan	44,451	47,582	+7.0
Campbell River	37,861	40,704	+7.5
Parksville	28,922	31,054	+7.4
Port Alberni	24,669	25,786	+4.5
Ladysmith	14,572	15,501	+6.4
Total	681,131	736,605	

Source: Statistics Canada (2022).

Likewise, Victoria leads in the per cent of external in-migrants (22.2 per cent of total in-migrants), followed by Nanaimo (13.1 per cent of total in-migrants) (see Table 12.3). It is understandable that external migrants, who may be less familiar with the area, would decide to move to the populous capital city of the province, rather than distant locations (such as Port Alberni) or less populated locations (such as Ladysmith).

Table 12.3. In-migration, Vancouver Island cities

City	Total in-migrants	Percent internal in-migrants[1]	External in-migrants[2]	Percent external in-migrants
Greater Victoria	72,750	77.8%	16,135	22.2%
Nanaimo	25,705	86.9%	3,360	13.1%
Courtenay	15,700	91.3%	1,360	8.7%
Duncan	10,565	91.6%	885	8.4%
Campbell River	8,525	92.4%	645	7.6%
Parksville	9,000	95.8%	380	4.2%
Port Alberni	5,000	95.0%	245	4.9%
Ladysmith	4,085	96.2%	160	3.9%
Total	151,330	84.7%	23,170	15.3%

Source: Statistics Canada (2023).

Figure 12.3 shows the locations of First Nation reserves on Vancouver Island.

Figure 12.3 Locations, First Nation reserves on Vancouver Island

Note: Contains information licensed under the Open Government Licence—Canada.

Source: Derived from data at Government of Canada (2023).

1 Migrants from British Columbia or other provinces.

2 Migrants from outside of Canada.

First Nation populations on Vancouver Island (see Figure 12.3) are listed in Table 12.4, in order of population change between the decadal British Columbia censuses of 2006 and 2016. High rates of population increase occur in locations within commuting distance from Victoria, or small towns with tourist jobs (such as Ucluelet). Lowest rates occur in northern or west coast locations.

Table 12.4 Population change, Vancouver Island First Nations

First Nation	Population 2006	Population 2016	Population change
Malahat	90	135	+50.0
Ucluelet	200	275	+37.5
Snuneymuxw	560	725	+29.5
Tseshaht	440	555	+26.1
Nanoose	195	230	+18.0
Cape Mudge	385	4/40	+14.3
Penelakut	N/A	555	NA
Cowichan	1,930	2,150	+11.4
Halalt	155	170	+9.7
T'Sou-ke	210	220	+4.8
Stz'uminus	810	840	+3.7
Tsawout	1,635	1,685	+3.1
Ehattesaht	90	90	0.0
Gwa'Sala-Nakwaxda'xw	435	430	-1.2
Quatsino	235	225	-4.3
Campbell River	380	360	-5.2
Kwaikiutl	270	255	-5.6
Homalco	220	205	-6.8
Hupacasath	160	135	-15.6
Didiaht	195	160	-18.0
Pacheedaht	105	85	-19.1
K'ómoks	275	220	-20.0
Tla-o-qui-aht	350	245	-30.00
Hesquiaht	110	50	-54.6

Note: Contains information licensed under the Open Government Licence — Canada.

Source: Derived from Crown-Indigenous Relations and Northern Affairs Canada (CIRNAC) (2021).

Actual percentages of First Nation populations may be higher than indicated in census enumerations, due to high mobility rates, transient populations, a historical mistrust of the national government that persists up until today and methodological problems such as geographical challenges and language barriers, in a wide range of circumstances and situations (Hubner 2007; Lujan 2014).

Population Projections

PEOPLE 2020 is a population projection by BC Stats covering the years from 2020 to 2041. A reduced growth rate of 1 per cent over 2020/21 is related to the impact of the COVID-19 pandemic on migration to and within Canada. The annual rate is expected to recover to 1.3 per cent before the growth rate gradually declines to about 0.9 per cent at the end of the projection period (Ip and Lavoie 2020).

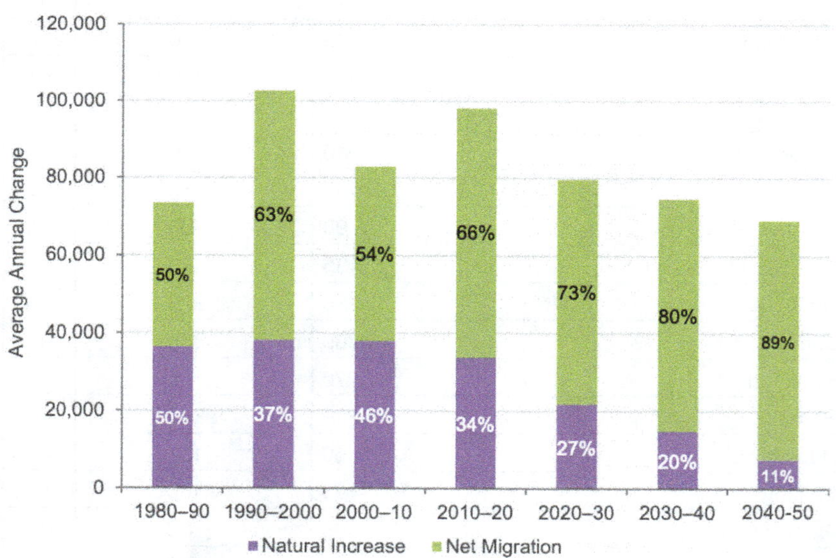

Figure 12.4 Population projection by decade, Washington
Source: OFM (2023c).

The State of Washington's Office of Financial Management (OFM) is charged with providing populations projections under the state's *Growth Management Act* (GMA). The projected decline in growth rates between 2020 and 2040 is presented in Figure 12.4 based on assumptions concerning fertility (lower among millennials and immigrants), mortality (ageing populations) and migration (holding steady).

The Human Footprint

The Salish Sea Ecosystem Conference provides a forum for numerous scientific reports related to current health of the Salish Sea region (SSEC 2014/2016/2018/2020). Available data have been summarised in overview publications such as Beamish and McFarlane (2014), Stewart (2017), Quinn (2010) and State of the Sound (2021). A brief summary of the current situation follows below.

When glaciers retreated and the seas invaded to form Puget Sound (and the Georgia Strait) the surrounding vegetation took over, climaxing in a coniferous forest of titanic and intimidating dimensions. Pioneers found an impenetrable temperate jungle, thick groves of trees that soared more than 200 feet, with trunks so thick early homesteaders made cabins out of their hollow stumps (Dietrich 1999: 20).

Coast Salish people are believed to have obtained as much as 90 per cent of their protein from marine sources for approximately 6,000 years prior to contact. Assuming Cole Harris' estimate of 50,000 people prior to the onset of the smallpox epidemics, consuming at the estimated per capita average rate, the Strait of Georgia would have supported a sustainable harvest of about five million sockeye salmon (Glavin 2014).

Earthquakes, volcanic eruptions, tsunamis, forest fires and clearances by Native people all left their mark, but these disturbances have been dwarfed by the impact of settlement. Over the past century and a half the biggest trees have been cut, fish runs imperilled, wetlands filled, fields ploughed and roads paved. Only the stumps of these old trees can still be seen today, surrounded by tree farms of cedar, hemlock, alder, spruce and Douglas fir (Dietrich 1999: 21).

Declines in bird populations have been tied to PCBs (polychlorinated biphenyls) in the effluents of pulp and paper mills, oiling of bird feathers following illegal discharge and catastrophic oil tanker spills, hunting and egg harvesting, entanglement in gillnets and lost fishing gear, pollution from landfills, plastic items which become concentrated in the crop of birds, boat and jet ski traffic and nesting site disturbance by dogs (Beamish and McFarlane 2014).

The most devastating loss took place in the region's once enormous salmon and steelhead runs, their habitat reduced by damming, siltation, agriculture, development and pollution. Mismanagement by a Department of Fisheries located thousands of miles away, allowing fish farms and unsustainable harvests, was also a major contribution. The basin's commercial salmon fishery has largely collapsed. Sports fishermen who switched from salmon to bottom fish noticed declines in these species as well (McFarlane and King 2014).

Reducing Our Footprint

It will take hundreds, and maybe even thousands of years for old growth forests to be restored in areas undergoing protective efforts. The insatiable quest for land and marine resources raises questions about the long-term viability of such efforts. The decline of fisheries due to damming, pollution and overharvest is proving difficult to reverse with hatchery production alone.

The importance of sustainability is especially evident to the inhabitants of small islands, given the cost of freight and ferry transport, and limited land, water and soil resources. Not surprisingly, sustainability is listed as a key goal of the Islands Trust Council (Islands Trust 2020):

- To foster preservation and protection of the trust area's ecosystems.
- To ensure that human activity and the scale, rate and type of development in the trust area are compatible with maintenance of the integrity of trust area ecosystems.
- To sustain island character and healthy communities.
- Efficient and collaborative governance.

The Islands Trust has helped preserve component islands as a 'heritage-scape', highlighting the islands' unique natural and cultural heritage as a way to promote tourism. However, with a surge in tourism and population, demand for new dwellings and potential exit of Salt Spring and perhaps others from the Islands Trust, a further transformation is in process, and conservation efforts are likely to weaken.

Locally specific sustainability plans have been developed by individual communities in Salt Spring (CESC 2017), and other Salish Sea Islands. Sustainability issues have been addressed in detail in Chapters 16 (on salmon), 17 (water), 18 (conservation), 19 (planning) and 20 (co-management), in the 'Environmental Management' section of this volume.

Given the observed declines in the effectiveness of government regulatory measures, in both fishing and logging, greater attention should be paid to Indigenous conservation methods. Turner and Berkes (2006) have shown how Indigenous food gathering, embodying belief systems, narratives, ceremonies and other measures, have been instrumental in sustainable resource use and conservation (see Chapter 13).

Concepts embedded in Indigenous food gathering systems include the kinship of all life forms, including mountains and plants; that all creatures have their own families and lives parallel to and as important as those of people; that food and other resources are gifts requiring appreciation; the avoidance of needless killing; the avoidance of waste; and allowing a portion of the catch to escape (Turner and Berkes 2006).

The role of ecological knowledge, preservation of ancestral sites, cultural and political boundaries, governments, ecocultural restoration and reconciling Indigenous and scientific views have been reviewed in Chapters 6 (on origin stories), 13 (Indigenous renaissance) and 20 (co-management) in this volume. The fate of the forests, salmon and other ecosystem components within and adjacent to the Salish Sea may well depend on harmonising these managerial systems.

Acknowledgements

I am most grateful to the generous advice, comments, suggested readings and referrals by Chris Arnett, Nancy Turner, Richard Kool, Russel Barsh, Jeff Corntassel and three anonymous reviewers.

References

Arnett, C., 1999. *The Terror of the Coast: Land Alienation and Colonial War on Vancouver Island and the Gulf Islands, 1849–1863*. Vancouver: Talonbooks.

Beamish, R. and G. McFarlane (eds), 2014. *The Sea Among Us*. Madeira Park: Harbour Publishing.

Belshaw, J.D., 2009. *Becoming British Columbia: A Population History*. Vancouver: UBC Press.

Boyd, R., 1994. 'Smallpox in the Pacific Northwest: The First Epidemics.' *BC Studies* 101: 5–40.

CBC News, 2021. 'B.C. First Nation Says More than 160 Unmarked Graves Found.' *CBC News*, 12 July. Viewed 31 May 2023 at: www.cbc.ca/news/canada/british-columbia/penelakut-kuper-residential-school-1.6100201

CESC (Community Economic Sustainability Commission), 2017. 'Salt Spring Island'. Viewed 31 May 2023 at: www.sustainableislands.ca/

Crown–Indigenous Relations and Northern Affairs Canada (CIRNAC), 2021. 'Welcome to First Nation Profiles.' Viewed 26 July 2023 at: fnp-ppn.aadnc-aandc.gc.ca/fnp/Main/index.aspx?lang=eng

Dietrich, W., 1999. 'Is Puget Sound in Peril?' *American Forests*, Winter.

FNHA-PHO (First Nations Health and Well-Being and Provincial Health Officer), 2018. 'Indigenous Health and Well-being.' Viewed 31 May 2023 at: fnha.ca/Documents/FNHA-PHO-Indigenous-Health-and-Well-Being-Report.pdf

Glavin, T., 2014. 'The Pre-Contact Era.' In R. Beamish and G. McFarlane (eds), *The Sea Among Us: The Amazing Strait of Georgia*. Madeira Park: Harbour Publishing.

Government of Canada, 2023. 'First Nations Location.' Viewed 24 July 2023 at: open.canada.ca/data/en/dataset/b6567c5c-8339-4055-99fa-63f92114d9e4

Harris, C., 1994. 'Voices of Disaster: Smallpox around the Strait of Georgia in 1782.' *Ethnohistory* 41(4): 591–626. doi.org/10.2307/482767

Hopper, T., 2021. 'How Canada Forgot About More Than 1,308 Graves at Former Residential Schools.' *National Post*, 13 July.

Hubner, B.E., 2007. '"This is the Whiteman's Law": Aboriginal Resistance, Bureaucratic Change and the Census of Canada, 1830–2006'. *Archival Science* 7: 195–206. doi.org/10.1007/s10502-007-9044-8

Ip, F., and S. Lavoie, 2020. 'PEOPLE 2020: BC Sub-Provincial Population Projections.' BC Stats. Viewed 31 May 2023 at: gov.bc.ca/assets/gov/data/statistics/people-population-community/population/people_population_projections_highlights.pdf

Islands Trust, 2020. 'State of the Islands Indicator Project: Final Report.' Viewed 25 June 2023 at: islandstrust.bc.ca/wp-content/uploads/2021/02/tas_2020-02_stateoftheislandsreport_final.pdf/

——, 2023a. 'Entire Region.' Viewed 29 July 2023 at: islandstrust.bc.ca/mapping-resources/mapping/entire-region/

——, 2023b. 'Islands Trust Area Census Profile 2021.' Viewed 25 July 2023 at: islandstrust.bc.ca/document/islands-trust-area-census-profile-2021/

Kluger, R., 2011. *The Bitter Waters of Medicine Creek: A Tragic Clash Between White and Native America*. New York: Knopf.

Lausanne, A., D. Fedje, Q. Mackie and I.J. Walker, 2019. 'Identifying Sites of High Geoarchaeological Potential Using Aerial LIDAR and GIS on Quadra Island, Canada.' *Journal of Island and Coastal Archaeology* 16(2-4): 482–508. doi.org/10.1080/15564894.2019.1659884

Lujan, C.C., 2014. 'American Indians and Alaska Natives Count: The US Census Bureau's Efforts to Enumerate the Native Population.' *American Indian Quarterly* 38(3): 319–41. doi.org/10.1353/aiq.2014.a552222

MacAulay, W., 2016. 'A Fit Judgement for their Intolerable Wickedness': Settler Responses to the 1862 Smallpox Epidemic in Victoria. University of Victoria (BA (Hons) thesis).

Marker, M., 2009. 'Indigenous Resistance and Racist Schooling on the Borders of Empires: Coast Salish Cultural Survival.' *Paedagogica Historica* 45(6): 757–72. doi.org/10.1080/00309230903335678

McFarlane, G. and J. King, 2014. 'The History of the Fisheries.' In R. Beamish and G. McFarlane (eds), *The Sea Among Us*. Madeira Park: Harbour Publishing.

Mosby, I., 2014. 'Administering Colonial Science: Nutrition Research and Human Biomedical Experimentation in Aboriginal Communities and Residential Schools, 1942–1952.' *Histoire Sociale* 46(91): 145–72. doi.org/10.1353/his.2013.0015

OFM (Office of Financial Management), State of Washington, 2023a. 'Components of Population Change.' Viewed 25 June 2023 at: ofm.wa.gov/washington-data-research/population-demographics/population-estimates/components-population-change

——, 2023b. 'Tribal Areas—2010 Census Detailed Demographic Profile.' Viewed 27 July 2023 at: ofm.wa.gov/washington-data-research/population-demographics/decennial-census/census-2010/2010-census-detailed-demographic-profiles/tribal-areas-2010-census-detailed-demographic-profile

——, 2023c. 'Forecast of the State Population: November 2022 Forecast.' Viewed 24 July 2023 at: ofm.wa.gov/sites/default/files/public/dataresearch/pop/stfc/stfc_2022.pdf

Quinn, T., 2010, 'An Environmental and Historical Overview of the Puget Sound Ecosystem'. In H. Shipman, M.N. Dethier, G. Gelfenbaum, K.L. Fresh, and R.S. Dinicola (eds), *Puget Sound Shorelines and the Impacts of Armoring—Proceedings of a State of the Science Workshop, May 2009*. Reston: US Department of the Interior and US Geological Survey (US Geological Survey Scientific Investigations Report 2010-5254).

SSEC (Salish Sea Ecosystem Conference), 2014/2016/2018/2020. Viewed 31 May 2023 at: cedar.wwu.edu/ssec

State of the Sound, 2021. '2021 State of the Sound.' Viewed 25 June 2023 at: stateofthesound.wa.gov/

Statistics Canada, 2022. 'Population and Dwelling Counts: Canada, Provinces and Territories, Census Metropolitan Areas and Census Agglomerations.' Viewed 29 July 2023 at: www150.statcan.gc.ca/t1/tbl1/en/tv.action?pid=9810000501

——, 2023. 'Census Metropolitan Area and Census Agglomeration Components of Migration 5 Years ago by Mother Tongue, Age and Gender: Canada.' Viewed 30 July 2023 at: www150.statcan.gc.ca/t1/tbl1/en/tv.action?pid=9810038101

Stewart, H.M., 2017. *Views of the Salish Sea*. Madeira Park: Harbour Publishing.

Suttles, W.P., 1951. Economic Life of the Coast Salish of Haro and Rosario Straits. University of Washington (PhD thesis).

Thom, B., 2005. Coast Salish Senses of Place: Dwelling, Meaning, Power, Property and Territory in the Coast Salish World. McGill University (PhD thesis).

Turner, N.J. and F. Berkes, 2006. 'Coming to Understanding: Developing Conservation Through Incremental Learning in The Pacific Northwest.' *Human Ecology* 34: 495–513. doi.org/10.1007/s10745-006-9042-0

VIEA (Vancouver Island Economic Alliance), 2023. *State of the Island Economic Report 2021*. Victoria: VIEA.

Weller, F.E., 2013. The 'How' of Transformative Change: Stories from the Salish Sea Islands. University of Victoria (PhD thesis).

Weller, J., 2017. 'Living on "Scenery and Fresh Air": Land-Use Planning and Environmental Regulation in the Gulf Islands.' *BC Studies* 193: 89–114.

Wikimedia Commons. 2021. 'San Juan Islands Map.' Viewed 29 July 2023 at: commons.wikimedia.org/wiki/File:San_Juan_Islands_map.png

Wonders, K. 2010. 'Coast Salish.' First Nations: Land Rights and Environmentalism in British Columbia. Viewed 25 June 2023 at: firstnations.de/development/coast_salish.htm

World Population Review, 2023. 'Victoria Population 2023.' Viewed 27 July 2023 at: worldpopulationreview.com/canadian-cities/victoria-population

13

Indigenous Political, Linguistic and Cultural Renaissance

Moshe Rapaport

> I think our people have to realize that they've become lost somewhere. We have come through a great disaster and we are like people in shock. We were almost destroyed. We are living in the wreckage of what was once our way of life. We have to look at this and try to do something about it. Now we are very much like the people who we say brought this upon us. This is a state of shock really—our memories have left us. Many of the young people don't know where they're coming from and where they are going. It's their future. We need to give them their past by telling them our history and we need to give them a future. (W̱SÁNEĆ Elder David Elliott Sr, 1990)

It is widely assumed that Indigenous peoples and cultures are under threat, rooted in landscapes and seascapes that have been radically transformed, and confronted by the forces of colonialism, racism and capitalism. Jeffrey Sissons (2005) suggests that Indigenous cultures today are in the midst of a renaissance, as 'diverse and alive as they ever were'. Instead of being absorbed into a homogeneous modernity, Indigenous cultures are actively shaping alternative futures for themselves.

Indigenous Nations of the Salish Sea Island region, spanning both sides of the US/Canada border, have developed ingenious methods of food cultivation and gathering on their lands—lands separated by sea and lacking the bountiful salmon-bearing rivers of continental landmasses such as the

Fraser or the Columbia. Traditional ecological knowledge, often encoded in stories and rituals, has been passed down from elders to younger generations over the millennia. Fortunately, this knowledge has not been lost.

This chapter begins with the struggle for sovereignty and territorial rights lost during Euro-American settlement and following periods. Next, selected examples of cultural revival and self-determination are presented: language, canoe journeys, reef net fishing (see also chapter on co-management of environmental resources) and clam gardens. The chapter concludes with questions pertaining to cultural revival in general: Why is this important? Can such efforts be successful?

Sovereignty

The loss of rights and title of Indigenous Nations in the Salish Sea area began with the Oregon Treaty of 1846, which set the boundary between the US and British North America at the 49th parallel, with the exception of Vancouver Island. In the decade that followed, treaties conducted by James Douglas on behalf of the British Crown and Isaac Stevens as governor of Washington Territory sought to extinguish Indigenous rights and title and replace them with treaty rights (Harris 2008).

Between 1850 and 1854, Douglas negotiated treaties with 14 Indigenous Nations on Vancouver Island. Verbal agreements were reached and each leader signed with an X. Around 930 square kilometres of land were exchanged for cash, clothing and blankets; the rights to retain village lands and fields; and to hunt and fish on surrendered lands. These have been followed by numerous disputes as blank documents were signed (Petrescu 2017; Price and Claxton 2021; see also Lutz 2017).

On the US side of the border, the Treaty of Point Elliott was concluded in 1855, with signatories including Chief Seattle, Territorial Governor Isaac Stevens and numerous Puget Sound Tribes including the Lummi, which claimed rights to the San Juan Islands. The treaty guaranteed fishing rights and reservations. However, some signatories were never granted reservations, and others did not sign at all, and these tribes continue to assert their self-determination and the rights associated with that (Petrescu 2017).

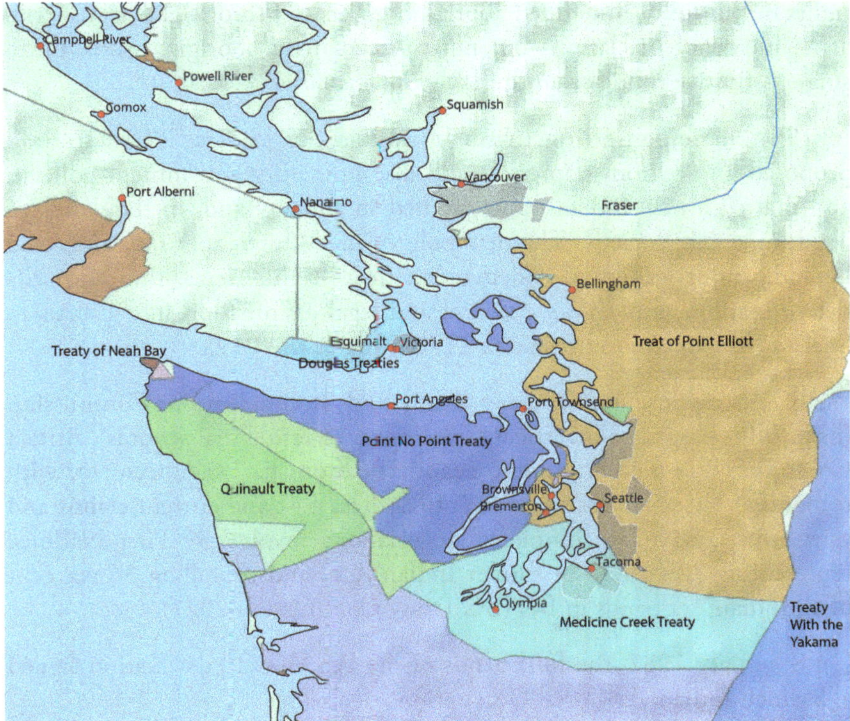

Figure 13.1 Territory ceded under Douglas and Stevens Treaties

Note: Stevens Treaties include Treaties of Point Elliott, Point No Point, Medicine Creek, Quinault, Neah Bay, Yakama and Nez Perce (not shown on the map).

Source: Derived from Native Land Digital (2023). Made with Natural Earth (www.naturalearthdata.com).

Fisheries rights have played out differently across the border. The Stevens treaties provided that 'the right of taking fish at usual and accustomed grounds and stations is further secured to said Indians in common with all citizens of the Territory'. In 1970, the US Government and several tribal governments sued the State of Washington for disregard for the fishing rights in the Stevens treaties, and under the Boldt decision of 1974 the Indian right to half of all the fish harvest each year was affirmed (Harris 2008).

The Douglas Treaties reserved to aboriginal peoples the right to 'their fisheries as formerly', but in practice offered only a patchwork of arrangements and agreements-in-principle, some affirmed and others in process, all with different fisheries provisions. For most of the province there are no treaty rights but, instead, ill-defined aboriginal rights to fish or rights to fisheries as an incidence of claimed but not-yet-recognised aboriginal title, and a single definitive ruling seems unlikely (Harris 2008).

In 1969, in response to activism by Indigenous leaders, Pierre Trudeau proposed a shift from oppressive and discriminatory government policies, and drafted a new Indian policy, rooted in equality and, in the words of Trudeau, 'a just society'. The new policy ('the White Paper'), was rejected by Indigenous leaders who demanded treaty rights, restitution and self-determination, rather than political and legal assimilation into Canadian Society (Indian Chiefs of Alberta 1970; King et al. 2019).

Partly in response to Canada's Truth and Reconciliation Commission (originally charged to investigate the impact of residential schools), British Columbia (BC) is attempting to negotiate Reconciliation Agreements with recognised First Nations and bands to address past and current claims and grievances. Active or completed negotiations involve 39 First Nations, representing 72 current or former Indian Act bands, totalling 36 per cent of all Indian Act bands in BC (BC Treaty Commission 2021).

On 8 January 1987, the four bands of the W̱SÁNEĆ First Nation signed a 'Saanich Indian Territorial Declaration', declaring:

> We do not recognize any past attempts to separate us from our homeland. We recognize that there were Treaties of Peaceful Co-existence entered into with the early settlers, but this did not involve the sale of rights or land …
>
> We will, from this day forward, take the necessary actions to govern our Saanich Territorial Homelands as outlined in our Territorial Map. (W̱SÁNEĆ Leadership Council n.d.)

Following the lead of the W̱SÁNEĆ, the Indigenous 'resurgence' school approached the reconciliation process with scepticism. Inspired by the writings of Frantz Fanon, Standing Rock Sioux scholar Vine Deloria Jr and others, Alfred (2005) and Corntassel (2012) propose turning away from the colonial state, and reclaiming and regenerating relational, place-based existence through daily or cyclical acts of renewal, such as speaking the language, honouring ancestors, and prayer, providing a way to 'power-surge against the empire with integrity' (Alfred 2005: 24).

A recent post to the 'Everyday Indigenous Resurgence' (EIR) Instagram account reflected on Indigenous responses to the COVID-19 epidemic, in the islands and beyond. By posting something each day, participants share the diverse ways they engage with cultural and familial practices during a pandemic (Corntassel et al. 2020).

Does resurgence require that the negotiations with the state be vacated altogether?

> Of course not. Settler-colonialism has rendered us a radical minority in our own homelands, and this necessitates that we continue to engage with the state's legal and political system. What our present condition does demand, however, is that we begin to approach our engagements with the settler-state legal apparatus with a degree of critical self- reflection, skepticism, and caution. (Coulthard 2014; see also Elliott 2018)

Is a 'Nation to Nation' relationship possible when each side demands sovereignty? Perhaps, writes Claxton (2003), citing the example of the Iroquois *Gus-Wen-Tah* (two row wampum): a bed of white wampum shell beads symbolising the sacredness of the treaty relationship, and two rows of purple wampum beads that represent separate but parallel paths by which the two parties travel down the same river. Each side travels side by side without 'steering the other's vessel'.

Tsawalk

Reconciliation between coloniser and colonised is not limited to apologies, treaties, land returns, and reparations. Important as material compensation may be, this cannot fully resolve radically divergent world views, long-standing grievances, and seemingly intractable political, social, economic, and racial hierarchies.

The theory of *tsawalk* (literally 'one') by Ahousaht hereditary chief E. Richard Atleo (*Umeek*), draws on analysis of traditional stories, social and political critique, and personal memoir. Assumptions about the nature of reality constitute the 'foundation upon which perspectives, laws, policies, and practices are situated'. Exploring these assumptions opens avenues towards ecologically just futures (Atleo 2011).

Atleo's work, interleaving insights from Nuu-Chah-Nulth tradition, Indigenous studies, anthropology, social philosophy, and environmental ethics, is an innovative, and unique achievement which continues to generate much discussion today in the Salish Sea area and beyond. As Atleo's oeuvre is voluminous I have based this section on an interpretive summary by Russell Duvernoy (2020).

Key concepts in Atleo's synthesis are *phase transitions* (such as global change and culture contact); *phase connectors*, such as the 1997 *Delgamuukw v. British Columbia* Supreme Court decision to make oral history permissible and relevant for sovereignty claims under Canadian law; *protocols* such as reciprocity for managing transitions; and *Oosumich* ('vision quest').

'Predation – by Wolf upon Deer, by colonizers upon colonized, by corporations upon the powerless and poor – is a common reality and is therefore a dimension of truth' (Atleo 2011: 122). The purpose of *protocols* is not to eradicate these potentials and polarities, but rather to manage them. In the words of Atleo, 'Life is dangerous if it is not appropriately managed, but beautiful if it is appropriately managed'.

Oosumich is defined as 'a secret and personal … spiritual activity that can inform varying degrees of fasting, cleansing, celibacy, prayer, and isolation for varying lengths of time depending on the purpose' and as a 'viable means for initiating a positive interaction with the spiritual realm' (Atleo 2004:17; Duvernoy 2020).

Oosumich experiences are individual and personal, but the practice is 'universal'. Interpretation is open to on-going revision based on outcomes in experience. This 'fallibilism' departs from stereotypes of a dogmatic priestly class leveraging power to maintain dominance. Such worries reflect historical patterns in some cultures, but *oosumich* is not confined to individuals of status (Atleo 2011:53; Duvernoy 2020).

Language

All 34 Indigenous languages in BC are considered endangered (FPCC 2018). First Nations communities are trying to revitalise their languages, but such efforts face numerous challenges: the few fluent speakers available to teach the languages, the passing of elders with specialised cultural and grammatical knowledge, limited language resources and barriers resulting from colonisation and assimilation policies and practices (Rosborough et al. 2017).

Reviving endangered languages can be particularly difficult in languages that are 'polysynthetic'—where morphemes (units of meaning) are conveyed in a single word, as in the examples shown below. Based on these and other examples, Nicolson (2013, cited by Rosborough et al. 2017: 431) has argued in her doctoral dissertation that 'replacing Kwak'wala with English shifts our relationship to our past, our histories, our lands and to each other as Kwakwa̱ka̱'wakw people'.

- *-axst(a)-* 'mouth'
- *ga'-axst-ala* 'to breakfast'
- *'i'k-axst-a* 'to speak nicely, good mouth'

Similar examples can be found in the *Twana Dictionary* (Thompson 1989), which includes ethnographic information from a variety of Puget Salish sources including different pronunciations and usage. The word for crow in Lushootseed is *Ka'ka*, to which Thompson adds the mythological being Crow and other cultural concepts associated with crows, providing information about linguistic aspects of the word as well as 'the world in which it lived and lives' (Buerge 2020).

The work Indigenous elders have often done to preserve their languages is extraordinary. Originally a custodian at the ȽÁU, WELṈEW̱ Tribal School, Dave Elliott purchased a used typewriter for $30 and set out to make the SENĆOŦEN writing system accessible to his people. The SENĆOŦEN alphabet is now used throughout the surrounding public schools of Saanich (School District 63), and the revitalisation of the SENĆOŦEN language is considered Dave's legacy (W̱SÁNEĆ School Board n.d.).

Kirsten Sadeghi-Yekta (2020) describes an innovative method to enliven language in Hul'q'umi'num'. Of an estimated 6,000 members of the Hul'q'umi'num' First Nation on Vancouver Island, around 40 fluent first-language speakers remain, and Hul'q'umi'num' is now considered an endangered language. Using traditional stories the Hul'q'umi'num' Heroes project aims to bring the language to the eyes and ears of the community, using the power of playfulness, laughter and exploration.

One of the stories being dramatised is 'The Monster at Lhap'qw'um', a young woman who saves her love interest and others from a sea monster. The idea of a female hero immediately aroused excitement. A call for participants in a theatre workshop was advertised and by autumn 2015, 30 participants were signed up. The story was told and listened to several times and broken down into scenes and characters, after which a script was developed, including songs for each participant (Sadeghi-Yekta 2020: 43).

A second play under development is based on a Ts'inukw'a' story 'Thunderbird Saves the People' in which a young boy gets a splinter in his eye and is transformed into a thunderbird. Initially ostracised by his community, he answers a call to defeat an orca who is eating up all the village's salmon. For the set design, the designers envision the creation of a traditional village—a plank house front, a weaving loom, a canoe and nets, and a seascape on the backdrop (Sadeghi-Yekta 2020).

Canoe Journeys

Canoe voyages have their roots in the most ancient times, as recalled by great flood origin legends among the Coast Salish and other Indigenous Nations. Canoes were highly valued, and often the only way of travelling between isolated settlements. Canoe carvers were highly esteemed, as were the canoes themselves. Neel (1995, quoted in Johansen 2012) goes so far as to say, 'a canoe, coming from a soul sometimes more than a thousand years old, is a spiritual being' (Johansen 2012: 133).

Canoeing can be seen as exercising self-determination and practising Nation-to-Nation protocols when arriving in another's territory (Jeff Corntassel, personal communication, 27 April 2021). I witnessed this practice during the welcome of an arriving canoe crew at a pole dedication potlatch at Old Masset several years ago. Permission was requested by the captain—standing tall in the canoe in full regalia—and granted, accompanied by drumming, songs and dances by the welcoming crowd.

Native builders retained their skills following the early period of Euro-American settlement, and there were even flourishing canoe racing competitions. By the 1920s, however, logging of old growth cedar, banning of the potlatch, forced attendance at residential boarding schools, and a shift to motorised boats, led to a decline and near complete disappearance of the old canoes. The revival which followed (discussed below) is summarised by Johansen (2012).

Renewed interest in traditional canoeing began in the 1980s. Haida artist Bill Reid completed work on the *Lootaas* based on a centuries-old canoe in the American Museum of Natural History, allowing Haida to visit distant relatives in Alaska. Similar voyages were made up and down the Inside Passage, sleeping at beaches or Big Houses on the way. On arriving at Bella Bella after a difficult voyage the crew were lifted out of the water 'still inside the canoe' and escorted into the Big House (Johansen 2012: 132).

Emmett Oliver of the Quinault Nation, who had retired from his job as Supervisor of Indian Education in Washington, is credited as originator of the Paddle to Seattle. An old growth cedar log was donated by the US Forest Service and trucked to the Quinault Reservation, and a craft was carved following traditional models. Travelling in the canoe along the North Olympic Peninsula coast and stopping at sites along the way, Oliver began promoting the idea of a 'paddle to Seattle' (Johansen 2012: 135).

Tribal canoe journeys are now well established, occurring on an annual basis (except during 2020/21 when COVID made this impossible), and ending at a different destination each time. Thousands of people from diverse tribes as well as 'tourists' follow the journey on land and by ferry, assisting the support canoe crews at designated stops for feasting, dancing and ceremonies, including the task of grilling hundreds of pounds of salmon to feed the multitudes (Johansen 2012).[1]

1 Multiple histories have been put forth for origin of the Canoe Journeys (Hundley 2022). These accounts are not in contradiction; seen together they provide an interwoven perspective on diverse journeys within the Salish Sea region and beyond, taking place at different times, and with different itineraries. The 'real beginning' of the Canoe Journeys has been ascribed to Tom Heidlebaugh and Philip Red Eagle, at the Washington coast in La Push, in 1970. Together they devised a ceremony in which participants discuss rules of the canoe and reflect on the meaning of the journey to themselves and their Tribal nations. A challenge by Frank Brown to Coast Salish nations during the 1989 Paddle to Seattle 'lit a fire across the region' and eventuated in the 1993 Qatuwas Festival Paddle to Bella Bella, involving 23 canoes from across the region. Qatuwas was the first Indigenous-led event connected with the Canoe journeys. Originally planned for every four years it now an annual event (Hundley 2022).

Reef Net Fishing

Reef nets (SX̱OLE) were used for salmon fishing by the W̱SÁNEĆ Straits Salish Nation since antiquity. As their territory lacked salmon-bearing rivers, the fishery relied on suspended nets strung across salmon-rich bays. In 1916 the Canadian Department of Indian Affairs banned the practice, despite the fact that J.H. Todd and Sons were permitted to fish using wire-fenced 'traps' attached to tall poles attached to the sea bottom (Elliott 1990).

The Lummi, on the US side of the border, likewise ceased to be competitive following competition by commercial traps by 1897 (Boxberger 1988). A 1934 ban on fish traps in Puget Sound gave Lummi fishermen renewed access to reef fishing but they were soon outcompeted by the entry of non-Indigenous fishermen and opening of a cannery in 1939. By 1974 there were only 43 reef nets, none of these operated by the Lummi (Northwest Treaty Tribes 2013).

Nick Claxton had been researching the practice of reef net fishing, tutored by his uncle Earl Claxton Sr and John Elliott Sr (STOLȻEŁ), a teacher of SENĆOŦEN language and culture at the ŁÁU, WELṈEW̱ tribal school. Interest soon grew among W̱SÁNEĆ community members interested in bringing back the practice, which had not been in use on either side of the border in generations (Claxton 2015: 155–9).

Earl Claxton passed away in 2011, after passing on much of his knowledge to his nephew Nick. STOLȻEŁ, assisted by Nick Claxton, began to focus his language and culture class on creating a model canoe and reef net, beginning each class with a prayer and acknowledging the spiritual and cultural significance of the project (Claxton 2015: 174). In acknowledgement for his work on the project, a ceremony was held in which Nick Claxton was given the title CWENÁLYEN—elder, captain and knowledge keeper.

The Lummi Nation had also begun reviving reef net fishing, had completed a model and full-scale net, culminating in a Reef Net Gathering and Salmon Barbeque on 28 August 2013 at Xwe'chi'eXen, Cherry Point, a centuries-old tribal village and traditional reef net site in Blaine, Washington. Invited by the Lummi, STOLȻEŁ and Claxton were treated with respect and hospitality, returning again in 2014 (Claxton 2015: 179–82).

On 9 August 2014 the W̱SÁNEĆ launched its first reef net fishing in about 100 years. Two canoes were launched from Tulista Park Boat Ramp in Sidney and paddled against the tide to Sidney Spit. From there they were towed by a Parks Canada boat to the mouth of Bedwell Harbor, South Pender Island. Approaching the harbour, the crews were met by a pod of orcas, circling counterclockwise within the harbour ahead of the boats, an event viewed as extremely rare (Claxton 2015: 190–3).

The reef net was set, but could only be left out for two hours, as the time was soon running out to get back to Tulista. While no salmon were caught the initiative was viewed as a success, viewing the effort as a 'proof of concept', in the sense that the technology worked as expected (Claxton 2015: 193–6). In recognition, Nick Claxton has been given the title of XEMŦOLTW̱. Trials continue among the W̱SÁNEĆ (Marelj 2020), and a modified form of reef netting is now well established among the Lummi (Karsten 2021).

Clam Gardens

The significance of clams as a food source for generations of Kwakwa̱ka̱'wakw, Coast Salish and other Northwest Coast peoples was often under-recognised within classic anthropological accounts, with an admittedly justified focus on salmon as a dominant source of nutrition (Moss 1993). Yet, much evidence suggests that clams contributed significantly to the stability and food security of coastal communities (Groesbeck et al. 2014).

Archaeological studies in the Salish Sea reaching back 11,500 years provide clues on long-term management of the intertidal rock-walled terraces known as clam gardens. Such records provide meaningful management targets and guidelines for applying traditional mariculture practices, such as the tending of clam gardens, to increase the productivity and sustainability of our foods today, and they supports claims of Indigenous peoples to their lands and seas (Toniello at al. 2019).

The size and growth of high-carbohydrate butter clams (*Saxidomus gigantea*) began to increase during the early postglacial warming period. By around 3,500 years ago, clams had become abundant and coastal populations began constructing clam gardens. Midden measurements suggest that clam gardens enhanced production despite increased harvesting pressure. Since European contact, clam indices have dwindled to that of the early postglacial period (Toniello at al. 2019).

'Listening to the Sea, Looking to the Future' is a five-year initiative led by Hul'q'umi'num and W̱SÁNEĆ Nations and Parks Canada aiming to restore and manage two clam garden sites according to the traditional practices of Hul'q'umi'num and W̱SÁNEĆ peoples. Elders and knowledge holders guide this work while Parks Canada scientists monitor changes to the intertidal ecosystem (Parks Canada 2018).

An evaluation by Augustine and Dearden (2014) supported by the Hul'qumi'num Treaty Group and Gulf Islands National Park Reserve suggests that restoration of clam gardens is possible without significant ecological impact, for the following reasons: intertidal regions recover quickly from disturbances, eelgrass areas will not be disturbed, the process has a limited footprint, the process retains existing trophic structure, and the clam gardens may potentially increase nutrient cycling.

Taking or Tending?

Archaeologists are now increasingly aware that plants, fields and other habitat can be as important in understanding pre-contact culture as cairns and non-living artefacts (Lepofsky et al. 2020). The revival of fishing, clam gardening and traditional resource management practices is gathering increasing attention worldwide, for ensuring the well-being both of populations, and also the species and habitats on which humans rely (Deur and Turner 2005).

Richard Hebda (Chapter 2, this volume), based on an earlier publication by Turner and Hebda (2012), has summarised the season cycle by month, from 'Elder Moon' (December), to 'Putting Your Paddle Away in the Bush' (November–December). This was in turn based on a 40-year collaboration with WSÁNEC Nation elders Elsie Claxton, Dave Elliott Sr, Christopher Paul and Violet Williams.

Nancy Turner (2020) has described the paradigm change she experienced as an aspiring botany student at the University of Victoria and University of British Columbia years ago. She came to learn that for Indigenous people, food gathering was not just a matter of sustenance. The relation between people and the environment were important in their own right and key to the way First People, living in place for generations, thought about the world (Turner 2020: 2473).

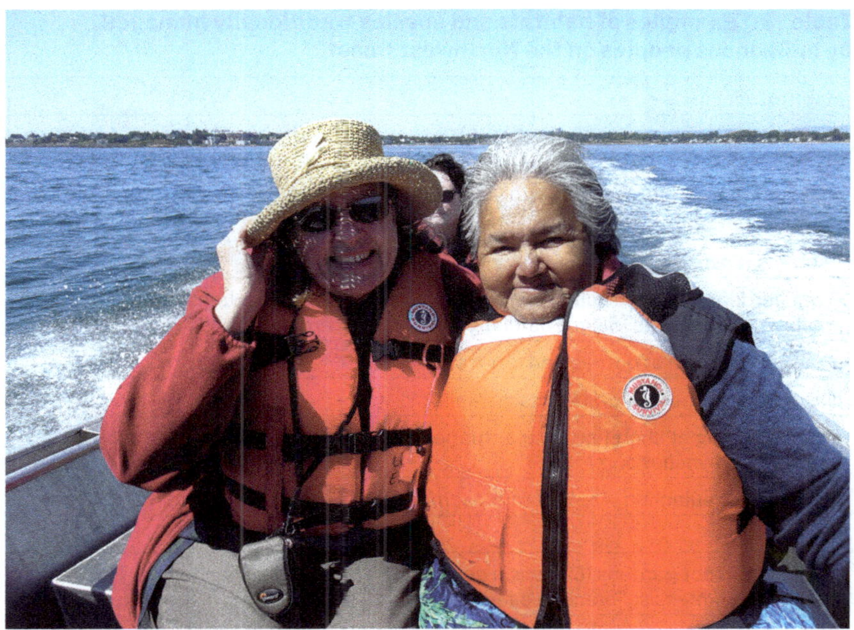

Figure 13.2 Joan Morris and Nancy Turner, near Tl'ches
Source: Darcy Mathews.

By the early 1990s Nancy had the fortune to meet Kwakwa̱ka̱'wakw scholar and cultural expert Kim Recalma-Clutesi of the Qualicum First Nation (on Vancouver Island), who in turn introduced Nancy to Clan Chief Adam Dick (Kwaxsistalla) and other trained experts in resource management and cultural protocols. It was then she came to realise what had been 'staring me in the face all along: *Indigenous Peoples of northwestern North America have been long-time cultivators*' (Turner 2020: 2476, emphasis in original).

In a related essay, Marlene Atleo (2020) reflects on her integration into the community of Ahousaht on the west coast of Vancouver Island, as the wife of Chief E. Richard Atleo (Umeek). Coached by her husband's paternal and maternal grandmothers, she provides an inside narrative on the variety of traditional food gathering and food tending practices, food preparation and the connections with daily life and ritual still extant in Nuučaan̓uł (Nuu-chah-nulth) culture.

Table 13.1 Examples of habitats and species traditionally managed by Indigenous peoples on the Northwest Coast

Habitat/species
Reef net salmon fishing (Straits Salish); sockeye (*Oncorhynchus nerka*) and other salmon species
Herring (*Clupea pallasi*), roe on kelp (*Macrocystis pyrifera*) and hemlock (*Tsuga heterophylla*) boughs
Eelgrass (*Zostera marina*) meadows
Clam gardens: littleneck clams (*Protothaca staminea*), butter clams (*Saxidomus gigantea*), razor clams (*Siliqua patula*), cockles (*Clinocardium nuttallii*), and other spp.
Seaweed (*Pyropia abbottiae* and other spp.)
Northern abalone (*Haliotis kamtschatkana*)
Salmon fishing at shoreline, river estuaries: stone fish traps, weirs and holding pools (*Oncorhynchus* spp.)
Oulachen, or eulachon (*Thaleichthys pacificus*)
Seabird eggs
Tidal marsh root gardens (*Camassia quamash*; *Fritillaria camscatchensis*; *Lupinus nootkatensis*; *Potentilla egedii*; *Trifolium wormskioldii*)
Crab apple (*Malus fusca*) 'orchards'

Source: Excerpted from Turner (2020). See original article for additional details and references.

Indigenous Guardians

Over the past three decades Indigenous guardian programs have emerged as institutions for environmental governance (Reed et al. 2020). A report by the Guardians of Mid-Island Estuaries Society (GoMIES) provides an example of one such program, in partnership with Wei Wai Kum (WWK), an Indigenous nation with a reserve at the mouth of the Campbell River estuary. The goal is to protect Carex (reed) marsh zones from overabundant Canada goose populations using wooden exclosures (Clermont 2021).

Over the past 100 years, the Campbell River estuary has been much impacted by logging, marinas, float plane docks, ship repair, barge loading facilities, gravel removal, and most of all by 3 hydroelectric dams wthin its watershed. Loss of estuary habitat, disturbed salmon spawning gravels, and high estuary flows disrupted the salmon cycle during critical life cycle stages.

Once considered the 'salmon capital of the world', and a major resource for the WWK food and ceremonies, large chinook salmon were no longer common. Most industrial activities had stopped over the last two decades and studies were launched to evaluate the possibility of restoration. Studies linked the salmon decline to loss of 90 per cent of Carex edge communities and herbivory by overabundant Canada Geese.

Beginning in 1981 Carex plugs were transplanted to the constructed islands. However, year round resident and visiting Canada Geese over-grazing and grubbing of channel edge vegetation led to the erosion of productive sediment and collapse of the marsh benches along most of Vancouver Island. Since 2017 GoMIES have been working with WWK First Nation to prevent overgrazing.

GoMIES developed an effective Carex transplant tool and restoration system that was subsequently implemented in five other estuaries. Exclosures perfected by WWK were built from alder poles and willow stakes, reinforced with hemp, willow branches, and twine. Carex channel edge communities now thrive in brackish tidal channels and provide optimal rearing and cover habitat for salmon fry (Clermont 2021).

These important achievements notwithstanding, a review by Reed et al. (2021) raises concern over a systemic lack of Indigenous control over funding and in some cases the design and implementation of guardian programs, and the implications such programs have for Indigenous decision-making institutions and knowledge systems when embedded within broader western environmental governance structures.

Conclusion

Isolated as they may be, Indigenous communities are embedded in the matrix of surrounding economies and governments. Yet cultures once perceived as endangered have undergone remarkable revitalisation in language, traditional food gathering practices, canoe voyaging and limited autonomy. This phenomenon, occurring throughout the world, has not gone unnoticed. The media, government publications and academic literature are replete with articles on what is variously described as Indigenous revival, renaissance, revitalisation or renewal.

The Salish Sea Islands, with environments and societies that are linked and have much in common, deserve to be recognised as an archipelagic region in their own right. The revitalisation of Indigenous cultures within this fragmented island world is worthy of special attention. Still trying to recover from the combined weight of colonisation, residential schools and socio-economic marginalisation, the contours of cultural revitalisation are being negotiated within Indigenous communities and with surrounding societies.

Beginning with epidemics introduced following European contact, Indigenous cultures have been decimated, subjugated, marginalised and displaced to reserves and reservations, losing their land, languages, religions and food gathering practices. Many Indigenous languages have completely disappeared. That Indigenous cultures survive at all is thanks to the help of elders, knowledge keepers and interested members of younger generations, and not least, women, who keep the traditions alive.

The development of a SENĆOŦEN orthography by elder Dave Elliott Sr and instruction in language and traditional food gathering at the ȽÁU, WELṈEW̱ tribal school have been remarkable achievements, as are language nests, theatre groups and Indigenous Studies and Language Programs at regional universities. Nick XEMŦOLTW Claxton's revival of reef net fishing, Emmett Oliver's work on Paddle to Seattle and Jeff Corntassel's teachings on everyday resurgence offer lessons in sustainable living and hope for the future.

Acknowledgements

This chapter could not have been written without generous advice, comments, suggested readings and referrals by Chris Arnett, Nancy Turner, Rick Kool, Sasha Harmon, Coll Thrush, Brian Thom, Russel Barsh, Jeff Corntassel and three anonymous reviewers.

References

Alfred, T., 2005. *Wasáse: Indigenous Pathways of Action and Freedom*. Peterborough: Broadview Press.

Atleo, E.R., 2004. Tsawalk: A Nuu-chah-nulth worldview. Vancouver: UBC Press. doi.org/10.59962/9780774851053

——, 2011. *Principles of Tsawalk: An Indigenous Approach to Global Crisis.* Vancouver: UBC Press. doi.org/10.59962/9780774821285

Atleo, M., 2020. 'Nuučaańuł Plants and Habitats as Reflected in Oral Traditions: Since Raven and Thunderbird Roamed.' In N.J. Turner (ed.), *Plants, People, and Places: The Roles of Ethnobotany and Ethnoecology in Indigenous Peoples' Land Rights in Canada and Beyond.* Montreal: McGill-Queen's University Press. doi.org/10.2307/j.ctv153k6x6.9

Augustine, S. and P. Dearden, 2014. 'Changing Paradigms in Marine and Coastal Conservation: A Case Study of Clam Gardens in the Southern Gulf Islands, Canada.' *The Canadian Geographer* 20(2): 1–10.

BC Treaty Commission, 2021. 'Negotiation Update.' Viewed 31 May 2023 at: bctreaty.ca/negotiation-update

Boxberger, D.L., 1988. 'In and Out of the Labor Force: The Lummi Indians and the Development of the Commercial Salmon Fishery of North Puget Sound, 1880–1900.' *Ethnohistory* 35(2): 161–90. doi.org/10.2307/482701

Buerge, D., 2020. 'Sleeping Languages Rising: NW Native Languages Fight for Survival.' *Post Alley Seattle: Marketplace of Ideas*, 8 November. Viewed 31 May 2023 at: postalley.org/2020/11/08/sleeping-languages-rising-nw-native-languages-fight-for-survival

Claxton, N., 2003. The Douglas Treaty and WSÁNEC Traditional Fisheries: A Model for Saanich Peoples Governance. University of Victoria (MA thesis).

——, 2015. To Fish as Formerly: A Resurgent Journey Back to the Saanich Reef Net Fishery. University of Victoria (PhD thesis).

Clermont, T.J., 2021. Eco-Cultural Restoration of the Campbell River Estuary. Guardians of Mid-Island Estuaries Society. Viewed 15 March 2024 at: a100.gov.bc.ca/pub/acat/documents/r59048/COA_F20_F_3090_1613501506484_F301B2B86F.pdf

Corntassel, J., 2012. 'Re-envisioning Resurgence: Indigenous Pathways to Decolonization and Sustainable Self-Determination.' *Decolonization: Indigeneity, Education & Society* 1(1) 86–101.

Corntassel, J., R. Edgar, R. Monchalin and C. Newman, 2020. 'Everyday Indigenous Resurgence During COVID-19: A Social Media Situation Report.' *AlterNative* 16(4): 403–5. doi.org/10.1177/1177180120968156

Coulthard, G., 2014. *Red Skin, White Masks: Rejecting the Colonial Politics of Recognition.* Minnesota: University of Minnesota Press. doi.org/10.5749/minnesota/9780816679645.001.0001

Deur, D. and N. Turner (eds), 2005. *Keeping It Living: Traditions of Plant Use and Cultivation on the Northwest Coast of North America.* Seattle and Vancouver: University of Washington Press and UBC Press.

Duvernoy, R.J., 2020. 'Life in Interregnum: Deleuze, Guattari, and Atleo.' In J. Hayes, G. Kuperus and B. Treanor (eds), *Philosophy in the American West: A Geography of Thought.* London: Routledge. doi.org/10.4324/9781003043621

Elliott, D. Sr, 1990. *Saltwater People, as Told by Dave Elliott, Sr: A Resource Book for the Saanich Native Studies Program,* Edited by Janet Poth (2nd ed.). First edition published in 1983. Saanich: Native Education School District 63. Viewed 9 June 2023 at: wsanec.com/wp-content/uploads/2019/03/saltwater-people-1983-delliot-sr-compressed.pdf

Elliott, M., 2018. 'Indigenous Resurgence: The Drive for Renewed Engagement and Reciprocity in the Turn Away from the State.' *Canadian Journal of Political Science* 51 (1): 61–81. doi.org/10.1017/S0008423917001032

FPCC (First Peoples' Cultural Council), 2018. 'Recognizing the Diversity of BC's First Nations Languages.' Viewed 31 May 2023 at: fpcc.ca/wp-content/uploads/2020/07/DiversityOfBCLanguages-February2018.pdf

Groesbeck, A.S., K. Rowell, D. Lepofsky and A.K. Salomon, 2014. 'Ancient Clam Gardens Increased Shellfish Production: Adaptive Strategies from the Past Can Inform Food Security Today.' *Plos One* 9(3): e91235. doi.org/10.1371/journal.pone.0091235

Harmon, A., 2008. 'Introduction.' In A. Harmon (ed.), *The Power of Promises: Rethinking Indian Treaties in the Pacific Northwest.* Seattle: University of Washington Press.

Harris, D.C., 2008. 'The Boldt Decision in Canada: Aboriginal Treaty Rights to Fish on the Pacific.' In A. Harmon (ed.), *The Power of Promises: Rethinking Indian Treaties in the Pacific Northwest.* Seattle: University of Washington.

Hundley, J.M. 2022. 'Histories of the Canoe Journey: Border Studies, Critical Indigenous Studies, and the Decolonization and Unsettling of Coast Salish Territory.' *Journal of Borderlands Studies,* 1–23. doi.org/10.1080/08865655.2022.2156373

Indian Chiefs of Alberta, 1970. 'Citizens Plus.' *Aboriginal Policy Studies* 1(1): 188–281.

Johansen, B.E., 2012. 'Canoe Journeys and Cultural Revival.' *American Indian Culture and Research Journal* 36(2): 131–41. doi.org/10.17953/aicr.36.2.w241221710101249

Karsten, C., 2021. 'The Reefnetters of Lummi Island.' Carmen Karsten's Imaginarium blog. Viewed 31 May 2023 at: cameronkarsten.blog/2021/01/30/the-reefnetters-of-lummi-island

King, H., S. Pasternak and R. Yesno, 2019. *Land Back: A Yellowhead Institute Red Paper*. Yellowhead Institute.

Lepofsky, D., C.G. Armstrong, D. Mathews and S. Greening, 2020. 'Understanding the Past for the Future: Archaeology, Plants, and First Nations' Land Use and Rights.' In N.J. Turner (ed.), *Plants, People, and Places: The Roles of Ethnobotany and Ethnoecology in Indigenous Peoples' Land Rights in Canada and Beyond.* Montreal: McGill-Queen's University Press. doi.org/10.1515/9780228003175-011

Lutz, J., 2017. 'Why did James Douglas's Treaty Era End after 1854?' *Times Colonist*, 19 February.

Marelj, B., 2020. 'W̱SÁNEĆ Youth Rise to the Challenge of Reef Net Fishing as they Reclaim the Formerly-Outlawed Practice.' *Indiginews*, 21 September.

Moss, M., 1993. 'Shellfish, Gender, and Status on the Northwest Coast: Reconciling Archeological, Ethnographic and Ethnohistorical Records of the Tlingit.' *American Anthropologist* 95(3): 631–52. doi.org/10.1525/aa.1993.95.3.02a00050

Native Land Digital, 2023. 'Treaties'. Viewed 24 July 2023 at: native-land.ca/

Northwest Treaty Tribes, 2013. 'Lummi Nation Holds Reef Net Fishery at Cherry Point.' Viewed 31 May 2023 at: nwtreatytribes.org/lummi-nation-holds-reef-net-fishery-cherry-point

Parks Canada, 2018. 'Listening to the Sea, Looking to the Future.' Viewed 31 May 2023 at: parks.canada.ca/agence-agency/bib-lib/rapports-reports/core-2018/ouest-west/ouest-west9

Petrescu, S., 2017. 'Lost in Translation: The Douglas Treaties.' *Times Colonist*, 19 February.

Price, J. and N. Claxton, 2021. 'From the Colonial Past to the Racist Present.' *Times Colonist*, 29 March.

Reed, G., N.D. Brunet, S. Longboat, and D.C. Natcher, 2020. 'Indigenous Guardians as an Emerging Approach to Indigenous Environmental Governance.' *Conservation Biology* 35(1): 179–189. doi.org/10.1111/cobi.13532

Rosborough, T., C.L. Rorick, and S. Urbanczyk, 2017. 'Beautiful Words: Enriching and Indigenizing Kwak'wala Revitalization through Understandings of Linguistic Structure.' *The Canadian Modern Language Review/La Revue canadienne des langues vivantes*, 73(4): 425–37. doi.org/10.3138/cmlr.4059

Sadeghi-Yekta, K., 2020. 'Drama as a Methodology for Coast Salish Language Revitalization.' *Canadian Theatre Review* 181: 41–5. doi.org/10.3138/ctr.181.007

Sissons, J., 2005. *First Peoples: Indigenous Cultures and Their Futures*. London: Reaktion Books.

Thompson, N., 1989. 'On the Need for Ethnographic Information in Native American Dictionary Construction.' Paper presented to the 24th International Conference on Salish and Neighboring Languages. Viewed 31 May 2023 at: lingpapers.sites.olt.ubc.ca/files/2018/03/1989_Thompson_N.pdf

Toniello, G., D. Lepofsky, G. Lertzman-Lepofsky, A.K. Salomon and K. Rowell, 2019. '11,500 y of Human–Clam Relationships Provide Long-Term Context for Intertidal Management in The Salish Sea, British Columbia.' *PNAS* 116(44): 22106–14. doi.org/10.1073/pnas.1905921116

Turner, N.J., 2020. 'From "Taking" to "Tending": Learning About Indigenous Land and Resource Management on the Pacific Northwest Coast of North America.' *ICES Journal of Marine Science* 77(7-8): 2472–82. doi.org/10.1093/icesjms/fsaa095

Turner, N.J. and R. Hebda, 2012. *Saanich Ethnobotany: Culturally Important Plants of the WSÁNEC People*. Victoria: Royal British Columbia Museum.

W̱SÁNEĆ Leadership Council, n.d. 'History and Territory.' Viewed 31 May 2023 at: wsanec.com/history-territory/#territory

W̱SÁNEĆ School Board, n.d. 'History of the Senćoŧen Language.' Viewed 31 May 2023 at: wsanecschoolboard.ca/sencoten-language/#main

14

Leaving Valdes, Staying Lyackson: Voices of the Indigenous Community of Valdes Island

Brian Thom

In 2021, few Lyackson First Nation members live on any of the reserve lands on Valdes Island, in spite of these having been ancestral village sites for millennia. The Lyackson First Nation's migration from their ancestral village sites reveals a particular response to twentieth-century policies that dramatically shifted the economies and livelihoods of Indigenous peoples in the Gulf Islands and beyond. In 2014, I was invited by Lyackson to make a study of the 'Lyackson migration', working with students to conduct oral history interviews, community meetings and an analysis of archival and archaeological sources.[1]

Our goal was to write a community-focused ethnography describing the historic context and circumstances around this profound change in occupancy of this Gulf Island landscape. This chapter is the result of this project, where undergraduate and graduate anthropology students from the University of Victoria and I worked to bring out the voices of the Lyackson people and their histories. One of the goals of the community was to give context and support to their work to establish a new land base for their community on Vancouver Island.

1 Further details in the Acknowledgments at end of this chapter.

Leeyqsun (Valdes Island) Overview

Located in the Strait of Georgia of British Columbia (BC), Valdes Island lies 40 kilometres southwest of Vancouver between Galiano Island and Gabriola Island. Valdes Island has a total area of 2,488 hectares and is the fourth largest of the Southern Gulf Islands. Valdes Island is the ancestral home of the Lyackson First Nation, one of numerous Hul'q'umi'num'-speaking communities of the Salish Sea. The ethnonym *Leeyqsun* is inspired by the place name Leeyqsun on Valdes Island (Rozen 1985).[2] The anglicised term 'Lyackson' has been adopted since the time of reserve establishment to refer collectively to the families and communities from Valdes Island (ibid.).

The gazetted names for Valdes Island (and its immediate neighbour Galiano Island) are in recognition of the 1792 survey by Cayetano Valdés y Flores and Dionisio Alacalá Galiano, who were among the first Europeans to survey this Gulf of Georgia area (Matson et al. 1999). In our conversations, it was clear that the Lyackson community would value officially recognising a name like Leeyqsun for their ancestral island.

The majority of Valdes Island land is now privately owned by a forestry company, with numerous small residential and recreational lots—also privately owned—dotting the coastline, particularly on the south and east. There are also several large parcels of unoccupied Crown lands. Despite extensive exploitation of natural resources through colonial settlement and industrial resource extraction on the private and Crown lands, Valdes Island remains the most 'undeveloped' Southern Gulf Island.

About a third of the island is held as Lyackson First Nation reserve lands. The Lyackson First Nation's three reserves have a total area of 791.6 ha. Lyackson IR 3, toward the north end of the island, is the largest with 710.6 ha. Lyackson IR 4 is a 79.0 ha parcel located at Shingle Point. Lyackson IR 5 consists of 2.0 ha and is situated at the south end of the island on Porlier Pass.

2 Quw'utsun Syuw'entst Lelum (2007) (Cowichan Teaching House) is the official source for the orthography (spelling) of place names in Hul'q'umi'num' appearing throughout this chapter. However, many other writing systems have been used in the past to write Hul'q'umi'num', and where an alternate orthography is used by a source cited in-text, this version will appear in parentheses next to the official orthography.

In a 1992 survey commissioned by Lyackson, 16.5 per cent (26 out of 158) of Lyackson band members lived on any of the Valdes Island reserves (Vickers and Associates Management Consultants and Lyackson First Nation 1992). When we did our work in 2014, there were no full-time residents on the island. As this chapter details, the distinctive circumstances of colonialism, along with regional economic changes, were instrumental in the reconfiguration of how people have lived on and experienced their ancestral villages on Valdes Island.

Place Names and Ancestral Villages

Prior to the introduction of the reserve system on Valdes Island, there were three permanent settlements on Valdes Island: T'eet'qe' (T'a'at'ka7), Tth'exul (Th'a'x̲el) and Tth'hwumqsun (Th'x̲weksen) (McLay 1999b: 11–2) (see Figure 14.1). These villages have an important relationship to another village, T'eet'qe', located on an island at the confluence of the Koksilah and Cowichan Rivers (George 2019: 1 footnote; Rozen 1985: 162, 278).

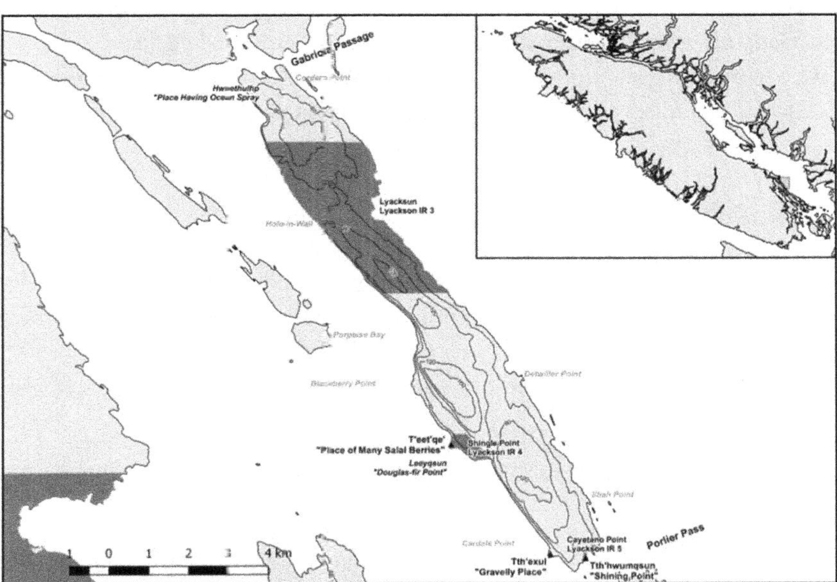

Figure 14.1 Valdes Island, with a selection of Hul'q'umi'num' place names

Source: Brian Thom.

T'eet'qe' was the largest village on Valdes Island and in the mid-nineteenth century hosted the second largest population of Coast Salish people in the Gulf Islands (DMCS 2011b: 7), estimated at up to 650 (McLay 1999a), and is today part of Lyackson IR 4.

Tth'exul is located at Cardale Point, on the southwest side of the island and was not set aside as reserve land, being privately owned today. The late Lyackson elder Agnes Thorne was born on Valdes Island and spent her childhood at Tth'exul, which is also where her father was born (ibid.). She recalled that there was considerable affinity between Tth'exul and T'eet'qe', however they retained their identity as distinct villages with distinct names (ibid.).

Tth'hwumqsun (Th'xwemksen), is located near Cayetano Point on the south end of Valdes Island (Rozen 1985: 69) and is part of Lyackson IR 5 today. In the 1980s the late Agnes Thorne estimated that in the previous century there may have been five long houses at the point, which housed between 100 and 150 people on a full-time basis (ibid.: 69–70). However, by 1915 only two or three houses were still standing at this smaller winter village site (DMCS 2011b: 7–8), as most people had moved to Shingle Point or other locations off-island (Rozen 1985: 70–1). Tom Peters, a Lyackson elder we spoke with who grew up on the southern end of Valdes Island, remembered five families and seven modern-style houses in the area.

Original Peoples

One of the earliest oral tradition recordings related to the origin of Valdes Island came from anthropologist Franz Boas' work in island Hul'qumi'num communities. The story, which was originally published in German in 1985 and translated and republished by Bouchard and Kennedy (2002: 148) said:

> A long, long time ago, Sqoē'te used to be a tree standing upright whose top reached up to the sky. Down this tree the people descended from the sky, as did deer with white backs and black legs and antlers which bent forward and covered the sides of their faces. When the people had arrived on earth, they pondered how they could bring down the tree. So two men called for the rats (?) and told them to gnaw through the tree. When the rats had gnawed for twenty days, they had almost reached the middle of the tree. The two men then told them to start from the opposite side, and here, too, the rats gnawed a deep hole. While the rats were gnawing, the people sang in order

> to keep their spirits up. The people were glad that the tree would fall soon and sang, 'Oh, let it fall and not break. Many deer will live on the trunk and we will build our houses on it.' When the rats had finished their task, they ran out of the tree and it fell over. But the top broke off and formed Ā'wik·sen Island. Many deer then lived on the islands. (The informant, an old man in S'ā̂menos, maintained that he once had seen one of these deer, but he didn't dare shoot it.)

Bouchard and Kennedy analyse Boas' imperfect representation of the term *Ā'wik·sen* and suggest that it refers to the name Leeyqsun 'Douglas Fir Point' (Bouchard and Kennedy 2002: 148 footnote; see also Rozen 1985: 74). This term has been more recently interpreted as 'Broken Point' or 'Tip of Douglas Fir' (DMCS 2011b: 9).

Archaeological evidence shows a continuous presence of Coast Salish peoples on Vancouver Island and the surrounding Gulf Islands, including Valdes Island, for at least the past 5,000 years (ex. Grier et al. 2009: 255, 262), with even earlier sites obscured by today's higher sea levels. Eric McLay's systematic archaeological survey of Valdes Island (one of the best surveys done in the entire region) has documented the presence of at least 56 additional archaeological sites (1999a). These sites span from the northern tip to the southern end of the island and include 47 shell matrix sites, two defensive sites, four cultural depressions, seven early pre-contact or early historical burials, three rock shelters, a petroglyph and other isolated 'utilization sites' (McLay 1999a: 80–94).

Valdes Island is archaeologically important, as many of the sites have remained largely undisturbed given the lack of urbanisation or other non-forestry development in the area. Though there have been surprisingly few controlled archaeological excavations at any of three village sites of T'eet'qe', Tth'hwumqsun and Tth'exul, their long-term occupation is evident from the impressive depth and aerial extent of shell deposits ('shell middens') (McLay 1999b). Excavations at T'eet'qe' by University of British Columbia archaeologist R.G. Matson revealed at least five ancient house floors (Matson et al. 1999), including one very large structure measuring 55 m long × 17 m wide (Matson 2003)

The monumentally large clam garden south of Shingle Point—spanning 4–6 meters wide and approximately 750 meters long—suggested long-term management of the site, requiring a significant and consistent investment of labour (Grier 2014). Grier has concluded that village sites like Shingle

Point throughout the Gulf Islands were the primary homes for groups based in the islands, rather than peripheral locations of interim occupation for communities based on Vancouver Island (Grier 2007).

Two defensive sites have been documented on Valdes Island, including the earthwork trench embankment at the Cardale Point, which is one of the most well-preserved sites of its kind in the Gulf of Georgia region (Angelbeck 2007: 266; McLay 1999a). In the nineteenth century T'eet'qe was the home of the powerful leader Tth'asiyten who helped establish the Fort Langley in the 1820s and continued to play an important role in the fur trade and in the historic Battle of Maple Bay (Suttles 1976; George 2019).

Population History and Census Data

It is likely that the first and most serious smallpox epidemic impacted the area around 1782, ten years prior to European contact, followed by smallpox, measles, influenza and tuberculosis outbreaks during the early colonial period (Galois 1995). Lyackson elder Sally Norris recounted a narrative shared with her by elders as a young girl, which detailed the impacts of smallpox and the role of Valdes Island as a refuge:

> It was something new to us, that smallpox … So, anybody that was okay was wrapped up and put on a canoe. I shall name it Leeyqsun, Valdes … they made tents there. (DMCS 2011b)

While there are no clear population records prior to the introduction of these diseases, the earliest census shows that T'eet'qe' was the second largest village in the Gulf Islands (after Penelakut), with 647 people living there in 1852 (Rozen 1985).

Following the devastating pattern of depopulation from these virgin soil diseases elsewhere in the Salish Sea, the population of Lyackson people living on Valdes Island plummeted in the 25 years following the 1852 census, hovering between about 50–80 people up until World War II. According to the Department of Indian Affairs annual reports, the population of Valdes Island dropped significantly again after 1885 as a result of a large forest fire that swept over the island (Lomas 1885, 1887).

The immediate post-war decades lack clear population statistics, but it was during this time a majority of Lyackson families migrated from Valdes Island to Vancouver Island (Thomas 2004). In the 1980s a few members tried to re-establish their reserve community on Valdes Island, but the lack of infrastructure and access to vital services limited these efforts. In the years that have followed, very few Lyackson people have lived full-time at their ancestral village sites.

Establishment of Reserves

While the everyday family situation of the shocking depopulation of nearly 90 per cent of Lyackson in the second half of the nineteenth century is hard to imagine, their autonomy in terms of land remained largely undisturbed until after confederation. There were a few notable exceptions, such as the granting of 307 ha of spectacular land at Gabriola Passage on the north tip of the Island in 1876 to Navy Captain Baldwin Wake. However, after BC joined the Canadian Confederation in 1877, the Joint Indian Reserve Commission (JIRC) undertook its work to establish reserve lands on Valdex Island, resulting in more drastic changes for the Lyackson First Nation (Sproat 1877).

Indian Reserve Commissioner Sproat recounted to the JIRC how Lyackson community members adamantly defended their title to Valdes Island and defended the significance of the land for their communities. However, the JIRC staunchly dismissed Lyackson First Nation's appeal to recognise Lyackson title over the entirety of their ancestral island. The JIRC justified their response by pointing out that the community had not attempted to cultivate the entirety of the island in the fashion of European agriculturalists (Sproat 1877)—a requisite for 'existing land use' by colonial authorities, but which did not reflect land use and occupation practices and patterns of Hul'qumi'num peoples. That year (1877), the JIRC designated three Indian Reserves on Valdes Island, opening up the remaining island lands to further grants and tenures of land and resources by the provincial Crown. The JIRC set aside 32 ha of land at Shingle Point was for the village site, graves and a 'small piece of rough timbered land around it for firewood' (ibid.: 12). The reserve at Porlier Pass was recognised as a dogfish fishing station, and the large piece of land comprising Lyackson IR 3 towards the north end of the island was intended for a 'cattle or sheep range' (ibid.: 12). Sproat wrote that the rough and poorly timbered area on the north of the island was not suitable for a white settler to make a living (ibid.: 13). In 1889, Samual Robins acquired 1,085 ha of land on Valdes Island through five Crown grants. After this, there was little remaining Crown land, as much of the rest of Valdes Island had been acquired as fee-simple private property.

Residential Schools

The shadow of Lyackson members' encounters at the Kuper Island residential school frequently came up in our discussions, making connections to its role in the alienation of people from their lands and resources. The Kuper Island Industrial School was established in 1890 and operated for 85 years until around 1975 (Indian Residential School Resources 2015). Generations of Lyackson children, like so many others across Canada, were forced from their families' homes on Valdes Island and isolated from their community (McLay 1999b). Lyackson scholar Robina Thomas (2004) asserts that the residential school experience has been the most devastating colonial impact that Indigenous peoples in Canada have encountered. Each former student's experience is unique, but the overall negative impacts of residential schools were made clear by Lyackson elders.

Abuses at the Kuper Island Industrial School were well known to the generation of Lyackson elders born in the early part of the twentieth century. Elders revealed that they were afraid to attend the school. They shared harrowing stories about children getting food poisoning and children who drowned while trying to escape from the school (Thomas Peters, Patricia Peters and Doreen Thomas, interview, 11 March 2014).

Some Lyackson children managed to evade the system, running away from the priests and nuns who came to collect them. Thomas Peters (interview, 11 March 2014) recalled how his father kept him from residential school,

> I was on Valdes. I didn't go to school because my dad had a bad thing that happened when he went to school. He said it was contaminated food, and part of his neck just dropped out, just dropped out right from along here. And there was scars in two places, one on his back. They thought he was going to die so they sent him home. But, my great-uncle knew all kinds of medicine, Indian medicine. He started giving it to him, and he survived. So when I was of school age he hid me from everybody. I didn't come to town too often, I stayed home. I was wondering why he did that, and after I met my cousins from Shell Beach [Thuq'min, near Ladysmith] there, and … I asked how did they learn how to read and they told me they went to school. And I asked him how come you didn't send me to school. Then he explained what happened to him. He almost died, and he didn't want me to go through the same thing. He was scared for me to go to school. I was the only son he has so he really cared a lot for me.

Chief Richard Thomas spoke about the negative impacts that residential school had in terms of interrupting the transmission of traditional knowledge and language. He identified residential school as a major part of his youth, which separated him from the island and from his family. As he reflected on the effect of residential school in his family, Chief Thomas recalled what his mother said about residential school:

> She really didn't share too much with me either. I think that's part of the effects of residential school for her too. The unfortunate part is that the punishment of the day for residential school students who were apprehended and brought over there and their only language was our Hul'q'umi'num' language and being punished for speaking it and not speaking English. So my mom, she told me, she apologized to me and she said 'I never taught you how to speak our language because of the punishments that were endured at the residential school.' (interview, 20 November 2014)

Chief Thomas went on to share how later in his life, he reconnected with his father and his cultural teachings by spending time together at Shingle Point, where he connected with hunting and fishing:

> a lot of things changed after residential school, that fishing was always in me … First things once I get to a hunting area is to have a sacred bath up there and do my prayers up there too. So that's the teachings that we get, so it's like so that's what I'm able to pass onto my kids, my grandkids. To pray in our own way. (interview, 20 November 2021)

Lyackson Migration History

In the post-war years, most of the people who had once lived on Valdes Island eventually moved away. Elders recalled family members moving to Penekalut Island and Vancouver Island, to the BC Lower Mainland and Washington State (Herman Norris, Catherine Sager, Henrietta E. Underwood and Thomas Peters, unpublished interviews). The last person from this era was Sally Norris' grandmother, who was buried in the cemetery at Shingle Point in the late 1950s or early 1960s (Sally Norris, interview).

A few people returned in the 1980s to try and re-establish the community at Shingle Point, building two modern homes and a Big House without the benefit of municipal infrastructure or amenities. These late twentieth-century residents of Valdes Island left for pragmatic and practical reasons, despite their attachment to their ancestral home (Thomas 2004).

Lyackson elders' narratives describe how remote island living in the late twentieth century limited their access to basic resources, health care, employment, education and security of property and possessions. The villages and reserve lands of Valdes are not particularly remote from urban centres—by boat Teet'qe' is merely 15 km from the town of Ladysmith, 25 km from downtown Nanaimo and 35 km from the historically important Fraser River port of Steveston in the Lower Mainland. However, the lack of transportation infrastructure, municipal infrastructure and services of any kind are significant pragmatic factors the community's departure.

Lyackson elder Thomas Peters described how Valdes Island has: 'No hydro, no running water. Everything was by kerosene lantern. No radio, no phone. You pretty well just had to have a boat to live out there' (unpublished interview).

Indeed, Valdes Island can only be reached by private boat. Because there are no wharves or docks on or near Lyackson reserve lands, vessels must be safely anchored, or beached and secured to the shores, which experience two low/high tide cycles every day. Many Lyackson First Nation members would be interested living full-time on Shingle Point reserves, according to a 2012–13 community survey, but to bring this about, improving accessibility and infrastructure on Leeyqsun would need to be 'very high' priorities (Lyackson First Nation 2013).

Doreen Thomas suggested, 'a breakwater there would have been nice so our boats could have been kept covered from the wind' (unpublished interview). Former Valdes Island resident Herman Norris identified the lack of water supply and power as being particularly prohibitive (unpublished interview). KPA Engineering Ltd's estimated costs of water supply facility construction was approximately CAD$250,000, and upwards of several hundred thousand dollars for a permanent dock (Vickers and Associates Management Consultants and Lyackson First Nation 1992: c-17). The high costs of infrastructure development are a major impediment to rebuilding the community on existing reserve locations.

These problems become more challenging when necessary access to emergency health care services are considered (Patricia Peters, unpublished interview). The absence of medical services on-site was spoken of again and again as a key reason why many interviewees no longer lived on Valdes. After many years living on Valdes Island, Doreen Thomas' family ultimately moved to Chemainus to be near their doctor. Chief Thomas explained that Victor and Frank Norris moved from the island to the Cowichan area in the late 1990s because of their older age and its accompanying medical demands. Chief Thomas's own father left his home at Shingle Point due to medical problems, which he could not care for alone (unpublished interviews).

Doreen Thomas highlighted the reality of economic challenges associated with acquiring provisions and preparing enough resources to live self-sufficiently on Valdes Island. Living expenses are high due to increased reliance on frequent motorboat transportation. '[When] we moved over there we thought, 'oh well it'll be cheap, we don't have to pay rent or anything like that." But it isn't that way', recalled Doreen in her interview.

> Like they say, you've got to haul your kerosene, your gas, your food, you've got to have everything prepared to take over, cause it's such a long run, you can't just come back and pick up what you forgot, you know? (unpublished interview)

Eight of the elders we spoke with identified lack of reliable transportation and access to resources as the predominant reasons for leaving the island. For those who had lived there, Chemainus—a minimum half-hour ride on calm waters on a small boat from Single Point—or Duncan (which by-and-large would have to be reached by car once ashore at Chemainus) were the primary resupply points for both fuel and food.

Elders also had concerns about environmental contaminants for the critical beach foods that sustain residents of the island. Two key environmental issues are ocean pollution from the nearby Crofton pulp and paper mill, which has been operating since 1957; and the poor management of habitats and fish catches which have made the fisheries in the Gulf Island precarious. In particular, shellfish closures and private foreshore enclosures in the Gulf Islands have jeopardised shellfish harvesting (Vickers and Associates 1992 Management Consultants and Lyackson First Nation: 11). Many men from Lyackson historically made their living as small-scale commercial fishermen, but the decline of locally available salmon as ended this as an income option (Thomas Peters and Herman Norris, unpublished interview). One Lyackson community member observed that in his lifetime, the availability of local fish has dramatically declined. In his youth, he would catch 20 to 30 fish in the mornings by trolling from his canoe; however, now people are lucky to catch one or two fish (DMCS 2011a: 30). Sally Norris remembered the impact of increasing numbers of recreational and commercial harvesters frequenting some of the last remaining open Valdes Island beaches for shellfish:

> Well the government at the time were taking a lot. The boats and the [non-Native] people were coming. There were lots of them coming over they were taking all the food. They would dig up all the clams. (unpublished interview)

Veronica Kauwell was also concerned with recreational and commercial shell fishing, saying that 'people came [and] they cleaned off most of the beach really and [the clams are] just coming back now and you can find oysters in certain areas' (unpublished interview). As shellfish and fin fish are vital to household food security, this shortage further pushed residents to seek out grocery stores and other off-island resources (DMCS 2011b).

After the closure of the Kuper Island Industrial School, students from Valdes Island were required to reach the next available school, on Vancouver Island, by small boat. In the 1995–96 new community survey, conducted on behalf of Lyackson First Nation, several respondents cited access to education as a reason for leaving Valdes Island. In our interviews, Tom and Pat Peters recalled choosing to move their family to Penelakut Island so their children could attend school there and elsewhere—as Penelakut Island is served by a ferry, unlike Valdes. In the same survey, comments indicated that improving access to educational facilities and opportunities would make Valdes Island a more desirable place to live (Lyackson First Nation 1996). Patricia Peters reiterated this point in her interview: 'the reason we really had to move off was for them [the children] to start school' (unpublished interview).

Current Situation

Lyackson scholar Robina Thomas concluded that limited access to Valdes Island, coupled with the assimilative policies of the federal government, has destabilised Lyackson's land-based cultural practices, created problems for passing on cultural teachings associated with dances, masks and songs and made the ongoing use of the Hul'q'umi'num' language challenging

(Thomas 2004). The lack of federal and provincial support for Lyackson peoples in their efforts to maintain connections with Valdes Island is seen by Thomas as denial of:

> not only [our] right to live on Valdes Island, but the right to protect our land, gather as a people, practice our culture and tradition, learn from our elders, and access all of the local resources. (Thomas 2004)

Lyackson elder Patricia Peters lamented:

> fifty years we've been off reserve. Never did move back to Valdes … It is a beautiful island. Sorry that we can't live in the life like that there. We wished we could though, but we can't. (unpublished interview)

Despite the ongoing difficult circumstances, every Lyackson member we spoke with expressed a strong connection to Valdes Island.

Being away from the island has made it difficult to protect their lands, resources and family homes. Troublingly, in recent years cemeteries on Lyackson reserves have been vandalised, and some ancestral burial grounds off reserve have been destroyed. Doreen Thomas was concerned about how even private Lyackson residences on Valdes Island were vulnerable to unwanted visitors, given the islands' isolation: 'If we could've had somebody live there, somebody staying there patrolling the place, that would've been nice' (Doreen Thomas, unpublished interview). Catherine Sager suggested that a wharf with a shared boat to take people back and forth from the island would help because potential thieves wouldn't know who has left their houses, as individual boats can easily be identified (unpublished interview). Veronica Kauwell noted that kayakers and residents from surrounding islands frequent Valdes Island and use the beaches without permission (unpublished interview). Non-Indigenous sport hunters and others have trespassed on reserve without permission, and foods and medicines that have long been cared for by Lyackson have been harvested without following local protocol and teachings.

Industrial forestry continues to fundamentally transform the landscape, and consultations over forest and community values are very limited on these private lands. Patricia Peters described her concerns about the scale of deforestation, easily visible on the island and on Google Earth[3]:

3 earthengine.google.com/timelapse#v=49.07077,-123.67396,10.863,latLng&t=3.83&ps=50&bt=19840101&et=20221231.

> the whole island is going bald—no more trees there, pretty well all gone now the trees ... Logged out in the whole island. There's going to be no future for the younger generations. (unpublished interview)

Growing up without living in their ancestral villages on the island, some Lyackson people have moved throughout their lives. Thomas Peters, for example, said,

> That's how I grew up, you know. I gill netted from ... six, seven years, and I went to work in the sawmill [on Vancouver Island]. I worked in the sawmill for about twelve years. And before that we were boom men in Kuper Island, and I knew how to boom. I worked there for about six years. The logging was finished, and then we moved to Shell Beach [Stz'uminus First Nation], that's where our kids went to school. ... That's about all I know, how I grew up from a child into a man anyways. Real rough life out there, not easy. My parents grew up there [on Valdes Island], my dad grew up there all his life. (unpublished interview)

Patricia Peters described the difficulty of living without accessible reserve land:

> The reason why we're off reserve is that, we've been trying to get land [to form a new reserve], 'Indian land' they call it. But we never get any [additions-to-reserve] and so we're off reserve. (unpublished interview)

For nearly 20 years, the Lyackson First Nation has been involved in internal discussions to acquire new land on Vancouver Island as an addition to reserve for their community. A community survey done in support of this work recorded 68 per cent of Lyackson respondents saying they would want to live in this new community with another 16 per cent saying 'maybe'. Respondents expressed their views of how a new reserve community on Vancouver Island would be beneficial:

> it would be nice to get our own land base and our own reserve to live on. All the other tribes have their own reserves, with their family around them, where our [Lyackson First Nation] members are all spread out, with no official place to call our own. (Lyackson First Nation 2013: 26)

Another respondent shared a similar sentiment, saying,

> I want my kids to be able to live among their family. I want them to have access to our Elders. I want access to my Elders on a daily basis. I want us to learn our language, our culture, and our traditions. I want us to know our family names. (ibid.: 27)

Indigenous and Northern Affairs Canada provides per capita funding that is contingent on the number of members living on reserve lands. However, with no members living on the reserves, Lyackson First Nation receives only a fraction of the available funding to develop and maintain their traditional lands (Thomas 2004). Lyackson First Nation has been using their reserve lands on Valdes Island to start a small eco-tourism business. As a largely undeveloped Gulf Island, Valdes Island is already a destination for eco-tourists, who take advantage of the beautiful beaches and a network of informal trails for hiking or mountain biking.

Lyackson First Nation has built yurts on their lands to support their business model, working to provide a characterful accommodation that is otherwise absent on Valdes Island. Veronica Kauwell explained how these yurts have also been used by members of Lyackson as 'a retreat for the children once a year in the summer and we have Elders come over. The kids usually pitch tents outside but the Elders stay in the yurts' (unpublished interview). This grassroots undertaking allows the Lyackson community to continue supporting people who wish to connect with their ancestral lands.

Future of Leeyqsun

Although the Lyackson First Nation faces considerable difficulties in re-establishing a full-time presence on Valdes Island, there is a strong feeling among community members that Valdes Island will continue to be a central feature in their lives. While reminiscing about her time living on the island, Doreen Thomas noted, 'we had fun. I wouldn't mind going back like if I was younger, and I wouldn't mind living back there again' (unpublished interview). Lyackson First Nation has a clear vision for their continued stewardship and governance over all of Valdes Island, including the protection of the island's cultural sites and distinctive landscapes and ecosystems. Lyackson First Nation continues to work with the Hul'qumi'num Treaty Group, and more recently the Cowichan Nation Alliance, to coordinate their goals and visions for the future with their neighbours and relatives in the region. Lyackson First Nation has been generous in sharing their island with other Coast Salish peoples who harvest animals or beach foods there with permission, and as a place of 'recreation, rest and renewal' with the broader community, when they respect Lyackson ownership of the land (Lyackson First Nation n.d.).

Acknowledgements

I would like to thank my incredible collaborators on this project. Chief Thomas, Veronica Kauwell, Herman Norris, Sally Norris, Thomas Peters, Patricia Peters, Catherine Sager, Doreen Thomas and Henrietta Underwood are members of the Lyackson First Nation who contributed directly to the oral histories in this chapter and inspired sharing these Indigenous histories of Valdes Island. Kathleen Johnnie is a member of Penekalut Tribe, and the principal of the Hul'q'umi'num' Lands and Resources Society, without whose dynamic facilitation this project could have never been completed. I also hold my hands up in thanks to my students in Anthropology 433 (2014), who worked closely with Lyackson elders to compile the oral histories and archival records for this report, as well as my research assistants Britney Oswell, Ursula Abramczyk, David Fargo and Jen Argan who made important editorial contributions. I also thank Eric McLay, whose long-term archaeological and ethnohistorical research for and with Lyackson First Nation has pulled together so many important threads.

References

Angelbeck, B., 2007. 'Conceptions of Coast Salish Warfare, or Coast Salish Pacifism Reconsidered: Archaeology, Ethnohistory, and Ethnography.' In B.G. Miller (ed.), *Be of Good Mind: Essays on the Coast Salish*. Vancouver: UBC Press.

Bouchard, R., and D. Kennedy (eds), 2002. *Indian Myths and Legends from the North Pacific Coast of America. A Translation of Franz Boas' 1895 Edition of Indian Sagen Von Der North-Pacifischen Küste Amerikas.* Translated by D. Bertz. Vancouver: Talonbooks.

DMCS (DM Cultural Services, Ltd.), 2011a. 'Lyackson First Nation Land Use and Occupancy Study: Final Report (Draft).' Prepared for Lyackson First Nation. Unpublished document.

——, 2011b. 'Report into Lyackson Historical Connections to Valdes Island and Tl'qutinus Visual and Annotated Bibliography of Maps Related to Lyackson History and Culture.' Prepared in Conjunction with the Lyackson Land Use and Occupancy Study (Archival Review) for the Lyackson First Nation. Unpublished document.

Galois, R.M., 1995. 'The Native Population of the Fort Langley Region, 1780–1857: A Demographic Overview.' A report submitted to Parks Canada, July.

George, H., 2019. Hul'q'umi'num Stories of Tth'asiyetun: The Last Coast Salish Warrior Chief. Simon Fraser University (MA thesis).

Grier, C., 2007. 'Consuming the Recent for Constructing the Ancient: The Role of Ethnography in Coast Salish Archaeological Interpretations.' In B.G. Miller (ed.), *Be of Good Mind: Essays on the Coast Salish*. Vancouver: UBC Press.

——, 2014. 'Landscape Construction, Ownership and Social Change in the Southern Gulf Islands of British Columbia.' *Canadian Journal of Archaeology* 38: 211–49.

Grier, C., P. Dolan, K. Derr and E. McLay, 2009. 'Assessing Sea Level Changes in the Southern Gulf Island of British Columbia Using Archaeological Data from Coastal Spit Location.' *Canadian Journal of Archaeology* 33(2): 254–80.

Indian Residential School Resources, 2015. 'Kuper Island Residential School.' Viewed 25 February 2015 at: web.archive.org/web/20151101170835/irsr.ca/kuper-island-residential-school

Lomas, W.H., 1885. 'Dominion of Canada. Annual Report of the Department of Indian Affairs. For the Year Ended 31st December, 1885.' Ottawa: Order of Parliament.

——, 1887. 'Dominion of Canada. Annual Report of the Department of Indian Affairs. For the Year Ended 31st December, 1887.' Ottawa: Order of Parliament.

Lyackson First Nation, 1996. 'Lyackson First Nation New Community Survey'. Unpublished documents, Lyackson First Nation.

——, 2013. 'Lyackson Community Profile.' Unpublished manuscript, Lyackson First Nation.

——, n.d. 'Lyackson First Nation on Vancouver Island and Valdes Island.' Viewed 12 October 2021 at: lyackson.bc.ca

Matson, R.G., 2003. 'The Coast Salish House.' In R.G. Matson, G. Coupland and Q. Mackie (eds), *Emerging from the Mist: Studies in Northwest Coast Culture and History*. Vancouver: University of British Columbia Press.

Matson, R.G., J. Green and E. McLay, 1999. 'Houses and Household in the Gulf of Georgia: Archaeological Investigations of Shingle Point (DgRv 2), Valdes Island, British Columbia.' Vancouver: University of British Columbia, Laboratory of Archaeology.

McLay, E.B., 1999a. Valdes Island Archaeological Survey. Unpublished Survey.

——, 1999b. The Diversity of Northwest Coast Shell Middens: Late Pre-contact Settlement Patterns on Valdes Island, British Columbia. University of British Columbia (MA thesis).

Quw'utsun Syuw'entst Lelum, 2007. *Qwu'utsun Hul'q'umi'num' Category Dictionary*. Duncan: Cowichan Tribes.

Rozen, D.L., 1985. Place-Names of the Island Halkomelem Indian People. Victoria: University of British Columbia (MA thesis).

Sproat, G.M., 1877. 'Report on Chemainus District Reserves.' Correspondence of Joint Reserve Commissioner to Department of Interior, Ottawa. RG 10 Volume 7537 Reel C-14809 File 27, 150-8-9, Library and Archives Canada.

Suttles, W., 1976. 'The Ethnographic Significance of the Fort Langley Journals.' In M. Maclachlan and W. Suttles (eds), *Fort Langley Journals*. Vancouver: University of British Columbia Press.

Thomas, R., 2004. 'Laayksen Mustimuhw and Snuw'uy'ul.' *ĆELÁNEN: A Journal of Indigenous Governance* 1(1). Viewed 1 June 2023 at: web.archive.org/web/20040326042347/web.uvic.ca/igov/research/journal/articles_thomas_p.htm

Vickers and Associates Management Consultants and Lyackson First Nation, 1992. First Nation of Lyackson: Physical Planning Study. Unpublished document. New Hazleton.

15

Place and Self-Governance: Gabriola and the Autonomous Local

Dyan Dunsmoor-Farley

This chapter[1] examines Gabriolans' efforts to govern through the impacts of globalisation and shows how the concept of place not only infused conventional local governance structures but made it possible and necessary for action-based, collaborative citizen self-governance. The expected resolution of the breached treaty between the Snuneymuxw First Nation and the Crown raises questions about the interface between settler self-governance ambitions and the expression of Snuneymuxw's inherent right to self-governance. By examining the localised regulation and self-governance strategies employed on Gabriola, outside of the normative governance structures and institutions, lessons can be learned that may benefit other small communities. Read alongside Richard Kool's chapter on the absence of 'island' thinking on Vancouver Island (Chapter 19), one can perhaps begin to see gestures towards new governance modalities and ways of living in water-bounded communities.

1 This study is part of a set of case studies examining how three communities—Tumbler Ridge, Tofino and Gabriola Island—were affected by ruptures associated with global recessions and how they responded. The research involved in-depth interviews; a survey to capture information about attitudes and values; historical, geographical and statistical information from documentary sources; and knowledge-exchange activities. The result is analytical insights into narrative accounts of transition, describing the experience of recovery from rupture and responses to globalisation. Statements should be understood as representing the perspectives of those surveyed and interviewed and are not generalisable to all Gabriolans. Unless otherwise noted, quotes throughout the chapter come from interviews conducted as part of the research.

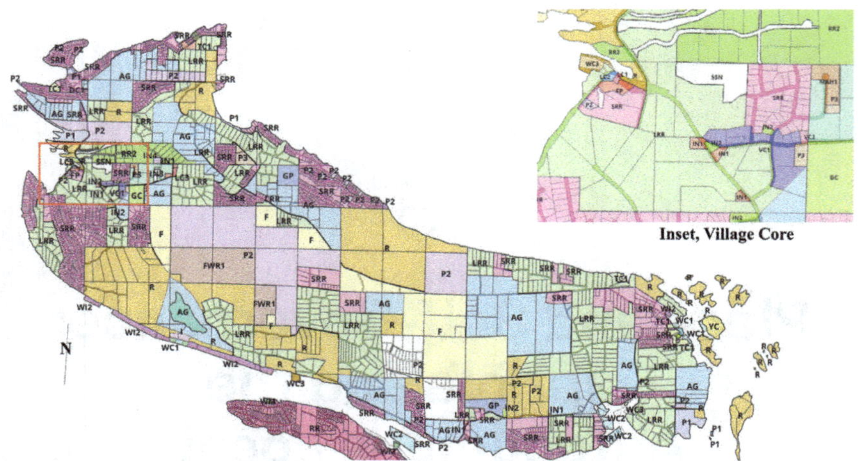

Figure 15.1 Gabriola Zoning Map

Note: SRR Small Rural Residential, LRR Large Rural Residential, AG Agriculture, F Forestry, FWR1 Forestry Wilderness/Recreation, IN Institutional, LC Land Commercial, R Resource, P Park, VC1 Village Commercial (Shopping Center), WC3 Marine Transportation (Ferry Landing), WP Water Protection. Inset (upper right) shows ferry landing and shopping center. Created with QGIS using open access cartographic data in cited website.

Source: Islands Trust 2024.

Gabriola is a rural, ferry-accessible island community situated 5 km from Nanaimo, British Columbia (BC). Although many of Gabriola's roughly 4,500 residents are older, well educated and own their own homes, residents' median income is lower than that of most British Columbians, as evidenced by a higher proportion of low-income households, a higher percentage of low-income children and high rates of homelessness (see Table 15.1).

Table 15.1 Population statistics on Gabriola

Number of residents	4,500
Median age	63
Per cent with post-secondary degree or diploma	57
Per cent homeowners	83
Per cent people in low-income households	7.8
Per cent low-income children	19
Proportion homeless	1/65

Sources: Statistics Canada (2021); 2019 Homelessness Count.

The island has a mixed economy of primarily low-paying, service industry jobs; the unemployment rate is slightly above average and self-employment levels are comparatively high (Gabriola Health and Wellbeing Collaborative 2020). Gabriola's earliest economy was the mixed economy of the Snuneymuxw which included hunting, fishing, food gathering and production, and trading with other Indigenous neighbours and with non-Indigenous traders. The incursions of the colonial government under Governor James Douglas saw the establishment of a treaty[2] covering not only Gabriola Island but much of present-day Nanaimo, Cedar and Lantzville, that confined the Indigenous population to small parcels of land but guaranteed their right to hunt and fish on their former territories.

That treaty was breached with the introduction of the *Land Registry Act* (1860), which allowed individuals (excluding the Snuneymuxw) to pre-empt up to 160 acres, conditional on making improvements. Within 20 years, Gabriola was carved up into 100 privately owned parcels (Gabriola Museum and Historical Society 2015).[3] Currently only 4 per cent of the population identifies as Indigenous and most Snuneymuxw reside in Nanaimo. Treaty negotiations to remedy the historic breach are currently under way between the Snuneymuxw, and the federal and provincial governments. Partial settlement for compensation for land in Nanaimo has been reached; 1,000 acres on Gabriola are currently under discussion. Chapter 13 (this volume) provides the context for understanding the potential of this transformation. A contrasting experience can be found in Chapter 14.

A defining feature of Gabriola is its lean governance structure. Rather than a municipal government, Gabriola is governed locally by two bodies, the Islands Trust[4] (see Figure 12.1) which sets land use policy (see Figure 15.1), and the Regional District of Nanaimo, which provides services such as building permits, recreation services, economic planning and environmental services like recycling and garbage removal.

2 This was one of several treaties negotiated on Vancouver Island and surrounding islands, commonly referred to as the Douglas Treaties.

3 The only exceptions were two tiny Indian Reserves representing the 'smallest reserve per capita land base of any First Nation in BC' (Gabriola Museum and Historical Society 2015).

4 In 1974, the New Democratic Party introduced the *Islands Trust Act* to protect the Gulf Islands from rampant development.

In practical terms, this means Gabriola has no social planning capacity and residents have no access to municipal water or sewerage services, no sidewalks and no streetlights, relying instead on individual wells, cisterns, septic fields and roadside pathways, forest trails and flashlights. More significantly, the Trust's primary planning tool, the Official Community Plan (OCP) is outdated in most communities, and time-consuming and costly to update. Moreover, OCPs do not provide the nimbleness necessary to respond to and anticipate changes to the planning context such as the climate crisis and reconciliation with First Nations.

The Islands Trust governance model primarily acts as a 'field of care'; it is introspective, as opposed to outward-looking, and is concerned with sustaining place, with little capacity to respond to the continuously changing political, social and environmental landscape. Necessarily, its expression on Gabriola goes beyond Tuan's 'fields of care' to what one might call 'fields of responsibility' (Tuan 1996).

Governing in the Gaps

Several theorists note that change happens at 'at the points of disjuncture and asymmetry and at the interstices of social relations' (Cutler 1999: 258), in the 'abyss' (Araujo 2016) or in the 'interstices of the dominant power structure' (White et al. 2016: 9). We are encouraged to 'decentre' the state (Magnusson 1992) and rather than using these abyssal and interstitial spaces to create another state, instead 'occupying myriad spaces in order to create other possibilities' (Araujo 2016: 81) . As you will see in the examples below, Gabriolans are not only prepared to act in spaces unoccupied by the state, but also to act in any sphere where it is necessary, whether it is occupied by the state or not. That is, they act in spaces occupied by the state but unenacted or ineffectively enacted, not as a form of resistance, but as an expression of self-governance.

The community has developed an extensive, robust social economy out of necessity in the absence of services, systems and infrastructure typically available in other communities through state institutions. The intensity of activities at the social and household level is partly due to the ways in which the state manifests itself and the spaces it occupies or leaves vacant.

Gabriolans surveyed are highly critical of the provincial and federal levels of government, seeing them as either indifferent to the needs of a small rural community (for instance, road maintenance policies meant for larger urban settings, a commuter ferry service based on cost recovery as opposed to being a form of public transportation, and inaccessible health services) or as acting in direct opposition to local interests (for instance, proposing freighter anchorages off the shore of the island). While Gabriolans can be critical of the Local Trust Committee (the local governing body for the Islands Trust) and Regional District of Nanaimo, they generally support the Islands Trust 'preserve and protect' mandate and appreciate government that is 'not in your face'.

Gabriolans manage their relationship to the global economy through a finely modulated, interconnected set of responses in the social and household economies which are shaped to either complement or supplement activities in the political realm. Former Islands Trustee Susan Yates and housing activist Tatha Cornish emphasise the household economy as the place where people can exercise some control through their choices. Cornish notes that 'the choices we make … impact our own family and personal life, but we have a huge impact on the health of our community and our planet' and that 'in finding ways to meet our needs for sustenance we're creating resilience in our ability to take care of our needs'.

Three initiatives demonstrate the variety of ways that Gabriolans have taken action to address community health, well-being and sustainability, including responding to the COVID pandemic. These include a health care initiative using a traditional non-profit approach, and two loosely structured networks, one addressing sustainability and the other health and well-being.

Gabriola Health Care Centre

In 2005, Gabriola was struggling to retain a doctor; many Gabriolans had to rely on doctors in Nanaimo. With fragmented primary health care, and the elimination of after-hours emergency evacuation ferry service, it was difficult for doctors to practice on the island and certainly difficult to recruit doctors and increase availability of local health care—a key consideration for a sustainable local economy. Every other Trust island, except Quadra and Salt Spring Island, which has its own hospital, received Island Health funding to provide an urgent care facility to treat, evaluate and stabilise patients (Gabriola Health Care Foundation 2006). On Gabriola, patients are evacuated by the Harbour Patrol Boat after ferry sailings end at 10:30 pm, an often unpleasant 6-kilometre crossing in any season, or by helicopter in emergency situations.

To ensure they would be able to recruit and retain physicians, the community came to together to raise funds to build a permanent health care clinic with urgent treatment capacity. With the creation of the Gabriola Health Care Society in 2007,[5] a community clinic with urgent treatment capacity was built using volunteers, donated land and donations of cash, goods and services. The new clinic which opened in June 2012, now hosts four doctors,[6] locums, a nurse practitioner and other health care services such as a mental health nurse, social worker, community care workers and a medical lab. It serves the resident population of 4,500 plus a large influx of tourists and seasonal residents. The immediate impact was a 90 per cent decrease in visits to Nanaimo Regional General Hospital emergency ward and an estimated saving to Island Health of $200,000 per year (McMartin 2012).

Even though Island Health espoused a 'geographic model of care' that responded to the needs of unique geographical areas, both the Health Authority and the Ministry of Health resisted supporting the community's efforts to create a clinic with urgent treatment capacity, refusing to provide full Medical On-Call Availability Program (MOCAP)[7] funding to the doctors for call out and only contributing $100,000 toward the completion of the Health Care Centre.

The development of the health centre follows a traditional pattern of formal organising and governance through the development of a non-profit foundation to manage the facility and attract funding. The following

5 The society transitioned to the Gabriola Health Care Foundation with a broad mandate to improve the primary health care services for the residents of and the visitors to Gabriola Island by: identifying the health care needs of those residents and visitors; working with donees and funders to develop and/or deliver programs to address those needs; providing medical facilities and/or medical equipment to be used for the benefit of residents of and visitors to Gabriola Island; and gifting funds to qualified donees. It currently has over 700 members.

6 This represents the first time since the Clinic was build that Gabriola has had a full complement of primary care providers. However, COVID and the unwillingness of the Health Authority to fully fund the Urgent Treatment facility has seriously impacted the physicians who provide service during office hours and by call-out after hours.

7 MOCAP provides compensation for doctors who are called out after hours. In 2021, Health Care Centre physician compensation issues were addressed by the Ministry of Health and Island Health.

initiatives take a different approach, aiming instead to be unfettered by external constraints such as the BC Society Act or the need to organise hierarchically.

Sustainable Gabriola

Sustainable Gabriola emerged in 2008 as a community-wide effort to plan a sustainable future. The organisation is a network of individuals 'who care about the long-term sustainability of the community and are committed to providing stewardship to achieve that goal' (Sustainable Gabriola n.d.). The participant make-up reflects a wide variety of ages (from 30s to 80s), education levels (high school graduates to PhDs), backgrounds (working class, professional), income levels (people of considerable means and people with limited means) and interests. Its purpose is to ensure 'future generations will have diverse, healthy ecosystems with access to renewable and non-renewable resources, and that individuals, and communities will experience economic prosperity, environmental quality, and social justice' (Sustainable Gabriola n.d.). They do this work by creating 'spaces for exploration and dialogue' through monthly meetings open to all, and regular community-wide planning opportunities, focusing on a broad array of sustainability topics beyond the physical environment—a 'biocultural approach' (Hanspach et al. 2019).

To date, Sustainable Gabriola has incubated six major initiatives: a community well-being survey; a cycling infrastructure plan which was adopted into the Official Community Plan (OCP); the Community Bus project—initially volunteer-operated and recently granted Regional District of Nanaimo funding through a referendum; the Heat Pump Social Enterprise providing heat pumps to Gabriolans at wholesale cost (plus a donation),[8] Gabriola Talks, a forum to discuss contentious community issues,[9] and the Mental Health and Substance Use Action Plan, aimed at addressing the high number of suicides between 2011 to 2014.

Sustainable Gabriola evolved out of interest from the Local Trust Committee to explore how certain aspects of sustainability planning embedded in the OCP could be supported using a planning model that operated outside

8 Almost 1000 heat pumps have been installed on Gabriola.

9 Two forums have been hosted: one to prioritise the issues Gabriolans wanted to work on (the climate crisis) and the second addressing housing affordability, biodiversity and freshwater conservation.

of the Local Trust Committee mandate. Former Trustee Malcolmson[10] describes how this shift came about, during an early planning workshop: 'a light went on that we didn't have to embed everything into the Official Community Plan', that some things could be covered off in 'a community sustainability plan which could be altered and changed without having to go to the Minister for approval'. Malcolmson concluded it would be better 'to create a structure that allows everyone to own it (the plan) collectively' (Trustee Sheila Malcolmson, personal communication, 21 April 2010). The result would be a separate Community Sustainability Plan outside of the OCP but with the capacity to influence and complement it.

Sustainable Gabriola resulted from an opening in the space normally occupied by the state, a willingness on the part of a local government representative to work outside of existing structures and a determination by many community organisations not to be constrained by an unresponsive planning model. Sustainable Gabriola's mandate emphasises linkages, articulating the role of Sustainable Gabriola as facilitative, connecting, educational, solution-seeking and action-taking, and its relationship to external entities as holding open a space for exploration and dialogue, strategising and integrating. This signifies a desire to not supplant the role of other organisations working on sustainability but to create opportunities to build on what others are doing.

Gabriola Health and Wellbeing Collaborative

The Gabriola Health and Wellbeing Collaborative (GHWC) uses a similar network structure. The Collaborative evolved from the success of Sustainable Gabriola's Mental Health and Substance Use Local Planning Group and a recognition that there were other critical health issues that needed to be addressed. The GHWC is made up of roughly 50 organisations and individuals concerned with community health. The group defines 'health' broadly, based on the social determinants of health. This is reflected in their membership, which includes representatives from health and social services including not only doctors and paramedics, but also alternative health practitioners, first responders, police, arts and culture groups, recreation and

10 Malcolmson, who was formerly the Member of Parliament for Nanaimo-Ladysmith, is currently the MLA for Nanaimo-North.

conservation groups, the ecumenical community, the business community and local government representatives. From the beginning, it fostered a strong, proactive relationship with the Regional Medical Health Officer, basing its planning on data and community input.

The Collaborative's mission is to create opportunities among the GHWC members for mutual support, shared advocacy and improved use of funding resources, in order to improve the health of all Gabriolans. Its objectives are:

- To improve access to primary health care
- To improve social and economic equity
- To improve access to affordable housing
- To increase environmental protection
- To improve community resilience and planning capacity, and
- To improve access to safe transportation.

The Collaborative's effectiveness rests in its ability to coordinate across resources, collaborate where joint action would be most effective and support member organisations where an individual approach would work best. The Collaborative has successfully launched multiple initiatives: the Gabriola Health Report, a repository of demographic, economic, health and wellness, and sustainability data available online to individuals and organisations (Gabriola Health and Wellbeing Collaborative 2020); the *Gabriola Talks Health and Wellness* planning initiative, and, in partnership with the Health Authority and People for a Healthy Community, the Palliative Care Project.

On 28 January 2020, BC had its first case of COVID-19. On 11 March, a pandemic was declared (CBC News 2020). The onset of the pandemic put the Collaborative model to the test. Within two weeks of the pandemic being declared, the Collaborative marshalled a working group comprising representatives from health care and social service organisations, local government, first responders, emergency services, transportation and business. Representatives from the two adjacent islands (Mudge and De Courcy) were added to the group to ensure planning addressed the needs of the smaller islands accessible to Gabriola only by private boat or water taxi.

The Gabriola Emergency Response and Recovery Committee (GERRC) recognised that the pandemic posed significant risks for a ferry-dependent community relying on supplies and specialised supports from larger urban centres. Meeting twice a week, the group focused on ensuring an effective, systemic response for all aspects of disease response and with a longer-term view to building recovery strategies. Because the participants already had a good sense of the capacity of the different member organisations, they focused their attention on a systems approach to risk management. Local government representatives ensured coordination with the Regional District's Emergency Operations Centre, and the physician and first responder representatives undertook coordination with the Medical Health Officer's directives.

In addition to ongoing capacity gaps such as not enough doctors, limited access to specialised services and supply chain vulnerability, the GERRC identified worst-case scenarios and response strategies in the following areas: access to acute care, transportation of patients, goods and services in the event of a ferry service interruption, social safety net coordination, management of COVID-related logistics and supplies, alternative communications mechanisms, and recovery planning including a business and food system recovery strategy. By 8 June, members were confident that they had a coordinated response plan that would be able to address all contingencies. The GERRC disbanded and COVID planning became part of the monthly Collaborative meeting. As of time of writing, there were no confirmed cases of COVID-19 on Gabriola, due in large part to a strong social fabric emphasising individual and community responsibility (Dunsmoor-Farley and Weller 2020).

The above examples demonstrate how, in the absence of comprehensive governance structures, Gabriolans have exploited the opportunity to govern in the interstices. These efforts have ranged from the more traditional organisational approaches authorised by the state such as non-profit societies, cooperatives and foundations, to what White and colleagues describe as more anarchic 'emancipatory praxis', focusing on mutuality, consensus decision making, self-organisation, neither dominating nor being dominated, horizontality and self-management (2016: 11). And while it would be tempting to see initiatives like Sustainable Gabriola as 'a repudiation of the role of government', the response is much more sophisticated than a binary categorisation of the state as good or bad.

Gabriolans are primarily focused on the local, whether through existing formal governance structures or through acts of self-governance—what political science scholar Warren Magnusson describes as 'particularist radicalism' aimed at self-transformation (1992: 97). In this case, the efforts are not aimed at resisting or transforming the state, although they may result in state transformation, rather efforts are aimed at forming judgements and acting as expressions of political will. But being 'islanded' can translate into insularity, limiting the appreciation of the relationship to that which is beyond the boundaries of the island, whether it is the global economy, or the existence of external interests which might impact Gabriola. When the external world intrudes on Gabriolans' sense of a circumscribed world, it is experienced as jarring and violent. Nonetheless, as the COVID response demonstrated, a collective community effort helped to transcend a potentially destructive event.

Possibilities and Opportunities

> sometimes the smallest act in the most limited circumstances bears the seed of … boundlessness, because one deed, and sometimes one word, suffices to change every constellation. (Hannah Arendt, *The Human Condition*, 1958: 190)

People have been attracted to Gabriola for many reasons: the Snuneymuxw for its abundant natural resources, and for early 'settlers' the availability of land, made possible by the dispossession of the local Indigenous population. More recently, demand for land has necessitated a law to 'preserve and protect' this unique place, resulting in a governance structure that prioritises place over economy. A culture of 'protection' is evident on Gabriola. 'Modernisation' efforts to change the governance structure to a municipality, or to connect the island to Vancouver Island by a bridge are met with strong resistance, suggesting that Gabriolans are satisfied with their governance arrangements and value being 'islanded' despite the hardships this may entail. The complex matrix of governance modalities, although an irritation to some Gabriolans, is also seen as an opportunity to exercise community control in addressing critical issues.

Many Gabriolans see the island as a place for testing out new ways of being and acting, including a willingness to live within a fragmented governance environment, recognising the opportunities for autonomous action in the interstices. As Chloe Straw observes: 'it's going to continue to be a demonstration community where we test things out and pilot different projects that will be taken up in other places'. The Snuneymuxw's efforts to redress breaches of their treaty provides an opportunity for Gabriolans to engage in reconciliation and forge new relations. Gabriolans' deep attachment to self-governance suggests that there is an opportunity to learn from a people who have been insisting on their right to self-governance for more than 150 years.

Gabriolans embrace a range of strategies, from individual autonomous action to networked autonomous action (like Sustainable Gabriola) to the creation of place-based governance entities as potential sites for autonomous action (for instance, the Islands Trust). The effectiveness of these strategies is determined by several factors that must be carefully considered by communities interested in strengthening the autonomous local. First, the exercise of autonomous action can be hampered if communities imagine the state as the locus of political action and the market economy as the primary agent for achieving community health and well-being; this in turn, has consequences for life control, self-determination and self-governance.

Second, the degree to which the community is willing to work outside of the normative governance structures affects their ability to imagine and create adaptive strategies capable of responding to the unpredictable mutations of global capital. Third, a lack of understanding of colonisation's ongoing impacts will hamper the community's capacity to create meaningful and productive relationships with the Snuneymuxw, relationships that will ultimately be critical to the economic and social well-being of all parties.

As the Gabriola example shows, self-governance is occurring in many forms and is often 'under the radar'. Gabriola has developed a robust self-governance infrastructure, not necessarily as an antidote to a 'fully constituted' state, but certainly as a form of self-regulation, in part because 'place', rather than simply being about the virtues of a particular geography, has been the determining factor in the form of local governance.

This governance arrangement creates a unique culture in which place is privileged, and, as the interviews show, citizens express a commitment to stewardship. It also serves as a constant reference point when contentious issues arise. On Gabriola, we see examples of individuals exercising judgement, taking responsibility and acting beyond their own narrow interests. Although some Gabriolans imagine the state as the container for politics and as the singular holder of power, there is ample evidence of a culture that respects the role of the state but does not expect it to address all concerns. But taking responsibility is not an unproblematic concept: in the absence of a concern about the common good, taking responsibility can simply reflect the desires and wants of the individualistic, atomised and thoroughly colonised citizen caring only for their own interests and needs, the citizen who acts within the law but without concern for their neighbour or the greater community well-being. Gabriolans' developing relationship with the Snuneymuxw demonstrates these challenges.

Post-contact, the availability of relatively inexpensive land pre-empted from local First Nations made it possible for non-Indigenous occupiers, in the form of early 'settlers', the following influx of 'hippies' in the 1970s looking for the freedom of an alternative lifestyle, and then waves of ageing baby boomers in the 1980s and 1990s, to be in this space. Despite the fact that those who have come here in the last 30 years have been attracted by a sense of sanctuary and community, it exists only through its denial of sanctuary to its original inhabitants. How the community reconciles its history with the Snuneymuxw will be a defining moment in its self-governance evolution. As treaty negotiations continue, present community efforts have focused on relationship building between organisations and Snuneymuxw leadership and a broader effort to educate the community by working with Elders and knowledge keepers.

For communities seeking to achieve greater control in addressing issues that matter, the fundamental principle is to be prepared to take responsibility and to act, not out of self-interest, but in the broadest interests of the community. Implicit in this is the understanding that responsibility must always be linked to accountability. The question is, can state structures adjust to take advantage of the potential that arises from community self-governance initiatives? The complexity of state and local relationships is well described in Chapter 20, 'Indigenous Co-management of Salish Sea Protected Areas'.

Conclusion

Fundamental to achieving large-scale systems change is the need to focus on self-governance; that is, to establish common objectives that shift the discourse away from narratives of inevitability and hopelessness towards a reimagined future. If we are not to succumb to neoliberal self-governance aspirations that focus on the individual, then our imagining of new concepts of economy and governance must be founded on principles of individual responsibility within the context of the common good. To do this we must avoid atomistic conceptions of the individual; individual responsibility is critical but insufficient on its own. It is only when individual action is linked with the actions of others that the possibility for transformative change emerges.

As the governance examples show, linkages are critical. But it is insufficient to imagine that linkages must be made only with formal structures of governance like the state and formal civil society entities.

Linkages must be formed between citizens and across the social, political and economic landscape. Focusing on the state and other formal structures limits our ability to imagine other ways of being and to exploit the cracks that exist within the state apparatuses. Although the health care initiative relied on the establishment of a formal structure—the Health Care Foundation—Sustainable Gabriola demonstrates the value of avoiding the tendency to valorise formal structures that reflect the state or the corporation over informal structures such as networks. Moreover, we must resist the temptation to formalise linkages and networks, thereby replicating the very structures that need to be subverted.

Autonomous action aimed at individual gratification is insufficient; the impetus for autonomous action may spring from an individual's need, but it must ultimately address that which is outside of the individual's interests and speak instead to common needs. This requires an engaged actor working across interests to imagine a multiplicity of governance capacities and strengthening those capacities that have previously been undervalued. It also means examining our own responsibilities as political actors with a duty not just to act, but to listen carefully, learn and build sustaining relationships across polities and perspectives.

For small communities, the task may feel daunting, and it is easy to feel a sense of hopelessness and futility. And yet community actors play a critical role in addressing the behaviours, processes and systems that are ultimately not in their collective self-interest. Rather than seeing this complexity as a problem, one might imagine it as a virtue, in that it provides multiple points of entry for adjusting, reconfiguring and even upending existing systems. Being strategic about how we target our efforts increases our chances of effectiveness: effectiveness increases our sense of efficacy and encourages us to do more.

Gabriola's next transformative challenge will be to embrace the necessary work of co-governance with the Snuneymuxw, flowing from the restoration of their inherent and unrelinquished treaty rights. All of Gabriola's self-governance attributes will need to be utilised in this work, if those displaced by global capital are to be restored to this 'sanctuary' presently known as Gabriola Island.

References

Araujo, E., 2016. 'What Do We Resist When We Resist the State?' In M. Lopes, R.J. White and S.D.S. Springer (eds), *Theories of Resistance: Anarchism, Geography and the Spirit of Revolt.* London: Rowman and Littlefield.

Arendt, H., 1998. *The Human Condition.* Chicago: University of Chicago Press.

CBC News, 2020. 'The COVID-19 Pandemic: A Timeline of Key Events Across British Columbia.' *CBC News*, 3 April. Viewed 20 June 2023 at: www.cbc.ca/news/canada/british-columbia/covid-19-bc-timeline-1.5520943

Cutler, C.A., 1999. *Private Power and Global Authority.* Cambridge: Cambridge University Press.

Dunsmoor-Farley, D. and F. Weller, 2020. 'Sustainable Gabriola Initiatives.' Viewed 31 May 2023 at: sustainablegabriola.ca/initiatives/

Gabriola Health Care Foundation, 2006. 'Archives.' Gabriola Health Care Foundation. Viewed 31 May 2023 at: ghcs.ca/archives/status.html

Gabriola Health and Wellness Collaborative, 2020. 'Gabriola Health Report: Taking the Pulse of Our Island.' Viewed 31 May 2023 at: ghcf.ca/2020_Gabriola_Health_Report.pdf

Gabriola Museum and Historical Society, 2015. 'Gabriola Roots: The Land Provides.' Viewed 31 May 2023 at: gabriolamuseum.org/exhibits/gabriola-roots-the-land-provides/

Hanspach, J., L.J. Haider, E. Oteros-Rozas, A.S. Olafsson, C.M. Gulsrud, M. Torralba, B. Martin-Lopez et al., 2019. 'Biocultural Approaches to Sustainability: A Systematic Review of the Scientific Literature.' *People and Nature* 2(3): 643–59. doi.org/10.1002/pan3.10120

Islands Trust, 2023. 'Islands Trust Area.' Overview of the Islands Trust. Viewed 7 June 2023 at: islandstrust.bc.ca/about-us/overview-of-islands-trust/

——, 2024. Mapping and Resources. Viewed 13 March 2024 at: islandstrust.bc.ca/mapping-resources/mapping/gabriola/

Magnusson, W., 1992. 'Globalization, Movements and the Decentred State.' In W.K. Carroll (ed.), *Organizing Dissent.* Toronto: Garamond Press.

McMartin, P., 2012. 'Gabriola Island Residents Take Health Care in Their Own Hands.' *Vancouver Sun*, 5 March. Viewed 31 May 2023 at: vancouversun.com/news/mcmartin-gabriola-island-residents-take-health-care-in-their-own-hands

Statistics Canada, 2023. '2021 Census of Population.' Viewed 18 July 2023 at: www12.statcan.gc.ca/census-recensement/2021/dp-pd/prof/index.cfm?Lang=E

Sustainable Gabriola. n.d. Viewed 31 May 2023 at: sustainablegabriola.ca/

Tuan, Y.-F., 1996. 'Space and Place: Humanistic Perspective.' In J.L. Agnew (ed.), *Human Geography: An Essential Anthology*. Oxford: Blackwell Publishing Ltd.

White, R.J., S. Springer and M.L. De Souza, 2016. *The Practice of Freedom: Anarchism, Geography and the Spirit of Revolt*. London: Rowman and Littlefield.

Wright, A.L., C. Gabel, M. Ballantyne, S.M. Jack and O. Wahoush, 2019. 'Using Two-Eyed Seeing in Research With Indigenous People: An Integrative Review.' *International Journal of Qualitative Methods* 18: 1–19.

Part Four: Environmental Management

16

Salmon and the Salish Sea: A Transboundary Approach to Salmon Recovery

Isobel A. Pearsall and Michael W. Schmidt

The Salish Sea is the inland sea that encompasses the Strait of Georgia, Puget Sound, the San Juan Islands and the Gulf Islands. The area spans from Olympia, Washington (WA) in the south to Campbell River, British Columbia (BC) in the north, and includes the cities of Seattle and Vancouver. It is one of the world's largest inland seas, with a total sea surface area of 16,935 km^2 and a maximum depth of 650 m. There are 419 islands within the Salish Sea, and it is home to over 8 million people, as well as orcas (*Orcinus orca*), Steller sea lions (*Eumetopias jubatus*), harbor seals (*Phoca vitulina*), humpback whales (*Megaptera novaeangliae*), Pacific salmon (*Oncorhynchus* spp.), and hundreds of other fish species, including forage fish species such as Pacific herring (*Clupea pallasii*) and Pacific sand lance (*Ammodytes hexapterus*). This rich ecosystem is estimated to support 37 species of mammals, 172 species of birds, 253 species of fish and more than 3,000 species of invertebrates (Gaydos and Pearson 2011).

The Salish Sea is an important nursery area for all species of Pacific salmon: sockeye (*Oncorhynchus nerka*), pink (*O. gorbuscha*), chum (*O. keta*), coho (*O. kisutch*) and Chinook (*O. tshawytscha*), as well as steelhead trout (*O. mykiss*). They are all anadromous, spending time in freshwater systems, estuaries and the open ocean. Adults lay their eggs in redds, or nests that the females create in depressions in river gravel. The eggs remain in the

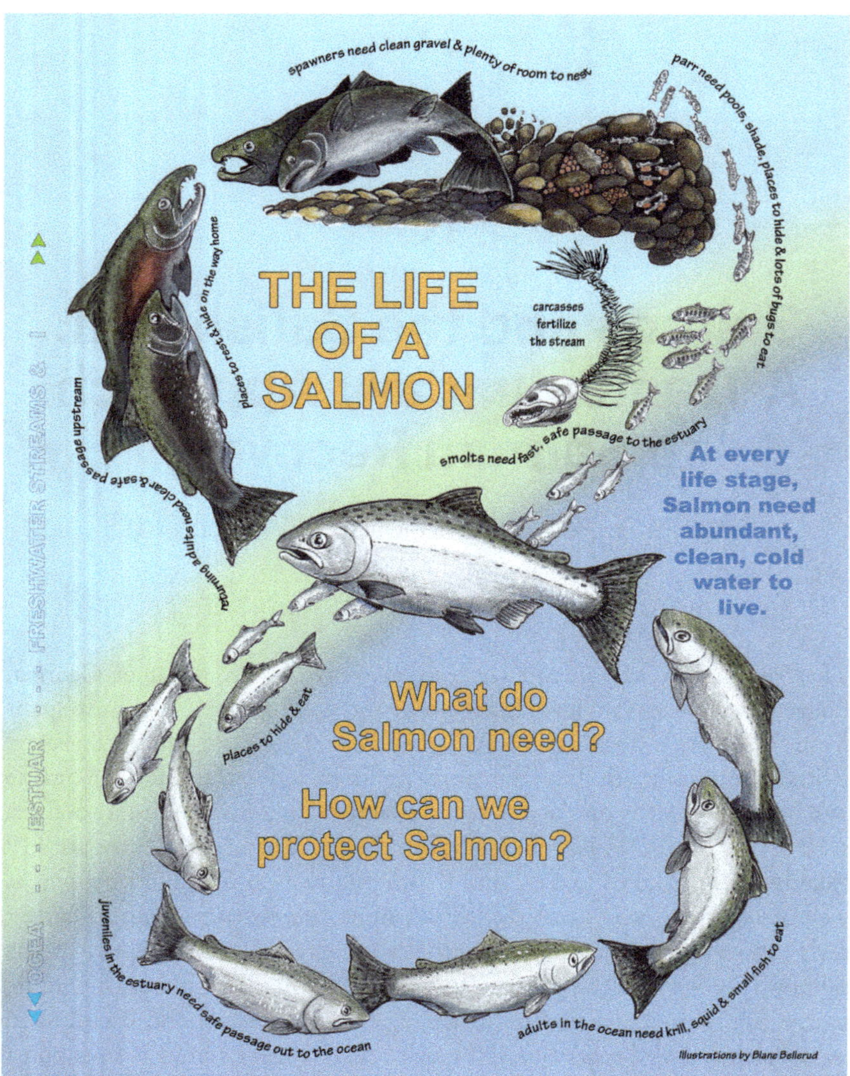

Figure 16.1 The life of a salmon

Source: NOAA (2019).

gravel through winter while the embryos develop, hatching as alevins in the spring. These tiny fish have the egg yolk sac attached, and once this is fully consumed, they emerge from the gravel as fry. Fry swim to the surface of the water, fill up their swim bladders with oxygen and begin to feed in their natal river or stream, remaining there for various lengths of time, depending on the species. After completing their freshwater stage, they go through a physiological change called smoltification which involves osmoregulatory changes that prepare them for entry into saltwater, and growth of silvery coloured scales. Chum and pink salmon are nearly smolts upon emergence from the gravel, going directly to estuaries and the ocean, while Chinook and coho may either go directly to sea the first spring or summer of their life (subyearling strategy), or remain in freshwater for a whole year before smolting (yearling strategy). Sockeye rear in fresh water, often in lakes, for one or two years before smolting, while steelhead may not smolt for two years or more. The young salmon imprint to their natal stream and are able to navigate back to the same systems when they return as adults (Groot and Margolis 1991).

Salmon may spend one to seven years in the ocean: some may stay resident for their entire lives within the Salish Sea, while others migrate out to the open ocean, either remaining in coastal areas, or migrating over 16,000 km throughout the north Pacific Ocean. During their time in salt water, they feed and grow, increasing their weight a thousand times and using the sun and earth's magnetic field to navigate. After their ocean migration, they return to their natal stream to spawn. In general, individuals of Pacific salmon typically die after spawning and are semelparous, while steelhead may sometimes exhibit repeated spawning, or iteroparity.

After adult salmon die, their nutrient-rich carcasses provide a critical resource for terrestrial ecosystems, increasing organic matter and nutrients to streams and rivers and the surrounding ecosystem. These nutrients are transferred to all levels of the food chain, and it is estimated that at least 137 different species depend on the marine-derived nutrients in the bodies of the salmon (Cedarholm et al. 2000). This includes predators and scavengers from aquatic insects up to bears, wolves and birds, which carry the carcasses into forests resulting in an input of up to 40 per cent of the nitrogen in riparian plants and trees (Bilby et al. 1996). Coastal bears can get from 33 to 94 per cent of their annual protein from salmon, resulting in larger bears and population densities up to 20 times greater in areas where salmon are abundant, as compared to areas where they do not occur (Hildebrand et al. 1999). The biomass of salmon in a stream also has been shown to be an indicator of the density and diversity of species of birds in the surrounding ecosystem (Field and Reynolds 2012), and is an important source of the carbon, nitrogen and phosphorus essential to rearing the next generation of salmon juveniles. Salmon also play a significant role in the survival of certain ocean species during their time in salt water. For example, Chinook salmon are the primary prey for the southern resident killer whale.

Given the influence of salmon on so many other species, anadromous salmon are considered a 'keystone' species in both aquatic and terrestrial environments. They serve as a bioindicator, and as such, the health of salmon populations can indicate general ecosystem health, including water and habitat quality, and ultimately, the integrity and persistence of the entire food web (Willson and Halupka 1995).

Salmon are at the heart of the BC and Washington State culture, economy and environment, and salmon and people have maintained a relationship in the Salish Sea region for over 10,000 years. They are particularly embedded in the culture, identity and existence of the First Nations people of BC and Washington State. Many First Nations communities are founded on traditional fishing grounds and Coast Salish people have depended on salmon for thousands of years as a staple food source as well as a source for wealth and trade.

Pacific salmon also support economically viable commercial and recreational fisheries in Canada, supporting tens of thousands of jobs and local economies and communities around the Pacific Rim. Commercial and recreational salmon fisheries in the US state and Canadian province bordering the Salish Sea were worth an average of CAD1.1 billion annually (GDP) from 2012 to 2015 (Gislason et al. 2017). Wildlife viewing of salmon also is a significant part of BC's tourism industry, contributing to the economy and providing employment.

In summary, salmon are not only iconic to our way of life in the Salish Sea; they are crucial to the health of our freshwater, estuarine and marine ecosystems. By protecting salmon and their habitats we can protect our economies, culture, communities and environment, including those on Vancouver Island, the Gulf Islands and San Juan Islands.

Gulf Islands and Salmon

Within the Salish Sea, the Gulf Islands and San Juan Islands are well known for their temperate, unique Pacific Northwest ecosystems and varied habitats, including rocky reefs and protected bays filled with eelgrass and kelp beds. Juvenile Pacific salmon rear around the islands, utilising the eelgrass and kelp nursery grounds for protection and feeding opportunities. Stocks derived from Strait of Georgia rivers are commonly found rearing around the BC Gulf Islands, with some such as Cowichan River Chinook showing extended residence in these areas (Beamish et al. 2012). Similarly, the marine areas adjacent to the San Juan Islands represent an important early marine rearing habitat for juveniles originating from the Nooksack, Skagit and potentially other rivers from Washington State (Davis et al. 2020). Recent studies have highlighted the importance of the San Juan Islands as rearing areas for Pacific sand lance (Davis et al. 2020), while many of the more northern BC Gulf Islands such as Hornby and Denman are well known herring nursery areas (Therriault et al. 2009). These forage fish species are crucial for both coho and Chinook, which shift to piscivory during their first summer rearing in the ocean. Subyearling Chinook caught in the marine areas surrounding the San Juan Islands are more likely to eat forage fish than elsewhere in Central or Southern Puget Sound, with up to three times fuller stomachs and distinct growth advantages (Davis et al. 2020). The population of sand lance in the San Juan Island wave field has been estimated at over 63 million fish (Selleck et al. 2015). Robinson et al. (2013) also found large schools of Pacific sand lance close to suitable sand burial habitats around the southern BC Gulf Islands.

State of Salmon Populations in the Salish Sea

The Salish Sea was once a very productive place for Chinook, coho and steelhead. However, things began to change abruptly through the late 1970s and 1980s when their marine survival rates[1] plummeted in a way not seen in other populations from coastal and Columbia River populations (Beamish et al. 2010). Although wild fish generally survive better than hatchery fish, this declining trend was observed among both groups (Figure 16.2).

1 Marine survival covers the period from release in fresh water at hatcheries or, for wild fish, from their downriver migration as juveniles and through their ocean phase to the point where these fish are either captured as adults in fisheries or return to rivers and hatcheries to spawn.

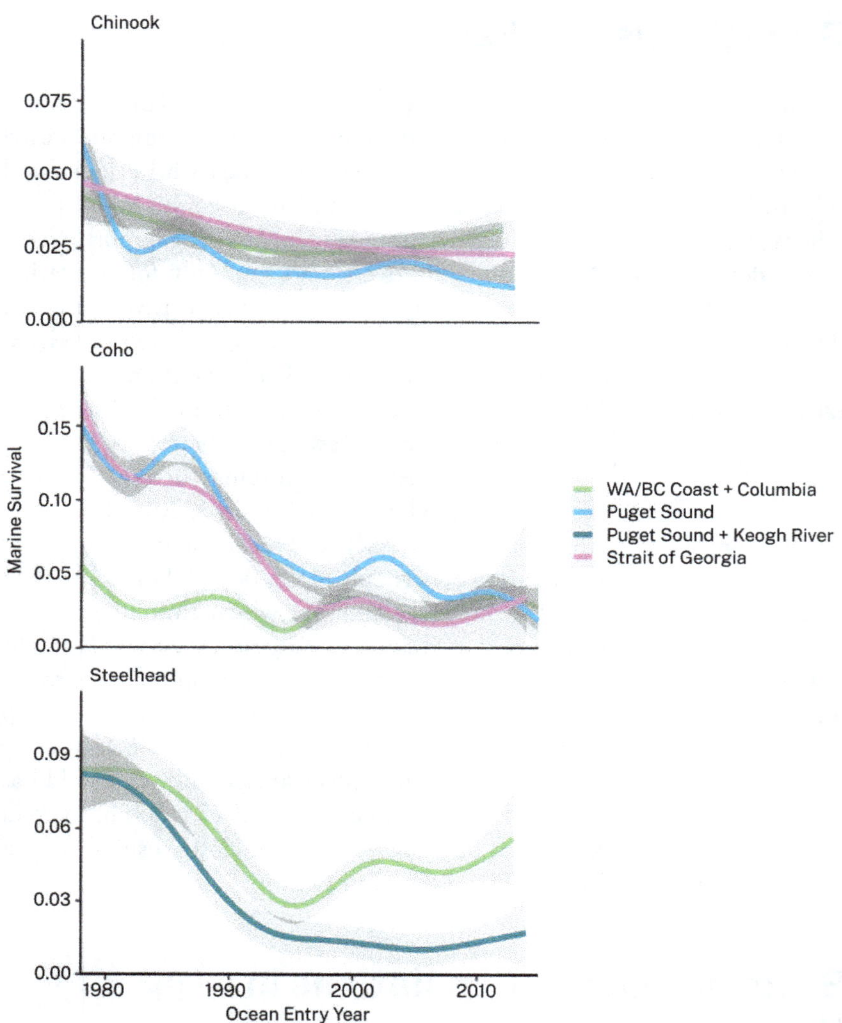

Figure 16.2 Marine survival trend for Chinook, coho and steelhead salmon

Notes: Chinook (top panel) and coho (middle panel) in Puget Sound (blue), Strait of Georgia (pink), and WA/BC Coast and Columbia River (green) and for steelhead (bottom panel) in Puget Sound and Strait of Georgia's Keogh River (teal). A smoothing function was applied to illustrate trend for ocean entry years 1978–2015.

Sources: Underlying data as described in Zimmerman et al. (2015); Ruff et al. (2017); Kendall et al. (2017); Sobocinski et al. (2021).

Both Puget Sound Chinook salmon and steelhead trout are listed as threatened under the US *Endangered Species Act*, while many populations of Chinook, coho and steelhead in the Strait of Georgia basin are listed as Species at Risk in Canada. These declines are having various negative impacts on the ecosystem; for example, it is believed that diminishing Chinook salmon populations have contributed to declines in endangered southern resident killer whales (Hanson et al. 2021). Current levels of Chinook, coho and steelhead are so low that it is suggested that these populations may be unable to adapt to any additional mortality that could result from continued climate change (Sobocinski et al. 2021).

In addition to declines in Pacific salmon, many researchers have documented other concomitant changes that have occurred within the Salish Sea (see Chapter 3 'Oceanography of the Salish Sea'). For example, seawater and river water temperatures have increased, deep-water oxygen has declined, sea level has risen and the timing of the Fraser River freshets has changed (Riche et al. 2014). We have also seen large increases in the number of harbor seals (Nelson et al. 2019), humpback whales and jellyfish (Greene et al. 2015), and communities around the Salish Sea have remarked on the losses of eelgrass and kelp habitats.

Salish Sea salmon populations have been negatively impacted by development, human population expansion and periods of historical overharvesting. Natural resource development, including logging and mining, has resulted in significant habitat degradation in freshwater and estuarine environments; dams, floodgates and other barriers have blocked upriver migrations, and human development and agriculture have resulted in reduced water quality and sedimentation.

Additionally, climate change continues to pose additional stressors. In fresh water, climate change is resulting in increased stream temperatures, which impact adult holding and upstream migration timing, deplete energy reserves and increase susceptibility to disease. Climate change–related changes to hydrology can result in flooding and the scouring and loss of eggs, or low flows resulting in thermal stress and increased stranding (Crozier et al. 2019). Coupled with increased winter flows is increased sedimentation of river and stream beds, which can increase or reduce the quantity and quality of gravel substrate available for spawning, with associated impacts on egg and alevin survival.

In the ocean, impacts of climate change include warming waters, increasing ocean acidification and increased numbers of destructive marine heatwaves, which impact ocean food webs (Grant et al. 2019). Changes in water temperature can reduce salmon growth, increase stress and reduce the availability of oxygen in the water. Climate change may influence prey availability for salmon by changing spring bloom timing (Allen and Wolfe 2013) or affecting plankton composition. It can also affect salmon and the ecosystem through many other pathways, including more acidic waters (Busch et al. 2013) and larger and more frequent harmful algae blooms (Tatters et al. 2013) that can impact salmon and their prey. Increased temperatures can also increase susceptibility to disease and contaminant impacts (Gouin et al. 2013).

In general, the salmon recovery efforts that began in the late 1990s in the Salish Sea resulted in significant investments in freshwater habitat restoration, harvest reduction and increased artificial production of salmonids in both hatcheries and aquaculture facilities. However, despite yearly releases of millions of juvenile salmon from enhancement facilities around the Salish Sea, and millions invested in freshwater habitat restoration, Pacific salmon continue to decline. There are also significant concerns about impacts of hatchery production on wild stocks including evidence for lower reproductive success of hatchery-origin fish in the wild, over-exploitation of wild populations harvested with more productive enhanced populations, and genetic and other ecological impacts of hatchery practices on naturally spawning fish, for example, as a result of interbreeding (e.g. Araki et al. 2008).

Until recently, little emphasis was put on assessing impacts on salmon in the marine environment. However, the elucidation of very different signals in marine survival of coho, Chinook and steelhead in the Salish Sea as compared to the outer coast highlighted the need to focus on factors impacting young salmon during their early life within the Salish Sea itself.

The Salish Sea Marine Survival Project: A Transboundary Approach to Identify Key Causes of Declines

The Pacific Salmon Foundation (Vancouver, BC) and Long Live the Kings (Seattle, WA) initiated a major transboundary program, the Salish Sea Marine Survival Project (SSMSP) in 2013 to identify the primary factors affecting the survival of juvenile Chinook salmon, coho salmon and steelhead trout in the Salish Sea marine environment. It was designed as an intensive, short-term (2014–18) study of the Salish Sea ecosystem to examine major components of the salmon ecosystem simultaneously. This international collaborative of over 60 federal, state, tribal, non-profit, academic and private entities implemented a coordinated research effort that encompassed all major hypothesised impacts on Chinook, coho and steelhead as they entered and transited the Salish Sea. Ultimately, several hundred scientists collaborated to implement over 90 studies.

Evidence suggests that for a short period after entering seawater, juvenile salmon must survive a critical period that plays a major role in dictating the overall number of adults that later return to spawn (Beamish and Mahnken 2001), and that high growth and attaining a large size at this time are crucial (e.g. Beamish et al. 2004). For this reason, experts have suspected that ongoing ecological changes in the Salish Sea may be largely to blame for the overall decline in Chinook, coho and steelhead in our local streams.

It was understood that numerous factors can affect salmon survival during the early marine phase, and these were assimilated into three key categories. Broadly, the primary hypotheses were:

- Bottom-up processes—annual environmental conditions determine the food supply for salmon and result in the variation in size and growth rate of juveniles.
- Top-down processes—predation is likely the direct cause of mortality, but fish condition may be compromised by other biological factors, increasing their susceptibility to predation (e.g. disease, harmful algae, competition).
- Additional indirect factors exacerbating these processes, including habitat loss and contaminants.

What distinguished SSMSP from previous efforts was the scope of topics considered simultaneously and the breadth of collaboration involved (Figure 16.3); as opposed to individual interests in separate species, years and locations.

In total, 21 hypotheses describing potential factors influencing juvenile salmon survival were addressed, including: changing water temperatures, reductions in food supply such as plankton and forage fish, increases in marine mammal abundance, contaminants and disease, as well as others. The ultimate goal was to determine the extent to which poor Chinook, coho and steelhead survival is driven by local factors, global processes or some combination and, based on results, to propose action-oriented research and management recommendations.

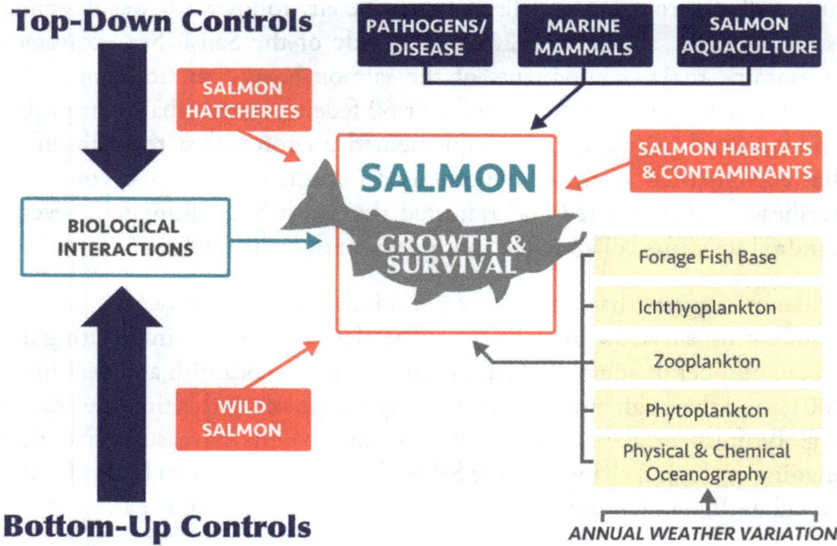

Figure 16.3 Schematic of factors addressed annually within the SSMSP

Source: Pacific Salmon Foundation.

Results showed that changes to Salish Sea food supply (availability, timing, quality and quantity) and levels of predation appeared to be the most critical factors impacting marine survival of Salish Sea salmon, although various other impacts (habitat loss, contaminants and disease) were found to be significant at population or sub-basin levels (see the summary/references in Pearsall and Schmidt et al. 2021).

Food Supply

SSMSP studies confirmed a connection between the availability of zooplankton and the survival of coho and Chinook salmon. Changes in weather may be affecting the timing or size of the spring phytoplankton bloom: though variable, the timing of the spring bloom has become earlier on average since 1990, which affects the subsequent zooplankton bloom that ultimately feeds salmon. Nutrients, primarily nitrogen from sewage, agriculture and stormwater, have also increased and could be affecting phytoplankton. Another concern is the increasing abundance of jellies in Salish Sea waters, which may be a result of warmer and more nutrient-rich waters. Jellyfish are voracious eaters but have little nutritional value, and they are therefore considered food web 'dead ends' in ecology. They also may consume phytoplankton and zooplankton, reducing overall productivity of the food web.

As young Chinook and coho live through the summer, they must consume increasingly energy-rich food to grow fast enough to avoid predators and fat enough to survive the coming winter. Forage fish, particularly herring, become extremely important to Chinook and coho once the salmon are large enough to eat them. Scientists found strong positive relationships between herring abundance in the Salish Sea and Chinook and coho growth and survival, but diet studies suggest that juvenile Chinook are currently eating fewer herring compared to the 1970s, when Chinook survival rates were higher. Herring populations have declined in many parts of the Salish Sea, with reduced diversity in spawning locations as well as spawn timing. Herring are on average spawning earlier in spring, which means that young Chinook and coho may not be able to find herring of the right size when they get large enough to eat fish.

The rapid outmigration of juvenile steelhead from the Salish Sea during a narrow time window suggests that foraging opportunities may be less important to their early marine survival.

Predation

Populations of harbor seals in the Salish Sea have increased concomitantly with declines in salmon marine survival; consumption estimates and other analyses indicated a stronger relationship between increased seal abundance and marine survival declines for coho and steelhead than for Chinook.

Overall, juvenile salmon and steelhead do not make up a significant portion of the seal diet, which prefer to feed on herring, Pacific hake (*Merluccius productus)* and Pacific cod (*Gadus macrocephalus*). However, the sheer abundance of seals appears to have a significant impact on juvenile salmon as they rear within the Salish Sea in the summer, likely as a byproduct of targeted feeding by seals on Pacific herring, a species with which salmon often co-locate. Declines in the overall abundance of forage fish may indirectly contribute to a greater consumption of young salmon and steelhead. For example, one study showed that more steelhead survived when large numbers of schooling anchovy were present in Puget Sound.

Seals are not the only animals that eat juvenile salmon. SSMSP scientists documented increased predation in the Cowichan River, BC, by raccoons, herons and other animals at times when streamflows were extremely low. This could be a significant concern as climate change alters flow patterns and causes more of these low-flow events.

Physical Habitat

Estuaries and the nearshore can be among the most productive marine habitats, providing food and shelter to young salmon as well as critical prey species, such as herring, sand lance and crab. However, estuary degradation, a result of development and climate change, is likely limiting the survival of some salmon populations, in particular the loss of estuary habitat for wild subyearling Chinook. SSMSP studies found that, within the Salish Sea, small Chinook rearing in healthy estuaries have a much greater chance of surviving to adulthood compared to those that rear in degraded estuaries.

Further, there may be linkages between lost kelp habitat, increased patchiness of eelgrass habitat and reductions in availability of salmon prey, such as herring and larval crab, which could be contributing to broader survival declines.

Disease

Studies of disease during the SSMSP led researchers to conclude that pathogens could be playing a substantial role in the survival of Chinook and probably coho and sockeye as well. Although there is no historic information to determine whether disease contributed to marine survival declines, experts noted that there is cause for concern because of known relationships between increasing water temperatures and higher disease susceptibility.

Researchers identified over 50 infectious agents in juvenile salmon in the Salish Sea, including 15 novel viruses and many agents never studied previously in BC salmon. The southern Strait of Georgia was identified as an infection hotspot in summer months, with more infection overall in the Strait of Georgia as compared to the outer Pacific coast. SSMSP scientists concluded that the length and intensity of exposure to disease agents are major determinants of impact on Pacific salmon, with highest loads associated with residency in specific areas of the Strait of Georgia.

Most of the disease work was conducted in the Strait of Georgia under the Strategic Salmon Health Initiative, launched in 2013 by the Pacific Salmon Foundation in partnership with Genome BC and Fisheries and Oceans Canada (DFO). Less is known about diseases among Chinook and coho salmon in Puget Sound.

Contaminants

Toxic chemicals are known to affect a large number of marine species, from plankton to killer whales. While Meador (2014) suggested that contaminant exposure may result in variation in marine survival among Chinook populations in Puget Sound, it is not known if contaminants were a primary contributor to declines in Chinook marine survival across the Salish Sea since the late 1970s. However, it is believed that contaminants are limiting the recovery of many Chinook populations, especially those that come from urbanised watersheds in Central and South Puget Sound and likely the Fraser River in the Southern Strait of Georgia (Figure 16.4).

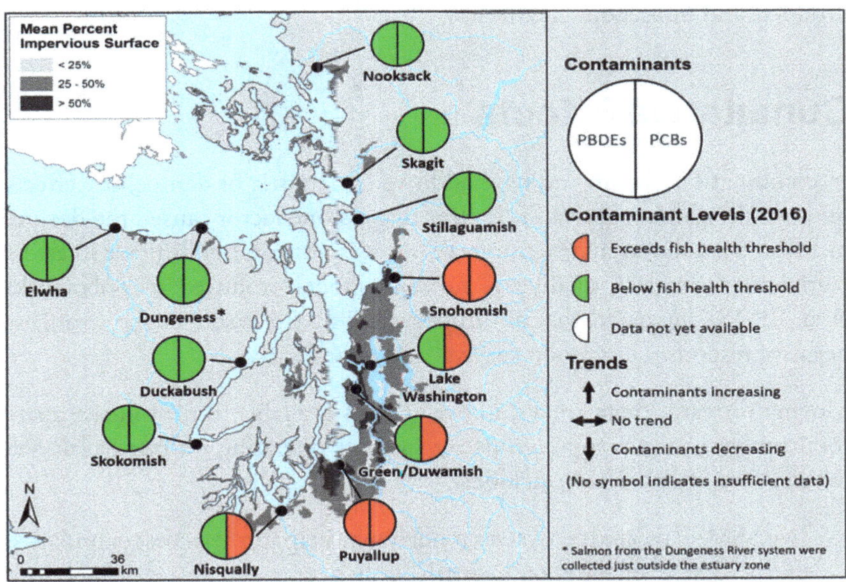

Figure 16.4 Contaminants in juvenile Chinook salmon

Note: Symbols on this map indicate sites where contaminants in juvenile Chinook salmon exceed (red) or remain below (green) adverse health effects thresholds.

Source: Puget Sound Partnership (2021): vitalsigns.pugetsoundinfo.wa.gov/VitalSignIndicator/Detail/49.

Two contaminants known to persist in the environment and negatively affect salmon growth and survival were broadly assessed in young and resident Chinook. They are PCBs (polychlorinated biphenyls), a group of oily compounds used in many commercial applications before being banned in the 1970s, and PBDEs (polybrominated diphenyl ethers), a group of chemicals used as flame retardants. Contaminant studies focused on the

US portion of the Salish Sea showed that juvenile Chinook captured in all urbanised rivers of Puget Sound contained PCBs at levels known to cause adverse health effects, with greater impacts on Chinook residing in central and southern Puget Sound for most of their lives. Further, there are very high levels of PBDEs in juvenile Chinook in several systems in Puget Sound that are of concern. Less is known about the impacts of PCBs and PBDEs on juvenile coho. PBDEs were found in steelhead in the Nisqually River at levels high enough to cause adverse effects.

Insufficient contaminant data currently exist for the Strait of Georgia, given the historic lack of a focused contaminant program in this region, as well as the impacts of Chemicals of Emerging Concern such as personal care products and household chemicals.

Cumulative Effects

In combination, factors can have additive, synergistic or dampening effects on juvenile salmon survival. It is clear that no one factor caused the decline in Salish Sea salmon marine survival, but it is notoriously difficult to assess cumulative effects. A qualitative network model produced by Sobocinski et al. (2018) suggested that anthropogenic impacts result in the strongest negative responses in salmon survival and abundance.

Moving forward, there are a number of tools available to try to tease apart the impacts of the many cumulative impacts operating on our Salish Sea marine ecosystems. These include:

- Two SSMSP-related ecosystem models may help further assess cumulative effects. These include an Ecopath with Ecosim model[2] developed by the University of British Columbia and an Atlantis model (Perryman et al. 2023) developed by the National Oceanic and Atmospheric Administration and Long Live the Kings.[3] These models are a critical part of the toolbox for supporting ecosystem-based management of fisheries and marine resources. Due to the uncertainty in understanding complex natural systems with limited data, using multiple models to evaluate and inform policy choices and management decisions is an emerging best practice.

2 See: goldford.github.io/
3 See: lltk.org/

- The Watershed Futures Initiative,[4] spearheaded by Dr Jonathon Moore, Simon Fraser University. This initiative aims to improve the management of cumulative effects by carrying out research on pathways of effects, establishing benchmarks and thresholds for harm to salmon, expanding on risk assessment and other cumulative effects assessment tools and improving networking among federal and provincial managers, First Nations community leaders, policy makers and regulatory agencies.
- Risk assessment methodologies such as the Committee on the Status of Endangered Wildlife in Canada Threats Calculator[5] and the Risk Assessment Methodology for Salmon developed by Dr Kim Hyatt[6] and colleagues at DFO. The latter uses life cycle modelling to rank the key limiting factors and assess cumulative impacts affecting the different life stages of Pacific salmon, as well as to prioritise possible actions.

Next Steps for Salmon Recovery?

Recovering our Salish Sea salmon is urgently required. Scientific research is ongoing and continues to further our understanding of impacts on salmon marine survival, but several recommendations from SSMSP provide a path forward for salmon recovery in the Salish Sea. Among others, recommendations include:

- Protect and restore estuary and nearshore habitat to benefit not only salmon but also their prey. This includes efforts to ensure connectivity within and among marsh, eelgrass and kelp habitats; accounting for climate change; and residential soft-shore initiatives that reduce artificial shoreline armour, promote healthy shorelines and improve forage fish spawning habitat.
- Recover herring populations, focusing on both abundance and diversity.
- Build resilience in salmon by promoting diversity. This may include restoration of habitat and experimenting with hatchery-rearing strategies that vary the timing when fish are released into the environment. These actions could make salmon more resilient to mismatches with prey and help reduce competition, disease and predation.

4 See: www.watershedfuturesinitiative.com/

5 See: www.cosewic.ca/index.php/en/reports/preparing-status-reports/threats-classification-assessment-calculator-status-report.html

6 Hyatt et al., paper in preparation.

- Investigate strategies to reduce seal predation, such as reducing barriers to salmon and steelhead migration that allow increased predation; removing log booms and obstructing the use of other seal haulouts; using predator deterrents; and restoring estuary and nearshore habitat that provides protection for salmon.
- Reduce toxic contamination in locations that have the greatest impact on salmon and steelhead.
- Optimise fish health in hatcheries. The Strategic Salmon Health Initiative[7] developed a salmon FIT-CHIP, a new genomic technology for rapidly assessing the health and physiology of young salmon which can be used in the hatchery environment to make sure that fish are released at the best time and in the best health.
- Protect and manage river flows to reduce predation of out-migrating salmon smolts and returning adults.

Determining the suite and sequence of actions required for successful restoration of Pacific salmon requires decision-making support tools that assist in prioritising and optimising the order of approaches to be taken. In this realm, decision support tools such as the Priority Threat Management Framework, developed by Dr Tara Martin and colleagues at the University of British Columbia, can be exceedingly useful (e.g. Carwardine et al. 2019). This framework is a decision tool that identifies the threat management strategies that will recover the most species for the least cost. The approach helps to answer questions around resource allocation and prioritisation of conservation approaches and management strategies.

Ultimately, actions taken are most likely to be successful when there is engagement of communities, local stewards and First Nations, as well as government agencies and academia. Much can be learned from ongoing research such as in Dr Anne Salomon's lab[8] at Simon Fraser University, where focus is on how to enable collaborative research and constructive dialogue among stakeholders and to find solutions which consider trade-offs between coastal conservation and resource use. A similar approach through the Puget Sound Partnership (2022) in Washington State encourages the collaborative efforts of regional, tribal and local partners to recover salmon populations throughout Puget Sound.

7 See: psf.ca/work/habitat/sshi/

8 See: www.sfu.ca/rem/about/people/salomon.html

Conclusion

The Salish Sea was once abundant with wild Pacific salmon, which supported fresh and saltwater ecosystems, thriving fisheries and Indigenous cultures. Beginning in the late 1970s, marine survival rates for Chinook, coho and steelhead sharply declined. This state of salmon within the Salish Sea continues to be a major concern on both sides of the border and throughout the island populations of the Salish Sea, who have noted major changes in the region over the past 30 years. However, past efforts to reduce harvest, restore habitat and improve hatchery practices have not led to recovery.

The Salish Sea Marine Survival Project was an international collaborative research project around the Salish Sea to understand a key question: what is limiting the survival of juvenile salmon and steelhead in the Salish Sea? The findings paint a complex picture of the interrelated factors at play in this critical early stage of the salmon life cycle. The body of evidence points to changes in the food web—both the availability of food for salmon, and the increasing impacts of salmon predators—as the largest contributors to marine survival, with estuarine habitat loss, pollution and disease also affecting local populations. There also are substantial concerns about the ongoing role of climate change in both the Salish Sea and north Pacific Ocean and how these changing conditions will continue to impact salmon.

The project provided an example of how a transboundary partnership can stimulate new investments into salmon research, result in the development of novel technologies, provide a platform for collaborative and integrative research and ultimately result in implementation of strategic management actions cross-border.

Another significant benefit of the SSMSP has been the encouragement and provision of a foundation for long-term monitoring of the Salish Sea and salmon health. Some of the outcomes include: an augmented DFO and new, collaborative Puget Sound zooplankton sampling programs; increased Washington State and DFO focus on seal and sea lion populations and their diets; and a significant expansion of nearshore habitat restoration, research and marine debris removal particularly around the Salish Sea Islands.

The SSMSP successfully built a strong salmon network of professional and community scientists to undertake the most comprehensive study of salmon in the Salish Sea marine ecosystem conducted to date. It is hoped that actions stemming from this program will ultimately benefit Chinook, coho, steelhead and the orca whales; Tribes and First Nations, and others who depend on and value salmon.

References

Allen, S.E. and M.A. Wolfe, 2013. 'Hindcast of the Timing of the Spring Phytoplankton Bloom in the Strait of Georgia, 1968–2010.' *Progress in Oceanography* 115: 6–13. doi.org/10.1016/j.pocean.2013.05.026

Araki, H., B.A. Berejikian, M.J. Ford and M.S. Blouin, 2008. 'Fitness of Hatchery-Reared Salmonids Fish in the Wild.' *Evolutionary Applications* 1(2): 342–55. doi.org/10.1111/j.1752-4571.2008.00026.x

Beamish, R.J. and C. Mahnken, 2001. 'A Critical Size and Period Hypothesis to Explain Natural Regulation of Salmon Abundance and the Linkage to Climate and Climate Change.' *Progress in Oceanography* 49: 423–37. doi.org/10.1016/S0079-6611(01)00034-9

Beamish, R.J., C. Mahnken and C.-E. Neville, 2004. 'Evidence That Reduced Early Marine Growth Is Associated with Lower Marine Survival of Coho Salmon.' *Transactions of the American Fisheries Society* 133(1): 26–33. doi.org/10.1577/T03-028

Beamish, R.J., R. Sweeting, K.L. Lange, D. Noakes, D. Preikshot and C.-E. Neville, 2010. 'Early Marine Survival of Coho Salmon in the Strait of Georgia Declines to Very Low Levels.' *Marine and Coastal Fisheries: Dynamics, Management, and Ecosystem Science* 2: 424–39. doi.org/10.1577/C09-040.1

Beamish, R.J., R. Sweeting, C.-L. Neville, K.L. Lange, T. Beacham and D. Preikshot, 2012. 'Wild Chinook Salmon Survive Better Than Hatchery Salmon in a Period of Poor Production.' *Environmental Biology of Fishes* 94: 135–48. doi.org/10.1007/s10641-011-9783-5

Bilby, R.E., B.R. Fransen and P.A. Bisson, 1996. 'Incorporation of Nitrogen and Carbon from Spawning Coho Salmon into the Trophic System of Small Streams: Evidence from Stable Isotopes.' *Canadian Journal of Fisheries and Aquatic Sciences* 53: 164–73. doi.org/10.1139/f95-159

Busch, D., C. Shallin, J. Harvey and P. McElhany, 2013. 'Potential Impacts of Ocean Acidification on the Puget Sound Food Web.' *ICES Journal of Marine Science* 70: 823–33. doi.org/10.1093/icesjms/fst061

Carwardine, J, T.G. Martin, J. Firn, R.P. Reyes, S. Nicol, A. Reeson, H.S. Grantham et al., 2019. 'Priority Threat Management for Biodiversity Conservation: A Handbook.' *Journal of Applied Ecology* 56(2): 481–90. doi.org/10.1111/1365-2664.13268

Cederholm, C., H. Johnson, R.E. Bilby, L.G. Dominguez, A.M. Garrett, W.H. Graeber, L.E. Greda et al., 2000. 'Pacific Salmon and Wildlife: Ecological Contexts, Relationships, and Implications for Management.' Olympia, Washington: Department of Fish and Wildlife (Special Edition Technical Report, Prepared for D.H. Johnson and T.A. O'Neil).

Crozier, L.G., M.M. McClure, T. Beechie, S.J. Bograd, D.A. Boughton et al., 2019. 'Climate Vulnerability Assessment for Pacific Salmon and Steelhead in the California Current Large Marine Ecosystem.' *Plos One* 14(7): e0217711. doi.org/10.1371/journal.pone.0217711

Davis, M.J., J. Chamberlin, J.R. Gardner, K.A. Connelly, M.M. Gamble, B.R. Beckman and D.A. Beauchamp, 2020. 'Variable Prey Consumption Leads to Distinct Regional Differences in Chinook Salmon Growth During the Early Marine Critical Period.' *Marine Ecology Progress Series* 640: 147–69. doi.org/10.3354/meps13279

Field, R.D. and J.D. Reynolds, 2012. 'Ecological Links Between Salmon, Large Carnivore Predation, and Scavenging Birds.' *Journal of Avian Biology* 44: 9–16. doi.org/10.1111/j.1600-048X.2012.05601.x

Gaydos, J.K. and S.F. Pearson, 2011. 'Birds and Mammals that Depend on the Salish Sea: A Compilation.' *Northwestern Naturalist* 92(2): 79–94. doi.org/10.1898/10-04.1

Gislason, G., E. Lam, G. Knapp and M. Guettabi, 2017. 'Economic Impacts of Pacific Salmon Fisheries.' Anchorage: Institute of Social and Economic Research, University of Alaska (Report prepared for the Pacific Salmon Commission by G.S. Gislason & Associates, Ltd).

Gouin, T., J.M. Armitage, I.T. Cousins, D.C.G. Muir, C.A. Ng, .L.R. Reid and S. Tao, 2013. 'Influence of Global Climate Change on Chemical Fate and Bioaccumulation: The Role of Multimedia Models.' *Environmental Toxicology and Chemistry* 32: 20–31. doi.org/10.1002/etc.2044

Grant, S., B.L. MacDonald and M.L. Winston, 2019. *State of Canadian Pacific Salmon: Responses to Changing Climate and Habitats*. Nanaimo: Department of Fisheries and Oceans.

Greene, C., L. Kuehne, C. Rice, K. Fresh and D. Penttila, 2015. 'Forty Years of Change in Forage Fish and Jellyfish Abundance Across Greater Puget Sound, Washington (USA): Anthropogenic and Climate Associations.' *Marine Ecology Progress Series* 525: 153–70. doi.org/10.3354/meps11251

Groot, C. and L. Margolis, 1991. *Pacific Salmon Life Histories*. Vancouver: University of British Columbia Press.

Hanson, M.B., C.K. Emmons, M.J. Ford, M. Everett, K. Parsons, L.K. Park and L. Barre, 2021. 'Endangered Predators and Endangered Prey: Seasonal Diet of Southern Resident Killer Whales.' *Plos One* 16(3): e0247031. doi.org/10.1371/journal.pone.0247031

Hildebrand, G.V., C.C. Schwatz, C.T. Robbins, M.E. Jacoby, T.A. Hanley, S.M. Arthur and C. Servheen, 1999. 'The Importance of Meat, Particularly Salmon, to Body Size, Population Productivity, and Conservation of North American Brown Bears.' *Canadian Journal of Zoology* 77: 132–38. doi.org/10.1139/z98-195

Kendall, N., G.W. Marston and M.W. Klungle, 2017. 'Declining Patterns of Pacific Northwest Steelhead Tout (*Oncorhynchus mykiss*) Adult Abundance and Smolt Survival in the Ocean.' *Canadian Journal of Fisheries and Aquatic Sciences* 74(8): 1–16. doi.org/10.1139/cjfas-2016-0486

Meador, J.P., 2014. 'Do Chemically Contaminated River Estuaries in Puget Sound (Washington, USA) Affect the Survival Rate of Hatchery-Reared Chinook Salmon? Pacific Northwest Steelhead Trout (*Oncorhynchus mykiss*) Abundance and Smolt Survival in the Ocean.' *Canadian Journal of Fisheries and Aquatic Sciences* 71(1): 162–80. doi.org/10.1139/cjfas-2013-0130

Nelson, B.W., C.J. Walters, A.W. Trites and M.K. McAllister, 2019. 'Wild Chinook Salmon Productivity is Negatively Related to Seal Density and Not Related to Hatchery Releases in the Pacific Northwest.' *Canadian Journal of Fisheries and Aquatic Sciences* 76(3): 447–62. doi.org/10.1139/cjfas-2013-0130

NOAA (National Oceanic and Atmospheric Administration), 2019. 'Pacific Salmon Life Cycle Poster'. Viewed 14 July 2023 at: fisheries.noaa.gov/resource/educational-materials/pacific-salmon-life-cycle-poster

Pearsall, I., M. Schmidt, I. Kemp and B. Riddell, 2021. 'Factors Limiting Survival of Juvenile Chinook Salmon, Coho Salmon and Steelhead in the Salish Sea: Synthesis of Findings of the Salish Sea Marine Survival Project. SSMSP Program Report.' Viewed 25 June 2023 at: marinesurvival.wpengine.com/wp-content/uploads/2021PSF-SynthesisPaper-Screen.pdf

Perryman H.A., I.C. Kaplan, J.L. Blanchard, G. Fay, S.K. Gaichas, V.L. McGregor, H.N. Morzaria-Luna, J. Porobic, H. Townsend and E.A. Fulton, 2023. 'Atlantis Ecosystem Model Summit 2022: Report from a Workshop.' *Ecological Modelling* 483: 110442. doi.org/10.1016/j.ecolmodel.2023.110442

Puget Sound Partnership, 2021. 'Contaminants in Juvenile Salmon.' Viewed 18 July 2023 at vitalsigns.pugetsoundinfo.wa.gov/VitalSignIndicator/Detail/49

——, 2022. 'Salmon Recovery in Puget Sound.' Viewed 25 June 2023 at: psp.wa.gov/salmon-recovery-overview.php

Riche, O., S.C. Johannessen and R.W. Macdonald, 2014. 'Why Timing Matters in a Coastal Sea: Trends, Variability and Tipping Points in the Strait of Georgia, Canada.' *Journal of Marine Systems* 131: 36–53. doi.org/10.1016/j.jmarsys.2013.11.003

Robinson, C., D. Hryncyk, V. Barrie and J.F. Schweigert, 2013. 'Identifying Subtidal Burying Habitat of Pacific Sand Lance (*Ammodytes hexapterus*) in the Strait of Georgia, British Columbia, Canada.' *Progress in Oceanography* 115: 119–28. doi.org/10.1016/j.pocean.2013.05.029

Ruff, C.P., J.H. Anderson, I.M. Kemp, N.W. Kendall, P.A. McHugh, A. Velez-Espino, C.M. Greene, M. Trudel, C. Holt and K.E. Ryding, 2017. 'Salish Sea Chinook Salmon Exhibit Weaker Coherence in Early Marine Survival Trends than Coastal Populations.' *Fisheries Oceanography* 26(6): 625–37. doi.org/10.1111/fog.12222

Selleck, R., C.F. Gibson, S. Shull and J.H. Gaydos, 2015. 'Nearshore Distribution of Pacific Sand Lance (*Ammodytes personatus*) in the Inland Waters of Washington State.' *Northwestern Naturalist* 96(3): 185–95. doi.org/10.1898/1051-1733-96.3.185

Sobocinski, K., C.M. Greene, J.H. Anderson, N.W. Kendall, M. Schmidt, M.S. Zimmerman, I.M. Kemp, S. Kim and C.P. Ruff, 2021. 'A Hypothesis-Driven Statistical Approach to Identifying Ecosystem Indicators of Coho and Chinook Salmon Marine Survival.' *Ecological Indicators* 124: 107403. doi.org/10.1016/j.ecolind.2021.107403

Sobocinski, K., C.M. Greene and M. Schmid, 2018. 'Using a Qualitative Model to Explore the Impacts of Ecosystem and Anthropogenic Drivers Upon Declining Marine Survival in Pacific Salmon.' *Environmental Conservation* 45(3): 278–90. doi.org/10.1017/S0376892917000509

Tatters, A.O., L.J. Flewelling, F. Fu, A.A. Granholm and D.A. Hutchins, 2013. 'High CO_2 Promotes the Production of Paralytic Shellfish Poisoning Toxins by *Alexandrium catenella* From Southern California Waters.' *Harmful Algae* 30: 37–43. doi.org/10.1016/j.hal.2013.08.007

Therriault, T.E. Hay and J.F. Schweigert, 2009. 'Biological Overview and Trends in Pelagic Forage Fish Abundance in the Salish Sea (Strait of Georgia, British Columbia)'. *Marine Ornithology* 37: 3–8.

Willson, M.F. and K.C. Halupka, 1995. 'Anadromous Fish as Keystone Species in Vertebrate Communities.' *Conservation Biology* 9: 489–97. doi.org/10.1046/j.1523-1739.1995.09030489.x

Zimmerman, M.S., J.R. Irvine, M. O'Neill, J.H. Anderson, C.M. Greene, J. Weinheimer, M. Trudel and K. Rawson, 2015. 'Spatial and Temporal Patterns in Smolt Survival of Wild and Hatchery Coho Salmon in the Salish Sea.' *Marine and Coastal Fisheries* 7(1):116–34. doi.org/10.1080/19425120.2015.1012246

17

Water: Sources, Supply and Governance

R.D. (Dan) Moore, Diana M. Allen, Oliver Brandes and Randy Christensen

The San Juan Islands in Washington and the Southern Gulf Islands in British Columbia experience similar climate and geological characteristics, and are not accessible by bridges from the mainland, lending these islands a unique socio-economic character. The chapter provides an overview of the hydroclimatic and geological contexts to provide a basis for describing the availability of water and issues associated with both limited quantity and impaired quality. The chapter concludes with an overview of water governance, highlighting the different institutional approaches between the Canadian Gulf Islands and the American San Juan Islands.

Hydroclimatic Context

Winter weather in the Salish Sea Islands is dominated by the passage of frontal storms that originate in the Pacific Ocean and bring cool, wet weather, while summer weather is dominated by high-pressure systems associated with fair weather and relatively light winds (Moore et al. 2010; see also Hebda, this volume). The Salish Sea Islands have a warm-summer Mediterranean climate, with mild, rainy winters and warm and relatively dry summers. The islands fall into the Coastal Douglas fir bio-geoclimatic zone.

Relatively few climate stations are maintained in the Salish Sea Islands. The longest record is for the Olga 2 SE station on Orcas Island, which extends back to 1891. Other records are shorter, typically beginning after the mid-twentieth century. All records contain gaps and/or missing observations.

As shown in Table 17.1, there is no clear spatial pattern for mean annual air temperature, but precipitation tends to increase from south to north. In addition to a south–north trend, precipitation is subject to substantial orographic influences. For example, the Washington Department of Ecology (1975) estimated that mean annual precipitation on the eastern portion of Orcas Island increases from about 760 mm near sea level to 1,140 mm at the highest points. Unfortunately, no long-term climate stations are maintained at higher elevations, so the magnitude of the orographic influence on different islands is unknown.

Table 17.1 Climate station locations and mean annual precipitation and air temperature

Station name	Latitude	Longitude	Elevation (m above sea level)	P (mm)	T (°C)
Gabriola Island	49° 09'	123° 44'	46	958	9.6
Salt Spring Is. St. Mary's Lake	48° 53'	123° 32'	46	987	10.7
Mayne Island	48° 50'	123° 19'	28	842	10.2
Salt Spring Is. Cusheon Lake	48° 49'	123° 28'	108	1,071	8.9
Saturna CAPMON	48° 46'	123° 1'9	178	812	10.2
Olga 2 SE	48° 37'	122° 48'	8	715	10.2
Friday Harbor	48° 31'	123° 1'	17	634	9.8

Notes: Values are normals for 1981–2010 for all stations except Friday Harbour, for which the normals are for 1971–2000. Stations are arranged from north at the top to south at the bottom.

Source: Authors' tabulation.

Air temperature remains above freezing most of the time, even through winter (Figure 17.1). As a result, snowfall accounts for less than 5 per cent of annual precipitation, except at higher elevation areas such as those on Orcas Island.

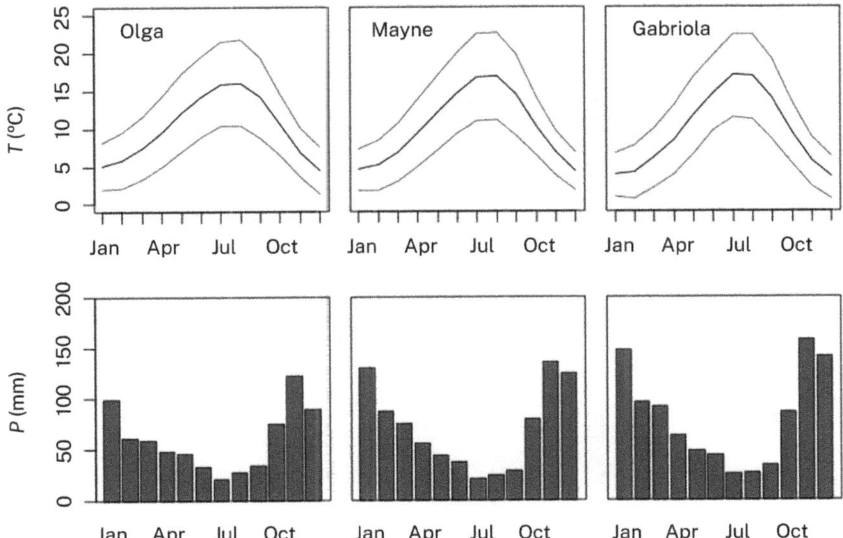

Figure 17.1 Climate normals based on the 1981–2010 period for three stations

Note: The top row shows mean daily maximum and minimum (upper and lower grey lines) and daily mean (black) air temperatures by month; the bottom row is mean total precipitation.

Source: R.D. Moore, Diana M. Allen, Oliver Brandes and Randy Christensen.

As seen in Figure 17.2, the smoothed lines through the data suggest a slight dip in winter precipitation in the 1920s and another drop toward the end of the 1970s, both of which are consistent with shifts in the Pacific Decadal Oscillation (Whitfield et al. 2010). Neither winter precipitation nor summer precipitation exhibits a statistically significant trend (using the Mann-Kendall test), but the increase in summer air temperature, especially evident since the 1960s, is highly significant.

Although the Salish Sea Islands generally experience the same weather systems, climatic conditions vary among sites due to topographic effects such as rain shadows, which vary with storm direction. Table 17.2 shows that summer precipitation variations are highly correlated among the Mayne Island, Salt Spring Island St Mary's Lake and Gabriola climate stations, with moderate correlations between these stations and those at Olga 2 SE on Orcas Island.

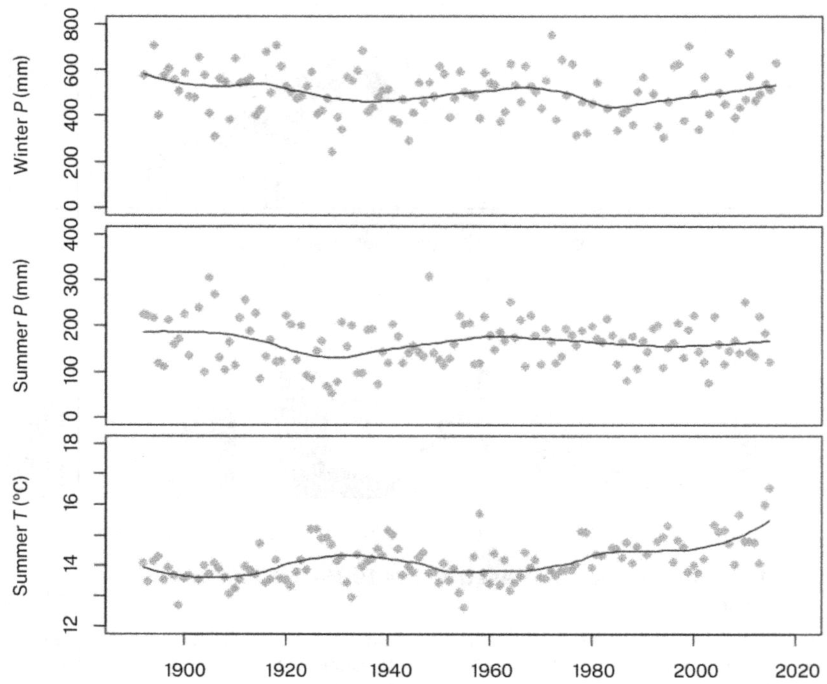

Figure 17.2 Time series of precipitation and temperature for Olga 2 SE, Orcas Island

Note: The black lines are smoothed versions of the time series generated using the loess algorithm.

Source: R.D. Moore, Diana M. Allen, Oliver Brandes and Randy Christensen.

Table 17.2 Correlation coefficients among the time series of summer (May–September) precipitation

	Mayne	Salt Spring	Gabriola
Olga 2 SE	0.72	0.71	0.64
Mayne	–	0.88	0.86
Salt Spring	–	–	0.89

Notes: Precipitation for climate stations on Mayne Island; Salt Spring St Mary's Lake, Gabriola Island; and Olga 2 SE on Orcas Island. A correlation coefficient of 1 indicates perfect correlation.

Source: Authors' tabulation.

Because the Salish Sea Islands receive little snowfall, their water resources are controlled by the quantity and timing of rainfall and its partitioning into soil moisture storage, evapotranspiration, surface runoff and groundwater recharge. Seasonal patterns of these water balance components were estimated for Olga 2 SE station using the Thornthwaite and Mather (1957) approach, with an assumed value of 100 mm for the maximum value of soil moisture storage (Figure 17.3). Although the data are from Olga 2 SE, the general seasonal pattern should be applicable throughout the Salish Sea Islands.

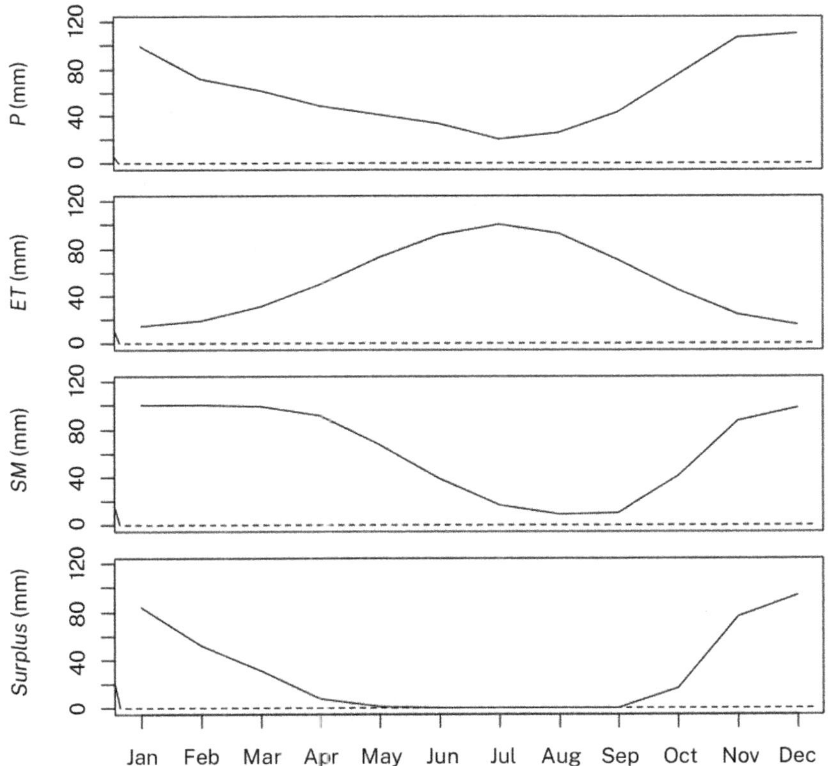

Figure 17.3 Mean monthly water budget components for Olga 2 SE, Orcas Island

Note: P is precipitation, ET is evapotranspiration, SM is soil moisture and Surplus is water that is available to become surface runoff or groundwater recharge.

Source: R.D. Moore, Diana M. Allen, Oliver Brandes and Randy Christensen.

During winter, abundant rainfall and low evapotranspiration maintain soil moisture at its maximum value. Up to about 80 per cent of the rainfall becomes surplus, which is available to become surface runoff or groundwater recharge. By March, with declining precipitation, the increasing evapotranspiration reduces the surplus and begins drawing down soil moisture storage. From May to September, evapotranspiration consumes all available precipitation and also draws down soil moisture, leaving no surplus. Through autumn, increasing precipitation and decreasing evapotranspiration returns soil moisture back to its maximum value and also increases the amount of rainfall that becomes surplus. Thus, water availability in the Salish Sea Islands is highly seasonal, with abundant surplus during winter and virtually no surplus during summer months, when demand increases.

Projections of future climatic conditions based on global climate models are subject to considerable uncertainty that is related to uncertainties in the models' ability to simulate climatic processes, as well as uncertainty about future changes to emissions of greenhouse gases and land cover change. The range of uncertainty can be expressed by comparing the outputs of multiple models and multiple future emissions scenarios, as in Table 17.3.

Table 17.3 Projected changes in mean air temperature and total precipitation for 2070–2099

Season	Precipitation change (%)	Air temperature change (°C)
Winter (DJF)	-5 to 32	0.8 to 4.6
Spring (MAM)	-7 to 36	0.7 to 3.7
Summer (JJA)	-46 to 5	1.6 to 6.3
Autumn (SON)	-5 to 42	0.8 to 4.6

Notes: Based on multiple climate models and emissions scenarios, relative to baseline conditions from 1961–1990. Values are the median changes listed for the 'PCIC (Pacific Climate Impacts Consortium) Planners Scenario' (PCIC 2016) based on the Intergovernmental Panel on Climate Change (IPCC) Fourth Assessment Report emissions scenarios for a custom polygon encompassing the Salish Sea Islands (IPCC 2007).

Source: Authors' tabulation.

Based on the projected changes for the region containing the Salish Sea, winters should be warmer and wetter than at present, while summers will likely be warmer and drier. The projections for summer are relatively consistent, indicating both warmer and drier conditions. Autumn and spring are both likely to be warmer and wetter.

Because snowfall is currently a minor portion of precipitation in the Salish Sea Islands, the main effect of future climatic changes would be primarily through changes in rainfall, with secondary effects associated with changes in evapotranspiration. Thus, it is likely that water surplus will increase in winter, and possibly also in spring and autumn, although increased air temperature and associated increases in evapotranspiration could offset that effect to some extent, especially in the shoulder seasons. Summer drought would become more extreme, resulting in further drawdown of soil moisture storage and increased evapotranspiration.

Surface Water

Many of the smaller islands lack perennial surface water bodies, with the only surface water being associated with seasonally wet areas located in depressions that dry out through the spring and summer. Most of the larger islands, however, contain lakes, streams and perennial wetlands. Beaver activity has resulted in higher water levels and/or greater extents of some wetlands, such as Laughlin Lake and the Great Beaver Swamp on Galiano Island.

Streamflow and water level records are scarce. The US Geological Survey (USGS) maintained three gauges on Lopez Island and one each on Orcas, San Juan and Shaw islands in the period 1996 to 1998. Water Survey of Canada (WSC) maintained gauges on five streams in the Gulf Islands, three of which only collected data from April through November. WSC also maintained gauges at six lakes on Salt Spring Island and one each on Pender and Gabriola. No gauges are currently maintained by WSC or USGS in the Salish Sea Islands.

Figure 17.4 illustrates streamflow for Cusheon Creek on Salt Spring Island, which has the longest and most complete record in the Salish Sea Islands. Consistent with the climatic water balance, relatively high flows and lake levels occur between November and April, with a decline through summer and autumn, often to zero-flow conditions. The general seasonal pattern at Cusheon Creek is broadly representative of streams throughout the Salish Sea Islands.

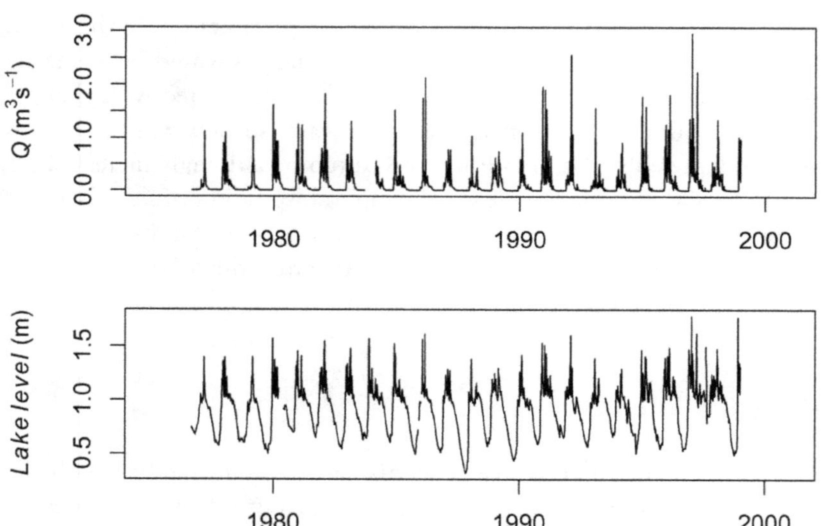

Figure 17.4 Streamflow (Q) and lake level at Cusheon Lake, Salt Spring Island

Source: R.D. Moore, Diana M. Allen, Oliver Brandes and Randy Christensen.

As shown in Figure 17.5, streamflow drops to summer baseflow levels by July each year and is negligible from August through October in most years, sometimes even through November. Only four years between 1977 and 1998 had continuous flow through the entire summer season, and zero-flow conditions were recorded 43 days each year, on average, with a maximum of 108 days recorded in 1985.

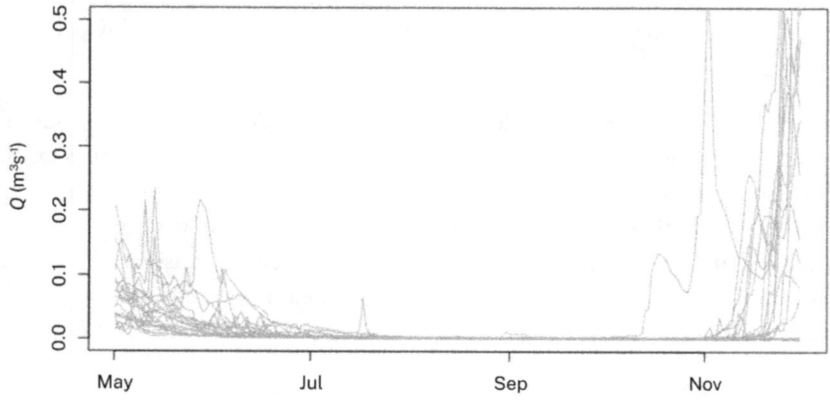

Figure 17.5 Warm-season streamflow (Q) for Cusheon Creek, 1977 to 1998

Note: Each line represents data for an individual year.

Source: R.D. Moore, Diana M. Allen, Oliver Brandes and Randy Christensen.

In addition to limited supply of surface water sources, there are increasing concerns about water quality. For example, St Mary's Lake on Salt Spring Island is increasingly impacted by cyanobacteria. The district, the Capital Regional District, the Ministry of Environment and Climate Change Strategy, the Ministry of Water, Land and Resource Stewardship, the Islands Trust, the Salt Spring Island Watershed Protection Authority (WPA), the Salt Spring Water Council, the Salt Spring Water Preservation Society and many concerned residents have worked and continue to work to improve water quality in St Mary and other island lakes.

One initiative of the Salt Spring Island WPA is the FreshWater Catalogue (FWC). The objectives of this citizen science project are to raise community awareness of the island's water diversity, help develop island watershed communities, support data gathering pertaining to the quantity and quality of the island's water resources and help inform island watershed preservation/management activities. The FWC currently includes over 3,000 monitoring points along streams where year-round measurement of flow, temperature and electrical conductivity are made by local volunteers across multiple sites. The dataset will inform planned water budget analyses across the island. The project also supports the ongoing Xwaaqw'um (Burgoyne Bay) watersheds restoration work, managed by the Quw'utsun Stqeeye' Learning Society, with the aim of characterising potential changes in surface flow conditions and groundwater contributions, and changes in flow seasonality and temperature that may result from watershed restoration efforts.

Despite the occurrence of extended low flows during summer and autumn, many streams support or have the potential to support a range of fish species. For example, records in the fisheries inventory maintained by the British Columbia Government indicate that Greig Creek on Galiano Island could maintain a run of coho salmon and that there are historical records of coho salmon in Laughlin Lake, which feeds Greig Creek (Government of British Columbia 2017). Kwiáht, a non-profit conservation biology research laboratory, has compiled observations of fish in streams on Orcas and San Juan Islands, including at least two significant populations of coastal cutthroat trout, along with a number of small, possibly transient occurrences of coastal cutthroat, rainbow trout and chum salmon (Kwiáht 2017).

Unfortunately, many stream habitats have been degraded through land development, water extraction and other human activity. Duck Creek on Salt Spring Island provides an example. Fish species observed in Duck

Creek include coastal cutthroat trout, coho salmon, chum salmon, three-spined stickleback, cutthroat trout, rainbow trout and steelhead. However, the provincial fisheries inventory records identify a number of impairments that have degraded habitat, including past agricultural use and removal of trees that provide shade. A number of enhancement activities have occurred, including construction of off-channel rearing areas, planting of riparian vegetation in agricultural areas, and the addition of boulder clusters, wood pieces and gravel. In addition, a continuous outflow from St Mary's Lake is maintained through the dry season to support a required minimum conservation flow (Kerr 2015).

Quw'utsun elders and knowledge keepers of the Stqeeye' Learning Society have taken an active stewardship role with research, monitoring and mapping within the Xwaaqw'um (Burgoyne Bay) watershed. Their leadership, teachings and work includes specialised ecological and educational values needed in a time of climate change and biodiversity collapse. This cultural work and witnessing has included workshops and learning days for multiple school classes.

Groundwater

Groundwater on the Salish Sea Islands is derived primarily from fractured bedrock, although localised deposits of sands and gravels may yield groundwater. Both the Nanaimo Group sedimentary rocks and the older igneous and metamorphic rocks have low primary porosity, and groundwater flows primarily through the network of fractures. Mackie et al. (2001) proposed a 'hydrostructural domain' conceptual model for representing the hierarchy of fracturing. Surrette and Allen (2008) later refined the domains and their relative permeabilities to include: (1) fracture zones, (2) highly fractured interbedded sandstone and mudstone, and (3) less fractured sandstone, as shown in Figure 17.6.

Recharge is generally higher in areas underlain by porous coarse-grained deposits (gravel and sand) than in areas underlain by fine-grained deposits (silt and clay), which limit water infiltration.

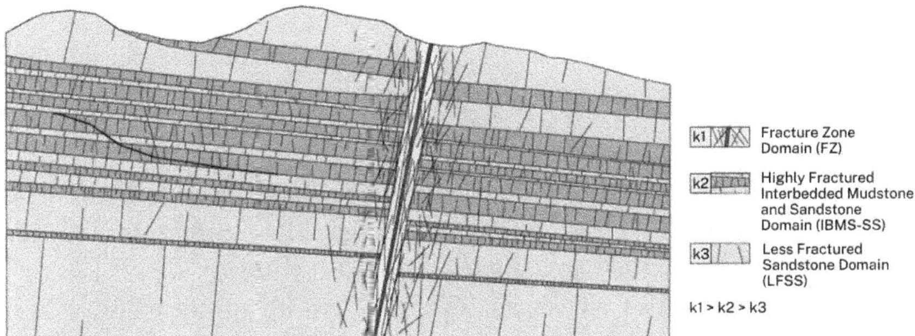

Figure 17.6 Hydrostructural domain conceptual model for the Nanaimo Group

Note: Figure shows, in order of decreasing permeability (k): fracture zone domain (FZ), highly fractured interbedded mudstone and sandstone domain (IBMS-SS), and less fractured sandstone domain (LFSS)

Source: Surrette and Allen (2008).

In bedrock aquifers, with the onset of the wet season (October or November), the low storage capacity of the fractures in the bedrock results in rapid filling, with a consequent rapid water table rise as shown in Figure 17.7. Between January and April, recharge and discharge are roughly in balance, resulting in a relatively steady water table. At the end of the wet season, the recharge rate declines, and the water table lowers over the summer (April to September), until the wet season begins again (Burgess 2017).

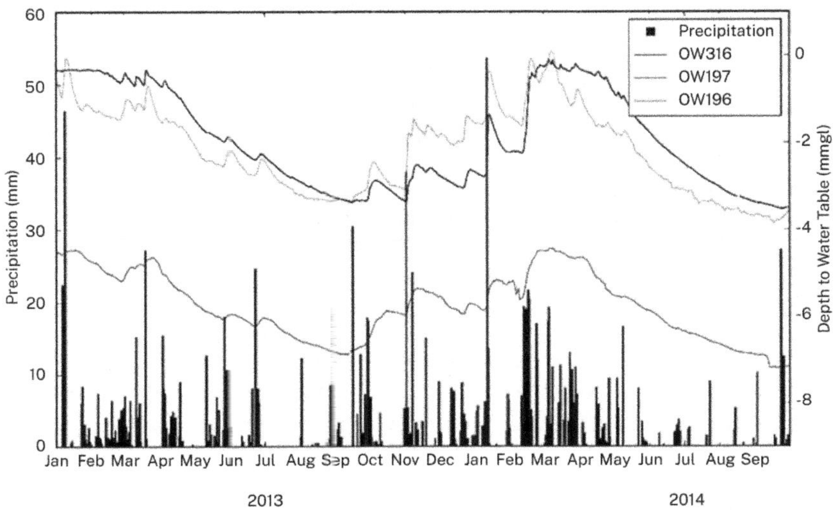

Figure 17.7 Seasonal groundwater level and precipitation variation for wells in British Columbia

Source: R.D. Moore, Diana M. Allen, Oliver Brandes and Randy Christensen.

Recharge and seepage areas are generally seasonally persistent in most areas. The magnitude of recharge is much lower during the summer as indicated by the near-zero values over a larger portion of the island. Seepage during the summer is also somewhat lower.

For Gabriola Island (Table 17.4), estimated recharge from a hydrologic model for the period 1995 to 2005 averaged 191 mm/year (20 per cent of the mean annual precipitation (MAP) (Burgess 2017). Actual evapotranspiration (AET) averaged 398 mm/year (40 per cent of MAP), and runoff averaged 357 mm/year (36 per cent of MAP).

The model was run for future climate conditions using downscaled[1] climate model data. For 2070–2099, MAP increased by 113 mm/year (11 per cent), evapotranspiration increased by 23 mm/year (6 per cent), recharge increased by 47 mm/year (24 per cent), and runoff increased by 91 mm/year (25 per cent) (Burgess 2017).

The increases in the annual water balance components in future projections reflect the expected increases in the fall, winter and spring. During summer, precipitation, AET and recharge are projected to decrease slightly.

Table 17.4 Water balance generated for Gabriola Island

Water balance component	Historical range mm/ year (% of MAP)	Historical average mm/year (% of MAP)	2080s change in average mm/year (% of MAP)
Precipitation	790–1,253	984	+113 (11%)
Evapotranspiration	359–435 (35 to 55%)	398 (40%)	+23 (6%)
Runoff	235–476 (29 to 42%)	357 (36%)	+91 (25%)
Recharge	188–215 (17 to 26%)	191 (20%)	+47 (24%)

Notes: Historical climate and future projected climate for the 2070–2099 (2080s) time slice based on downscaled results from multiple climate models and the A2 emissions scenarios. The range reflects the spatial variability across the island and the average is the spatial average.

Source: Authors' tabulation.

Groundwater quality of the islands varies spatially and temporally due to a complex interplay between the geology and geological history of the islands, groundwater use, land use and proximity to the ocean.

1 Deriving patterns and changes from large scale climate projections to the local climate.

All islands have a natural 'freshwater lens' in which the pores or fractures are filled with fresh water. This freshwater lens essentially floats on top of saline water at depth. There is often a transition zone between the fresh water and the saline water, although what this transition zone looks like in the fractured bedrock setting of the Salish Sea Islands is unknown. Certainly, wells that are drilled too deep, particularly near the coast, may intersect groundwater in this transition zone or indeed the saline groundwater beneath. The water quality of these wells would indicate salinisation to some degree.

The position of the fresh water–saltwater interface on the Salish Sea Islands has not remained stable over time. During the Pleistocene, the islands were depressed by as much as 300 m below present sea level due to the weight of the overlying glaciers (Clague 1983), leaving most islands fully submerged. During this time, seawater entered the bedrock through the fractures (Allen and Liteanu 2006) and the freshwater lens all but disappeared. It wasn't until the islands re-emerged that fresh water began displacing the saltwater (freshening) to re-establish a new freshwater lens. The current chemical and isotopic composition of the groundwater provides evidence of this freshening process (Allen 2004; Allen et al. 2015).

Groundwater samples collected on various Gulf Islands show a diverse chemical composition, ranging from fresh water to saline water. Most deep wells and wells located near the coast are more saline compared to shallower wells and wells at higher elevations. Chemically, it is difficult to distinguish between salinity originating from remnant seawater from when the islands were submerged, saline water present in the natural saltwater wedge underlying the island, or perhaps (although unlikely) connate water.[2] The Nanaimo Group bedrock is of marine origin, and there may be pockets of trapped marine (connate) water that is saline. There is, however, a distinct chemical composition of groundwater impacted by seawater intrusion.

Seawater intrusion can arise from (1) gradual landward encroachment of the fresh water–saltwater interface, (2) inundation by overtopping of ocean water during storm overwash events (or tsunamis), or (3) excess pumping. Thus, salinisation of the freshwater lens can occur from a variety of directions (over the top, from the side or from below) and can impact the quality and quantity of fresh water.

2 Water trapped in the pores of a rock during the rock's formation.

Seawater intrusion has occurred locally in some areas on the Salish Sea Islands, particularly in coastal wells that have been drilled too deep and may have intersected the fresh water–saltwater interface, or in wells where pumping has resulted in upconing (Kelly 2005; Klassen et al. 2014).[3] Seawater intrusion has also occurred laterally along discrete fractures that connect a well to the ocean (Allen et al. 2002). During summer months, when precipitation is low and water demand is high, groundwater levels decline and the quality of the water further deteriorates (Allen and Suchy 2001). Continued development on these islands is leading to increased concern about the quantity of groundwater available for use, and the potential impacts of pumping and climate change on seawater intrusion.

Salt springs have been identified on several islands. These springs have a chemical composition similar to seawater. Many of these springs have been developed by resorts; however, to date, no studies have explored the origin of these springs in any detail.

Apart from salinity, fertilisers in agricultural areas, septic systems, sea spray and well disinfectants are other sources of contamination, which, in addition to chloride, may add other chemical constituents, such as nitrate, to groundwater. Iron and manganese concentrations (above the aesthetic[4] limits) are often elevated and hydrogen sulphide gas (and occasionally methane gas) is also reported. Finally, arsenic has been found above drinking water limits in isolated areas of the islands.

Microbiological contamination of groundwater on the Gulf Islands has not been extensively studied. Provincial water quality studies in British Columbia have occasionally tested for total coliform, faecal coliform and E-coli, which are found to range from <1 to >2,000 CFU (colony forming units)/100 ml (total coliform) on Salt Spring Island, with the majority of wells tested being <1 CFU/100 ml. But other than limited scope provincial sampling programs and local consulting studies, microbiological sampling has not been carried out with any rigour.

Shallow dug wells and drilled wells that do not have a surface seal (surface grout) are particularly at risk of contamination. A particular problem identified across all islands is the potential for inadequately constructed septic fields, given the shallow fractured bedrock. Homeowners are encouraged to

3 Upward movement of salt water in a cone-shaped manner from below a fresh water/saltwater interface, under the influence of freshwater pumping above the interface.

4 Meaning 'palatable'.

test for bacteria and retrofit wells. Iron and manganese concentrations are also elevated on many islands. While the aesthetic limits may be exceeded, the high concentrations are not considered a health threat. Rather, bacteria associated with iron and manganese often cause well clogging problems.

Water Supply

Water supply on the islands comprises a mix of surface water, groundwater and rainwater harvesting. Salt Spring Island contains several lakes, and Coomes (1979, cited in Barnett et al. 1993) estimated that 72 per cent of freshwater demand was supplied from surface water sources at that time, with the rest supplied by groundwater. In contrast, Galiano Island depends almost entirely on groundwater, although there are licensed withdrawals from several streams, mainly for irrigation or livestock purposes. Several market gardeners on Galiano have excavated ponds to supply irrigation needs. In the Gulf Islands as a whole, the majority of residents use groundwater from the bedrock aquifers as their primary source of fresh water (Denny et al. 2007). Many of these wells are private and serve a single household, although some wells serve two or more households and are managed as community water supplies in the Gulf Islands.

In the San Juan Islands, surface water is currently very limited (Washington Department of Ecology 2016), and groundwater is the dominant source of water. However, locally important supplies of fresh water are also found in the large and elevated lakes on Orcas, Blakeley and San Juan Islands (Washington Department of Ecology 1975). For example, Mountain Lake and Cascade Lake on Orcas, and Thatcher Lake and Blakeley Lake on Blakeley Island, are capable of supplying a moderately large quantity of water at all seasons of the year. Numerous streams issue from all sides of Mount Constitution Range on Orcas.

To the authors' knowledge, there are no statistics available on the numbers of households that harvest rainwater as their primary water source. However, anecdotal evidence indicates that rainwater harvesters are a significant minority and are especially prevalent in areas affected by seawater intrusion or where aquifer permeability is limited. Considering the seasonality of rainfall in the Salish Sea Islands, it is important that rainwater harvesting systems include sufficient storage to provide water supply through the dry summer months.

Water Governance and Management

British Columbia's Gulf Islands and Washington State's San Juan Islands are similar in their geologic and climatic features. However, the legal regimes and governance arrangements for fresh water differ between the two countries, despite both being heavily influenced by Western water law.

Water Law Regimes — British Columbia

British Columbia's colonial (Crown) water law regime is a direct result of European settlement, which emphasised settlement of land for development, agriculture and mining. Recent legislative reform—including the 2016 *Water Sustainability Act*—moves toward a more sustainably oriented water law system. However, the following foundational principles of Western water law continue to dominate the governance approach (Percy 2004; Brandes and Curran 2017):

- The Crown asserts ownership over water resources[5]
- Water is allocated and diverted on a first in time, first in right system[6]
- A requirement exists to 'use it or lose it' (beneficial use principle)
- Water is prioritised as a resource for economic development

The Canadian Constitution places primary responsibility for fresh water with the provincial governments (*Constitution Act 1867*). While surface water and groundwater are constitutionally vested in the provincial Crown, this ownership is potentially subject to aboriginal rights and title claims protected under s. 35 of the Canadian Constitution. In British Columbia, aboriginal rights and title claims to water have not yet been specifically recognised in court decisions and are not effectively factored into the existing water allocation and governance regime. Many commentators believe that

5 Assertion of ownership by the Crown is likely the most controversial aspect of British Columbia's water law regime. Lack of recognition of aboriginal water rights or effective acknowledgement of Indigenous water laws remains a significant outstanding concern. Indigenous people have been using water the longest and yet are not entitled to the oldest water rights, making the recent new *Water Sustainability Act* a missed opportunity to address this inequality. The province continues to assert Crown ownership over all water in British Columbia even with numerous outstanding claims over land and water in First Nations traditional territories throughout the province.

6 Although slightly different than the US concept of *prior appropriation*, which is predicated on the date of first use (or when water was first appropriated), the Canadian system is based on a *prior allocation* system, where the priority date is based on the date at which senior government granted permission (date of allocation).

it will only be a matter of time before this recognition is required through a combination of government policy or court decisions. And, although Canadian provinces play a primary role in freshwater management, water is often viewed as a multijurisdictional responsibility, stemming from overlaps and gaps in jurisdiction and the shared nature of this flow resource (Percy 2004; Brandes and Curran 2017).

In Canada, the federal government holds constitutional powers over a number of areas that directly or indirectly affect fresh water. The formal federal role largely deals with things that 'touch' water (e.g. boats, fish, bridges and dams) rather than water itself. These powers include sea coast and inland fisheries, navigation and shipping, and international or interprovincial matters including transboundary treaties or agreements (*Constitution Act 1867*). However, over the last three decades the federal government has retreated quite substantially from any meaningful role in water management or governance in Canada (Brandes 2005; Pentland and Wood 2013; Curran 2015).

Because of the unique attributes of the Gulf Islands and the fragile availability of water, distinct legal and governance arrangements have been created for the region. The key local governing body in the Gulf Islands is the Islands Trust,[7] which consists of 12 local trust committees (for individual islands or groups of islands) and the Bowen Island Municipality (situated outside the Gulf Islands). The Islands Trust is a federation of local governments that plans land use and regulates development in the trust area. It also manages land trust assets on behalf of the 25,000 permanent residents and 10,000 non-resident landowners (Government of British Columbia n.d.).

Local governments in British Columbia play an important role in water management, including conservation, source protection and drinking water provision. Local trust committees in the Gulf Islands also play a specific role in some aspects of water governance and groundwater protection. These committees control development through official community plans, land use planning, zoning development permit areas, bylaw enforcement and subdivision servicing bylaws.

7 The Islands Trust was established under the *Islands Trust Act* (1974): 'to preserve and protect the trust area and its unique amenities and environment for the benefit of the residents of the trust area and of British Columbia generally, in cooperation with municipalities, regional districts, improvement districts, other persons and organizations and the government of British Columbia' (s. 3).

Gulf Islands' residents generally recognise water as an important issue: a recent survey indicates that over 40 per cent of Islanders are concerned about household water supply failures. Local trust committees' planning and priorities reflect these concerns, particularly with respect to groundwater. To protect groundwater, the Islands Trust (directly or through local trust committees) can investigate, regulate, control land use for and engage in long-range planning for water use, water source protection and development location and intensity (Nichols 2014). Some islands address water protection through development permit areas that control for local circumstances including requirements for water storage (ibid.).

A recent initiative is the development of the Islands Trust's Freshwater Sustainability Strategy. The process included the creation of a cultural knowledge holders advisory group. Currently, three cultural knowledge holders representing different parts of the Islands Trust region are involved, with the goal of integrating what is learned about the connection between water and the indigenous cultural heritage in the area into the historical and cultural context of the strategy.

A number of key provincial policies and laws directly or indirectly influence water in British Columbia, including: Living Water Smart—British Columbia's Water Plan; British Columbia Drought Response Plan; Action Plan for Safe Drinking Water and *Drinking Water Protection Act*; and the *Water Protection Act*. The most significant freshwater legislation is the *Water Sustainability Act* (WSA), which came into force through an initial set of regulations in 2016.

The WSA primarily deals with water allocations and planning. While it retains many elements of the previous water law regime (going as far back as the original *Water Act of 1909*), it also makes several significant and substantive changes and offers an array of new sustainability-oriented tools. Key provisions in the Act that generally apply across British Columbia (and thus also in the Gulf Islands) include:

- Groundwater use regulation for all non-domestic uses;
- Requirement for statutory decision makers to consider environmental flows in all future water licensing decisions;
- Legal protection for minimum (critical) flows and fish populations during periods of shortage; and,
- The possibility for existing licences to be reviewed (including changed or cancelled) in 2046 (30 years from when the Act came into force for existing licences—or 30 years after date of new licence approval).

As shown in Table 17.5, the WSA contains a number of innovations that support better water management and governance. However, it is difficult to comment in detail as very few of the new sustainability tools have been deployed to date—implementation of the WSA and a broader shift towards a new water management regime in British Columbia will take years.[8]

Table 17.5 Comparison of water governance, British Columbia and Washington State

Issue	British Columbia	Washington State
Jurisdictional context	Federated state with water primarily managed by the province: • authority and legitimacy of government action being challenged by aboriginal water rights claims and assertion of Indigenous water laws.	Water management is primarily a state function, but federal legislation and reserved water rights impact water management.
Key principles of Western water law in effect	• Government ownership of water; • First in time, first in right; • Use it or lose it/water as economic good for development.	Same as British Columbia.
Environmental flow protection	Consideration required under the WSA (s. 15) – but also possible through: • beneficial use considerations (s. 7); • Water Objectives (s. 43); • various planning mechanisms including Water Sustainability Plans (ss. 64–85).	The Dept of Fish and Wildlife can request the Dept of Ecology to establish an instream flow rule. Once established, it functions as a water right, with a priority date corresponding to its date of adoption.
Watershed planning within the legal framework	Enabled through the WSA with opportunities including: • Area Based Regulations (s. 124); • Sensitive Stream Designations (s. 128); • Water Reservations (s. 39); • Water Sustainability Plans (s. 64–85). No formal plans exist in the Gulf Islands.	Watershed planning may be required under Washington State law but has not been required for the San Juan Islands.

8 For a discussion of priorities for WSA implementation and a blueprint towards a broader watershed governance regime in British Columbia see Brandes and O'Riordan (2014) and for a specific discussion of WSA in comparison to water law developments in California to inform further development and implementation in British Columbia see Christensen and Brandes (2015).

Issue	British Columbia	Washington State
Groundwater use	Licences required but domestic water uses (i.e. households) are exempt. Potential to regulate domestic (household) wells in water stressed regions (such as the Gulf Islands) via Area Based Regulations (but none currently exist).	Licences required but some uses are exempt (e.g. non-commercial gardens under ½ acre, residential use under 5,000 gallons per day).
Drought response	British Columbia's Provincial Drought Response Plan provides a progressive 5-stage system that emphasises voluntary cutbacks and preparedness, but culminates in possible regulatory action including Critical Flow and Fish Protection Orders available under the WSA (ss. 86–88).	The state can declare a drought emergency in a region if it meets two criteria: 1. The region is receiving, or is projected to receive, less than 75 per cent of its normal water supply 2. Water users in the region will likely incur undue hardships as a result of the shortage.
Rainwater use	Governed by the common law 'rule of capture'. Local Trust Committees can limit development or require cisterns for the storage of rainwater.	In 2009, the Dept of Ecology issued a rainwater interpretive policy, which clarifies that you do not need a water right permit to use water collected from your rooftop. Some counties in Washington also allow rainwater collection as a main drinking water source, although many counties do not. In some areas of the state, such as the San Juan Islands, rainwater may be the only viable water supply for new construction. Rainwater collection may be limited if it is shown to interfere with prior water rights.
Water closures	St Mary's Lake (Salt Spring Island) has a moratorium on new connections for water supply.	Two lakes permanently and one seasonally.
Expanded role for local governments/ local authorities	The WSA offers possibilities for shared or co-governed water through formal advisory councils or delegated authority provisions (ss. 115 and 126).	The state primarily manages water resources through 62 Water Resource Inventory Areas, which largely track watershed boundaries.
Saltwater intrusion	Serious concern — but no legal or governance response identified.	Water permit applications may be denied if the use will exacerbate saltwater intrusion and the impacts cannot be mitigated.

Source: Authors' tabulation.

It is clear that the new Act, complemented by the provincial Living Water Smart Policy, ushers in a new emphasis on partnerships and a shared approach to water management in British Columbia. The new partnership model emphasises a creative mix of local, Indigenous and government actors working together through shared decision making and transforming how communities interact with water and how nature might be fundamentally included. However, much of this potential remains unfulfilled in the Gulf Islands—as is the case across much of the province.

Water Law Regimes—Washington State

Water law in Washington is based on 'common law' (law based on custom, tradition and court decisions) as well as on state legislative statutes. Washington court decisions impacting water rights date to the early 1900s, supplemented by detailed regulations. Washington's water law regime shares many features with British Columbia:

- Water is considered publicly owned. Individuals may obtain a right to use water through water permits or governance arrangements that exempt certain users from permit requirements (e.g. right of landowners to drill a well for a single residence).
- Washington State regulates water rights on a first in time, first in right basis.
- Washington State follows the 'use it or lose it' approach to water rights, which discourages efficient use of water.
- Washington State's water governance regime developed out of the 'prior appropriation' approach, which recognised that persons would gain right to use water upon taking the water and putting it to 'beneficial use' (which historically was agricultural, industrial or household).

Washington State has created an innovative 'Trust Water Rights' program to temper the impacts of the 'use it or lose it rule' (Washington Department of Ecology n.d.a). The program allows water users to temporarily place water rights with the Department of Ecology without the risk of relinquishment. While held in trust, the water rights contribute to streamflows and groundwater recharge, while retaining their original priority date.

To protect instream flows for aquatic habitat, Washington State has adopted a regime where the Department of Fish and Wildlife can request the establishment of an instream flow rule by the Department of Ecology. Instream flows adopted as rules are considered water rights and have the date of rule adoption as their priority date.

The federal government does not play a direct role in water management in Washington State, but its laws and funding may affect water management. A key example is the *Endangered Species Act* (ESA). The ESA has certainly impacted water management decisions and, based on court decisions, some scholars argue that the ESA pre-empts state water law in certain circumstances (Craig 2014).

While there have been past land claims by Indian Tribes in Washington State, current land claims do not have a significant impact on water management. Washington State does recognise federal reserved rights, which were validated by a US Supreme Court case that held when the United States sets aside an Indian reservation, it impliedly reserves sufficient water to fulfil the purposes of the reservation, with the priority date established as of the date of the reservation.

Washington State divides water resources into Water Resource Inventory Areas (WRIAs). The San Juan Islands are part of WRIA #2, which is not subject to an instream flow rule. The lack of an instream flow rule is explained, at least in part, due to the lack of major rivers within the WRIA.

Some areas within Washington State have been required to complete watershed planning, but the San Juans (WRIA #2) are not among those watersheds (Washington Department of Ecology n.d.b).

Saltwater intrusion is a concern in some areas, meaning that the Department of Ecology cannot issue a new water right if pumping will cause saltwater contamination, unless the effect of the intrusion is mitigated.

In WRIA #2, only two lakes are permanently closed and one lake is seasonally closed to future water rights licensing. In other areas, new surface and groundwater users licences may be issued. As well, the 'groundwater permit exemption' allows users of small quantities of groundwater (e.g. owners of a single residence) to construct wells and develop their water supply without obtaining a permit. Both licences and exempt wells are subject to the seniority system meaning that use may be prohibited to guarantee water availability to more senior uses despite the possession of a licence.

Washington State has specific criteria for determining drought. First, a geographical area is receiving or expected to receive less than 75 per cent of normal water supply as the result of natural conditions; and second, the deficiency causes or is expected to cause 'undue hardship'. If these two conditions are met, the state can issue a drought declaration, including: facts leading to issuance; the statutory authority; descriptions of possible actions under the order; and provisions for the termination of withdrawals if essential minimum flows are jeopardised. 'Curtailment' of water rights is only available against junior water rights holders and may be imposed to protect instream flows. During the 2015 drought, the Department of Ecology issued 883 curtailments statewide.

Similarities

Many similarities exist between the two jurisdictions in the Salish Sea, both of which are governed by same building-block principles of Western water law. Unsurprisingly, they have not created dramatically different management and legal systems on the San Juan/Gulf Islands. A previous analysis, conducted in the mid-2000s, concluded that the difference in governance arrangements between British Columbia and Washington State also did not result in dramatic differences in water management (Cohen 2003). As is often the case with water law and governance, there is an ongoing period of flux as new laws, management regimes and governance arrangements are developed and implemented. Provisions to deal with drought, rainwater and groundwater will be particularly important in Salish Sea Islands, given the existing issues and challenges noted in this chapter.

Concluding Comments

The Salish Sea archipelago experiences a Mediterranean climate, with wet, cool winters and warm, dry summers. Demand for water peaks in late summer, when part-time residents and tourists augment the islands' populations, and both groundwater and surface water supplies become limited. Furthermore, water quality can become impaired—such as by seawater intrusion for groundwater and cyanobacteria in surface water sources. Evidence of increasingly sophisticated and comprehensive approaches to water are emerging on both sides of the border in the Salish Sea in response to significant changes to hydrology, amplified by a changing climate and social expectations, which challenge the historical law and governance regimes. However, more work remains as the new laws, policies and governance are implemented over time. In addition, these formal management and governance arrangements are increasingly being augmented by community-based initiatives focused on monitoring, watershed restoration and other activities.

References

Allen, D.M., 2004. 'Sources of Ground Water Salinity on Islands Using Stable Isotopes of ^{18}O, ^{2}H and ^{34}S.' *Ground Water* 42(1): 17–31. doi.org/10.1111/j.1745-6584.2004.tb02447.x

Allen D.M., D.G. Abbey, D.C. Mackie, R.D. Luzitano and M. Cleary, 2002. 'Investigation of Potential Saltwater Intrusion Pathways in a Fractured Aquifer Using an Integrated Geophysical, Geological and Geochemical Approach.' *Journal of Environmental and Engineering Geophysics* 7(1): 19–36. doi.org/10.4133/JEEG7.1.19

Allen, D.M., D. Kirste, J. Klassen, I. Larcoque, S. Foster and G. Henderson, 2015. 'Research Monitoring Well on Salt Spring Island. Final Report Prepared for BC Ministry of Forests, Lands and Natural Resource Operations, and BC Ministry of Environment.' Viewed 31 May 2023 at: sfu.ca/personal/dallen/Risk_to_Coastal_Bedrock_Aquifers.html

Allen, D.M. and E. Liteanu, 2008. 'Long-term Dynamics of the Saltwater–Freshwater Interface on the Gulf Islands, British Columbia, Canada.' In G. Barrocu (ed.), *Proceedings of the First International Joint Saltwater Intrusion Conference*, SWIM-SWICA, Cagliari, Italy, 25–29 September 2006.

Allen, D.M. and M.S. Suchy, 2001. 'Geochemical Evolution of Groundwater on Saturna Island, British Columbia.' *Canadian Journal of Earth Sciences* 38: 1059–80. doi.org/10.1139/e01-007

Barnett, L., B. Blecic and W. van Bruggen, 1993. *Salt Spring Island Water Allocation Plan.* Victoria: Ministry of Environment, Lands and Parks. Regional Water Management.

Brandes, O.M., 2005. 'At a Watershed: Ecological Governance and Sustainable Water Management in Canada.' *Journal of Environmental Law and Practice* 16(1): 79–97.

Brandes, O.M. and D. Curran, 2017. 'Changing Currents: A Case Study in the Evolution of Water Law in Western Canada'. In S. Renzetti and D.P. Dupont (eds), *Water Policy and Governance in Canada.* Switzerland: Springer International Publishing (Global Issues in Water Policy). doi.org/10.1007/978-3-319-42806-2_4

Brandes, O.M. and J. O'Riordan, 2014. *Decision Makers Brief: A Blueprint for Watershed Governance in British Columbia.* Victoria: University of Victoria (POLIS Project on Ecological Governance).

Burgess, R., 2017. Characterizing Recharge to Fractured Bedrock in a Temperate Climate. Simon Fraser University (MSc thesis).

Christensen, R. and O.M. Brandes, 2015. *California's Oranges and B.C.'s Apples? Lessons for B.C. From California Groundwater Reform.* Victoria: University of Victoria (POLIS Project on Ecological Governance).

Clague, J.J., 1983. 'Glacio-isostatic Effects of the Cordilleran Ice Sheet, British Columbia; Canada.' In D.E. Smith and A.G. Dawson (eds), *Shorelines and Isostasy*. London: Academic Press.

Cohen, A., 2003. What Role for Regulation? The Case of Groundwater Governance on the Gulf and San Juan Islands. University of British Columbia (MA thesis).

Craig, R,K., 2014. 'Does the Endangered Species Act Preempt State Water Law?' *Kansas Law Review* 62: 851–91. Viewed 31 May 2023 at: kuscholarworks.ku.edu/bitstream/handle/1808/20259/1_KLR_Site_Craig_Final_Press.pdf;sequence=1

Curran, D., 2015. 'Water Law as Watershed Endeavor: Federal Inactivity as an Opportunity for Local Initiative.' *Journal of Environmental Law and Practice* 28(1): 53–88.

Denny, S.C., D.M. Allen and J.M. Journeay, 2007. 'DRASTIC-Fm: A Modified Vulnerability Mapping Method for Structurally Controlled Aquifers in the Southern Gulf Islands, British Columbia, Canada.' *Hydrogeology Journal* 15(3): 483–93. doi.org/10.1007/s10040-006-0102-8

Government of British Columbia, 2017. 'Ministry of Environment and Climate Change Strategy. Viewed 26 June 2023 at: www2.gov.bc.ca/gov/content/governments/organizational-structure/ministries-organizations/ministries/environment-climate-change

——, n.d. 'Islands Trust.' Viewed 26 June 2023 at: gov.bc.ca/gov/content/governments/local-governments/improvement-districts-governance-bodies/islands-trust

IPCC (Intergovernmental Panel on Climate Change), 2007. 'Climate Change 2007: Synthesis Report. Contribution of Working Groups I, II and III to the Fourth Assessment Report of the Intergovernmental Panel on Climate Change.' Edited by R.K Pachauri and A. Reisinger. Geneva, Switzerland: IPCC.

Kelly, D., 2005. 'Seawater Intrusion Topic Paper (Final), Island County: WRIA 6 Watershed Planning Process 1–30.' Viewed 26 June 2023 at: apps.ecology.wa.gov/publications/publications/1203271.pdf

Kerr, W.L., 2015. 'St. Mary Lake Watershed Water Availability and Demand—Climate Change Assessment. Report prepared for the North Salt Spring Waterworks.' (KWL Project No. 2932.004).

Klassen, J., D.M. Allen and D. Kirste, 2014. 'Chemical Indicators of Saltwater Intrusion for the Gulf Islands, British Columbia. Final report Submitted to BC Ministry of Forests, Lands and Natural Resource Operations and BC Ministry of Environment, June 2014'. Viewed 31 May 2023 at: sfu.ca/personal/dallen/Risk_to_Coastal_Bedrock_Aquifers.html

Kwiáht, 2017. 'Coastal Cutthroat Trout.' Viewed 31 May 2023 at: kwiaht.org/coastalcutthroat.htm

Mackie, D.C., D.M. Allen, P.S. Mustar and M. Journeay, 2001. 'Conceptual Model for the Structurally Complex, Bedrock-Controlled Groundwater Flow System on the Southern Gulf Islands, British Columbia.' Calgary, Alberta: Proceedings of the 2nd Joint CGS-IAH Groundwater Specialty Conference, September.

Moore, R.D., D.L. Spittlehouse, P.H. Whitfield and K. Stahl, 2010. 'Weather and Climate.' In R.G. Pike, T.E. Redding, R.D. Moore, R.D. Winkler and K.D. Bladon (eds), *Compendium of Forest Hydrology and Geomorphology in British Columba*. Victoria: BC Ministry of Forest and Range.

Nichols, K., 2014. *Gulf Islands Groundwater Protection: A Regulatory Toolkit*. Victoria, BC: Islands Trust. Viewed 31 May 2023 at: islandstrust.bc.ca/wp-content/uploads/2020/05/groundwater-toolkit.pdf

PCIC (Pacific Climate Impacts Consortium), 2016. Regional analysis tool. Viewed 31 May 2023 at: tools.pacificclimate.org/tools/regionalanalysis/

Pentland, R. and C. Wood, 2013. *Down the Drain: How We Are Failing to Protect Our Water Resources*. Vancouver, Canada: Greystone Books Ltd.

Percy, D.R., 2004. 'The Limits of Western Canadian Water Allocation Law.' *Journal of Environmental Law and Practice* 14: 313–27.

Surrette, M. and D.M. Allen, 2008. 'Quantifying Heterogeneity in Variably Fractured Sedimentary Rock Using a Hydrostructural Domain.' *Geological Society of America Bulletin*, 120(1): 225–37. doi.org/10.1130/B26078.1

Thornthwaite, C.W. and J.R. Mather, 1957. 'Instructions and Tables for Computing Potential Evapotranspiration and Water Balance.' *Publications on Climatology* 10: 185–231.

Washington Department of Ecology, 1975. 'Geology and Water Resources of the San Juan Islands, San Juan County Washington.' In R.H. Russell (ed.), *Water Supply Bulletin* (No. 46). Washington Department of Ecology Office of Technical Services.

——, 2016. 'Focus on Water Availability. San Juan Islands Watershed, WRIA 2.' Washington Department of Ecology, Water Resources Program (Publication Number 11-11-007). Viewed 26 June 2016 at: web.archive.org/web/20170131193801/https://fortress.wa.gov/ecy/publications/documents/1111007.pdf

——, n.d.a. 'Trust Water Rights Program.' Viewed 31 May 2023 at: ecology.wa.gov/Water-Shorelines/Water-supply/Water-rights/Trust-water-rights

——, n.d.b. 'Watershed Plan Archive.' Viewed 31 May 2023 at: ecology.wa.gov/Water-Shorelines/Water-supply/Streamflow-restoration/Watershed-plan-archive

Whitfield, P.H., R.D. Moore, S.W. Fleming and A. Zawadzki, 2010. 'Pacific Decadal Oscillation and the Hydroclimatology of Western Canada—Review and Prospects.' *Canadian Water Resources Journal* 35: 1–28. doi.org/10.4296/cwrj3501001

18

Culture Change and Conservation

Russel L. Barsh and Madrona Murphy

It is not unusual to overhear first-time visitors to the San Juan and Gulf Islands remarking, from the deck of the ferry, on the 'pristine' appearance of the landscape—in particular, the coniferous forests that seem to extend from the shoreline to the hilltops. In comparison to the streets of Seattle or Vancouver the islands are indeed experienced as shades of green and blue, with relatively few homes or businesses easily seen from the water. The San Juan and Gulf Islands terrestrial ecosystems are relatively young, assembled only within the last 12,000 years on rocks and thin soils scraped bare by glacial ice. But pristine they are not. Humans have travelled, camped, cleared, hunted, gardened and built homes in the islands for 8,000 years or longer, modifying soils and landscapes, deleting and adding to species diversity, and changing goals and tools with changing cultures, while attaining population levels comparable to the present long before the arrival of Europeans in the 1790s. Island ecosystems reflect multiple human cultures played out in a very small space, just separated enough geographically from the mainland of Northwest North America, and for long enough, to be distinctively different.

Differences between the San Juan group and Gulf group of islands extend beyond the historically arbitrary international border that separates them today. The two island clusters differ geologically, and thus in topography, hydrology and soils. Their locations relative to the mountainous spine of Vancouver Island and the Olympic mountain range results in significant differences in weather, including the 'rain shadow' effect on mean precipitation. There is more open sea (fetch) between the San Juan group and prevailing southwesterly winds. The proximity of the Gulf group to Vancouver Island is associated with a similarity of plant species, whereas the San Juan group is more similar (as a whole) to the nearby mainland of western Washington State—the Puget Lowlands and Cascade Mountains. The impact of different post-contact settlement patterns and land use regulations between the Gulf and San Juan groups should not be underestimated, either; Euro-American cultures and laws have made the two ends of the archipelago more alike in some ways and accentuated their differences in others.

Humans have modified habitats and (wittingly or unwittingly) facilitated the transfer of species between mainland habitats and the islands for millennia, under changing cultural frameworks. Considering the relatively small geographic area of the islands, the number of pre-contact settlements and the islands' role as an early point of post-contact extractive development, it would be difficult to imagine any part of the islands' ecosystems that are not, to some degree, cultural artefacts.

Coast Salish Management

The pre-contact archaeological record begins with several thousand years of large terrestrial and marine mammal remains and scattered fire rings, suggesting hunting and seasonally shifting settlements throughout the islands (Carlson 1960; refined by Stein 2000). From roughly 4,000 to 2,500 years ago, a period of cooling weather and growing coniferous forests, food gathering increasingly focused on fish and shellfish, while settlements grew larger and more permanent, as evidenced by thick accumulations of food waste and household debris (shell middens) and post moulds formed by the decay of rows of cedar house posts. Knapped (flaked) stone tools persist but are increasingly accompanied by ground stone artefacts, a product of time and wealth, and there is an increase in the use of exotic materials, a reflection of growing long-distance trade. Triggered by climate change (Lepofsky et al. 2005), the rise of a maritime civilisation was facilitated by a number of technological innovations, the chronology of which remain uncertain.

One innovation unique to the islands was reef net fishing, which deployed a barricade of floating nets, suspended from pairs of anchored canoes, at strategic points in the homeward migration of sockeye salmon through the islands to the Fraser River and smaller lake-fed streams (Suttles 1972). According to Coast Salish reef netters interviewed in 1895, a battery of a half-dozen 'gears' could impound thousands of salmon per day, or more than 25 tons in the course of a single summer season, requiring large summer processing camps at each fishing site and a high degree of coordination (Barsh 2021). A pilot study at the Watmough-Colville reef net shore processing site on Lopez Island found evidence of 3,000 years of salmon fishing, with a shift away from Chinook and coho salmon to mostly sockeye roughly 1,800 years before the present, determined by use of species-specific genetic markers (Hatch 2005). On Lopez Island, at least, reef net gears began replacing hook-and-line fisheries nearly two millennia ago.

Production of an annual surplus of dried salmon made large year-round settlements more secure, contributed to trade with peoples in the interior (where salmon can be harvested but generally in poorer, less fatty condition), and made it feasible for Salish Sea villagers to raise large flocks of specially bred dogs for shearing, spinning and weaving (Barsh 2016). Salmon was therefore necessary for producing textiles until the introduction of machine-spun sheep yarns by the Hudson's Bay Company and American Fur Company. By the 1850s, Coast Salish sold surplus salmon to the trading companies for bobbins of yarn, rather than feeding salmon to dog flocks. The antiquity of woolly dog breeding, and of textile production, remains speculative since neither woollens nor wooden looms persist in the conditions typical of shell middens in the Salish Sea. But maintaining large flocks of dogs would undoubtedly have affected the vegetation and wildlife of smaller islands that served as seawater-bounded corrals, such as Guemes in the San Juan group, which was known as 'dog island' in the Northern Straits language and was so marked on early nautical charts (Richardson and Galloway 2011).

Another important technological innovation (with potentially greater direct impacts on terrestrial ecosystems in the islands) moved beyond the organised harvesting and processing of wild *Camassia* to a true cultivation system that included demarcated family gardens, seasonal hoeing and weeding, periodic use of flashy fires to suppress woody succession, and selection and transplanting of bulbs to maintain the vigour and productivity of families' camas stocks (Barsh and Murphy 2016). Multiple species were combined with camas in gardens, such as Columbia lily (*Lilium columbianum*), Indian carrot (*Perideridia gairdneri*) and bracken fern (*Pteridium aquilinum*), which has starchy rhizomes. Ancient gardens in the islands, where they have remained undisturbed by subsequent Euro-American ploughing and pasturage, are distinguished by deep, dark, charcoal-rich anthropogenic soils that differ significantly from the sandy glacial till and thin brown clay loams typical of the islands. Ancient garden soils often support plants rarely seen elsewhere in the San Juan and Gulf Islands, among them Indian celery (*Lomatium nudicaule*), of traditional medicinal importance.

There is growing evidence for pre-contact construction of rock jetties along shorelines to build up artificial substrates for growing bivalves such as *Saxidomus gigantea*, the 'butter clam' of the Salish Sea (Groesbeck et al. 2014; Cox et al. 2019; Lepofsky et al. 2021). Shoreline modification on this scale may have slowed shoreline erosion, created or expanded salt marsh habitat and contributed to shorebird and waterfowl populations as well as intertidal mollusc, crustacean and fish populations. Coast Salish peoples and their neighbours on Vancouver Island and the Pacific Coast were also very fond of coastal marsh plants such as springbank clover (*Trifolium wormskioldii*), Pacific silverweed (*Potentilla anserina*), and northern rice-root (*Fritillaria camtschatcensis*) (Turner and Kuhnlein 1982; Turner and Kuhnlein 1983), which benefit from cultivation and drainage management, and may have been the focus of a third gardening system in the Salish Sea region involving dikes or check-dams along shorelines (Deur 2005; Grier 2014). This, too, would have expanded salt marshes and enhanced the populations of birds, mammals and invertebrates rearing and foraging in the nearshore.

Building the cooperative plank houses of Coast Salish villages, each large enough to accommodate hundreds of residents, required significant quantities of western red cedar for posts, beams and boards. Bark from straight cedars was also harvested for basket-making, and the long straight stems of ocean spray (*Holodiscus discolor*), a major component of the understory in humid coniferous forests, were valued for lean-to poles, fishing gear, garden hoes and other tools. Possessing only fire, ground stone axes, antler wedges and stone sledgehammers to fell and process trees, and lacking powered mechanical means of transporting woody materials, Indigenous Islanders would presumably have cut only high-value trees, not very far from where they planned to use them. They would have had a strong incentive to use fire lightly to clear the understory from paths, gardens and berry patches, without burning up their sources of lumber or firewood. Light burning and selective logging would have produced parkland-like landscapes. Sadly the earliest photographs of San Juan and Gulf Islands landscapes, taken by C.B.R. Kennerly in the 1850s, have disappeared, but a number of dendrochronological studies indicate that small, flashy fires maintained park-like conditions two to five centuries ago where there is a dense, closed conifer canopy today (Spurbeck and Keenum 2003; Sprenger and Dunwiddie 2011).

Clearing around villages and gardens with axes and fires may well have increased forest diversity. The pollen record suggests that for the past 5,000 years, the trees in open canopy habitats were mainly Douglas firs (Lucas and Lacourse 2013). In the 1850s, Garry oaks were restricted to discrete patches such as the Old Oak Prairie on San Juan Island (Kennerly 1860: 44), much of which has disappeared into farm fields and residential lawns. Like Pacific madrone (*Arbutus menziesii*), oaks establish today chiefly as pioneers on steep rocky slopes following landslides or clearing. Pioneer species such as Pacific madrone and Garry oak would have benefited from selective pre-contact logging and clearing around villages and gardens, such that the region's iconic Garry oak ecosystems are in actuality a cultural relic (Vellend et al. 2008; McCune et al. 2013; Pellatt and Gedalof 2014).

A 10-metre-deep peat core collected by the authors at Hummel Lake (Lopez Island), albeit not yet analysed completely, contains evidence of two fires hot enough to leave thick layers of charred woody debris and carbonised seeds. One took place approximately 150 years ago, coinciding with the early Euro-American clearing and settlement of the island. The other fire took place about 900 years ago and appears to have resulted in the replacement of conifers by shrubby Labrador tea (*Rhododendron groenlandicum*), which is valued by Coast Salish peoples for brewing tea. This site may be evidence of sporadic large-scale clearing exercises to promote or protect cultural resources. Notably, the Hummel peat record contains sprinklings of very fine windblown ash at every level to at least 8,000 years ago, suggesting that the island was peopled throughout that time.

One other consequence of large cedar plank houses in the islands may have been an increase in the abundance and diversity of bats, and possibly other synanthropic animals such as rodents (squirrels, mice), crows or deer (attracted to gardens). Over a period of more than ten years, the authors have identified dozens of bat maternity colonies in visits to more than 200 island homes and barns, and the evidence from these visits and bioacoustic surveys is that seven of the nine species of bats documented in the San Juan group of islands live primarily in occupied human structures.

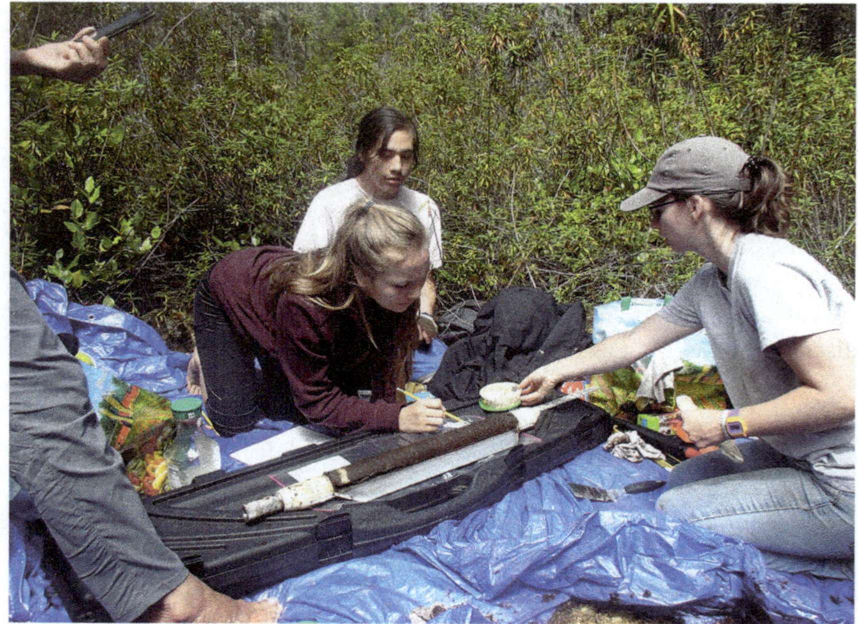

Figure 18.1 Peat drilling team with ash core
Source: Russel Barsh.

It is difficult to estimate the pre-contact population of the islands, not least because there is archaeological evidence of a relatively brief but significant contraction of that population roughly 200 years prior to contact, probably due to indirect contact with European diseases through trade as well as intensification of warfare with Pacific coastal peoples that were already securing firearms from the British and Spanish empires (Mitchell 1984; Stein 2000). An approximation for the Indigenous population of the San Juan group at the time of contact can be derived from the number of villages that were remembered and named by Suttles' (1972) elderly informants in the 1940s, and the size of plank house villages seen in the islands by early explorers such as William Warren (1860): no fewer than 2,400 to as many as 12,000. This is also roughly the range of population of San Juan County in the twentieth century.

Plainly, the Indigenous inhabitants of the San Juan and Gulf Islands were habitat builders and landscape transformers, in particular by maintaining and enhancing herbaceous meadows; and by promoting coastal marshes and tide flats. Cessation of Coast Salish cultivation and shoreline modification by the mid-nineteenth century probably resulted in the expansion of coniferous forests into long-cultivated landscapes, and erosion of undefended anthropogenic beaches and marshes, reducing diversity. Pre-contact logging may also have played a role in the development of the landscapes first described by Europeans two centuries ago.

Euro-American Settlement and Change

The arrival of European settlers in the mid-nineteenth century introduced a number of dramatic changes in landscape management.

At first, farmers could settle where Indigenous people had previously cleared and maintained gardens: these 'prairies' had deep, anthropogenic soils and could be tilled and planted with potatoes and grains with little or no additional clearing. Indeed, many Coast Salish gardens were already growing with potatoes acquired through eighteenth-century indirect trade with Mexico (Barsh and Murphy 2016). The low cobblestone walls that Coast Salish reportedly used to demarcate their garden plots were dismantled to open up larger spaces for sheep. Eurasian grasses were sown because sheep and cattle could not subsist on the sparse, tough native grasses that Coast Salish had weeded from their gardens. Settlers set fire to old, mature forests in the 1850s in an effort to clear them—unsuccessfully, in the view of C.B.R. Kennerly, staff scientist with the US Boundary Commission (Kennerly 1860: 8, 11, 16, 25, 58). These unmanaged conflagrations were followed by a century without controlled fires. Indeed, in the mid-twentieth century, US officials insisted that landowners in the San Juan Islands dig ponds as fire reservoirs, with profound impacts on the islands' fragile watersheds.

While sheep dispersed and fertilised Eurasian grass seeds, European settlers concluded that the willows, alders, aspens, native crab apples and wild cherries that grew around vernal pools and streams were 'trash trees', whereas Douglas fir had resale value for railroad ties, poles and sawn lumber. As a result, commercial loggers took out several generations of coniferous forest in the islands between the 1880s and 1860s, while farmers and homebuilders cleared deciduous woodlands. By the 1930s, Royal Canadian Air Force aerial photos of the islands show very little remaining forest. Clearcut logging and grading disrupted thin postglacial soils and altered drainage patterns. An unusually large cohort of pioneer species such as arbutus and Garry oak was established, and by the 1950s was being replaced by Douglas fir and other fast-growing conifers, altering the composition of woodlands that regrew in the mid-to-late twentieth century. Overall, then, fire frequency decreased, fire-sensitive species such as western red cedar have increased and the mean age and mean size of trees has fallen (Bjorkman and Vellend 2010).

Recent research on Indigenous 'clam gardens' suggests another dimension of ecosystem change since the 1850s. Coast Salish expanded salt marshes and beaches, which often protected shorelines from erosion. Euro-Americans tried to protect beaches and soft bluffs from erosion by impounding them with bulkheads, concrete seawalls and rock piles, which tend to 'starve' beaches of sand and promote scour. Euro-Americans also went to great lengths to clear shorelines and estuaries of drift logs, which are now understood to play an important role sheltering fish eggs and juvenile fish from predators (see generally Augustine and Dearden 2014; Martin 2015).

Settlement patterns may have inflicted the greatest impacts on the islands' biodiversity. Coast Salish lived in a small number of populous villages. Euro-Americans have lived in widely distanced family farms or dispersed individual residences surrounded by lawns and outbuildings. The cleared, built and domesticated portion of the landscape has increased significantly, both in total and per capita terms. More of the landscape is paved, impermeable or converted to some kind of enforced monoculture—mainly Eurasian grasses. Reduction of habitat diversity, and fragmentation of remaining wildlands, would almost certainly have resulted in a significant loss of species in the islands.

Post-Agricultural Island Economies

Changes in the post-contact economy have resulted in several boom–bust cycles affecting land use and disturbance in the archipelago (Cocheba et al. 1973). Sheep were associated with the conversion of herbaceous meadows and camas gardens into grasslands. Logging was followed by commercial orchards, planted mainly between 1880 and 1910, then gradually abandoned or uprooted after irrigation works on the Columbia River shifted regional fruit production from the Salish Sea lowlands to the interior. Peas and oats briefly surpassed sheep and apples as exports after the Great Depression, but were in decline by the mid-1950s, when the population of the San Juan group fell by more than half. The mid-twentieth century, much like the early nineteenth century, was a period of shrinking settlements and expanding coniferous forests.

The rapid growth of the Vancouver and Seattle metropolises as international financial, trade and technology centres in the 1970s–80s resulted in a surge of new, largely seasonal and recreational uses of the islands, accompanied by the construction of retirement homes and summer cottages where there had been farms, pastures and orchards (Eis and Craigdallie 1980; Kondo et al. 2012). Population regrowth peaked in the first decade of the twenty-first century and development activities—clearing, building, leveling and planting lawns and digging ponds—shifted to the edges of the islands, overlooking the sea, much of which had been left undisturbed since the nineteenth century. From marginal agriculture the island economy shifted to services for a wealthier, ageing resident population and recreational visitors. The ecological consequences of these changes are significant: more roads and road runoff, several times as much built surface area, an increase in consumption of limited water supplies and disposal of solid waste, more Eurasian grass, fewer trees and hedgerows and very heavy wear on public access lands, especially the fragile coastal meadows where much of the archipelago's remaining native plant and animal diversity is concentrated.

The 2020 COVID-19 pandemic resulted in another surge of residential construction and year-round occupation of existing structures: a 'COVID refugee' phenomenon that increased the resident population of the San Juan group of islands by 25 per cent or more in a matter of months, overwhelming the capacity of wells, electronic communications, ferries, schools and public parks, while driving up real estate prices sharply. As the pandemic waned in summer 2021, recreational travel boomed, filling campgrounds and increasing human disturbance of sensitive habitats significantly.

While the construction of more homes and businesses is regulated in the islands, with varying degrees of wisdom and efficacy, recreational activities have long enjoyed the privilege of being regarded as low-impact and a net economic benefit. As tourist numbers continue to grow, however, the impacts of hiking and sightseeing on island ecosystems have grown more conspicuous, and public land managers in both the Gulf and San Juan island groups face challenges protecting the archipelago's most unusual and fragile habitats—that is, habitats that are publicly held, and popular with visitors, precisely because they are rare and fragile.

Recreational impacts on glacially sculpted rocky outcrops ('balds') are conspicuous and provide an example. As they are liberated from a protective blanket of soil or shingle by landslides, shoreline erosion or other forces, balds are colonised by diverse lichens that gradually—over centuries—

liberate enough minerals from the exposed rock, and accumulate sufficient decomposing organic matter and water, for mosses and the seeds of some vascular plants to establish. Pioneer soil-building lichens are brittle, especially in dry summer weather, and grow barely a millimetre per year. Hikers quickly remove this lichen crust irreversibly, halting successional processes (Sadler and Bradfield 2010). In recognition of this threat, San Juan County (2016) has explicitly called on recreational visitors to avoid hiking or picnicking on balds as a part of the county's 'Leave No Trace' policy.

Recreational visitors increase the risk of wildfire in natural areas (e.g. McMorrow et al. 2009). Recreational visitors (and their pets) can also be significant vectors of invasive species since plants easily disperse via seeds inside trouser cuffs or adhering to socks, shoes or pet fur. Ticks and other ectoparasites may be introduced by visitors and their pets as well. A recently launched study by the authors has so far identified an exotic tick species, *Dermacentor variabilis*, on Lopez Island that presumably arrived by these means. Exotic tick-borne pathogens could have significant impacts on wildlife and human health.

Residents are also potential vectors of invasive plants and animals, of course. Lawns and gardens are frequently seeded ignorantly with exotic, aggressive plant species; two of the most invasive shrubs in the islands, English hawthorn and Himalayan (or Armenian) blackberry, were introduced as a rootstock for pears and a commercial berry variety, respectively. The list of invasive animals deliberately introduced to the San Juan group with mistaken good intentions over the last century includes the European red fox, Belgian hare, Eastern grey squirrel, Appalachian fox squirrel, American bullfrog, Louisiana red swamp crayfish and red-eared slider ('dime store') turtles. In the islands' lakes and streams, introduced species of fish now outnumber native species and include large-mouth bass, bluegills, pumpkinseeds and brook trout, all of which can compete with native coastal cutthroat trout and reticulated sculpins for prey—and devour their eggs. The European praying mantis and Asian multicoloured ladybeetle, sold to gardeners as natural pest controls in recent years, are well established in natural areas; while native ladybird beetles are now rare.

One of the most pervasive impacts of twentieth-century human agency in the islands was introduction of non-migratory Canada geese to satisfy the interests of waterfowl hunters. Canada geese benefit from and help disperse Eurasian grasses, building on ecological changes wrought by the introduction of sheep and Eurasian pasture grasses in the mid-nineteenth century (Isaac-Renton et al. 2010). Canada geese prefer small, uninhabited islands for their nests, hence they become the main vector for dispersing exotic grasses to isolated islands that might otherwise have maintained their native herbaceous vegetation intact. Geese affect small island ecology to a greater extent than rats or rabbits, which, if they succeeded in landing on isolated islands, would remain trapped there and quickly exhaust their food resources.

Disturbance and Succession

In the highly modified landscapes of the San Juan and Gulf Islands, with their complex histories of intensive, changing human cultures and economies, it is not surprising that native plant diversity is greatest in relatively small patches, isolated from human activity by seas or surrounded by dense, oligarchic Douglas fir forests (Bennett and Arcese 2013). Some important native species such as camas are well adapted to disturbance, however, and may be over-represented in culturally dynamic landscapes (Bennett 2014). At the same time, there is evidence that many exotic species have only just begun to establish footholds in the islands and have not yet dispersed widely, facilitating their control or eradication (Bennett et al. 2013; Murphy 2016). Examples include spurge laurel (*Daphne laureola*) and tickseed (*Coreopsis lanceolata*), each of which is concentrated mainly on a single small island in the San Juan group.

At present, succession in island landscapes reflects the behaviour of native and exotic plant species. Treeless meadows today are dominated by Eurasian grass species, intermixed with both native and exotic forbs. Rosette-forming non-native forbs and the tallest, tiller-forming Eurasian grasses are highly effective at out-competing native species and reducing patch diversity. A long-term study of a wildflower meadow on Lopez Island found that the colonisation of herbaceous meadows by aggressive native shrubs such as trailing blackberry, snowberry and Nootka rose is facilitated by tall Eurasian grasses (Barsh et al. 2006). Douglas fir seedlings also establish in tall grasses and are sheltered from browsing deer by growing shrubs. The Lopez study estimated the time from initial shrub encroachment to Douglas fir canopy as 25 years.

While it is important to protect the archipelago's xeric coastal meadows, it is also essential to recognise that this must be a dynamic process. Protected meadows will gradually become woody and in a generation or two, forested.

Periodic disturbance promotes nutrient recycling, structural change and diversity in ecosystems. The key management question is how to suppress woody succession or, alternatively, how to create disturbances in woodlands that foster re-colonisation by native herbaceous species as one stage in the successional cycle.

Fire Regimes

One of the most widespread sources of disturbance is fire, which can result from natural ignitions such as lightning and volcanic eruptions, or human activities intentional or accidental. Fires periodically 'wipe the slate clean' and restart the processes of ecological assembly and succession. Species gradually return to burnt ground, but not necessarily all of the same species that were there before the fire, and not always in the same order. In this way, fires can stimulate adaptation and evolution. Successful, widespread, long-lived organisms like trees—the 'winners' of past selection and competition—are reduced to raw materials, and the game starts over. Some of the planet's most productive, species-rich landscapes are products of frequent, patchy fires that result in complex, ever-changing mosaics of vegetation in varying stages of regrowth, from moss carpets to wildflower meadows, parklands of shrubs and young trees to close-canopy old forests with deep organic litter and diverse fungi.

Natural ignitions are infrequent in the Salish Sea, however. James Brown and Jane Kapler (2000: 6–7) concluded that western Washington woodlands would burn down naturally only once every two to five centuries, as compared to three or more times per century in most other regions of the United States. Thus while frequent wildfires are 'natural' in eastern Washington and the Rocky Mountains (Baker 2009), fires in the Salish Sea Islands are typically anthropogenic and, for more than a century, largely accidental. Although First Nations routinely burned garden trash and underbrush, their fires were small and flashy, barely scorching nearby trees (Lepofsky et al. 2003; Spurbeck and Keenum 2003; Sprenger and Dunwiddie 2011). Evidence from tree-ring studies on Lopez and Waldron islands indicate that pre-1850 clearing fires rarely extended over more than a fraction of a hectare. The Pender core contained ash from only four large fires—possible wildfires—over the last 1,300 years (Lucas and Lacourse 2013).

Figure 18.2 Landscape, Iceberg Point, Lopez Island
Source: Russel Barsh.

Settlers burned hundreds of hectares of mature forest, however, as Kennerly witnessed, and they were followed by commercial loggers who felled much of what big trees remained, and by sheep ranching, which prevented conifer seedlings from re-establishing in logged-over ground. In point of fact, it was not fire suppression that led to the reforestation of much of the San Juan and Gulf Islands after the 1930s, but rather the decline of logging and ranching. An excellent example is Iceberg Point in the San Juan group; sheep were grazed there until the 1980s when the landscape was re-designated as an Area of Critical Environmental Concern (Barsh et al. 2006). By 2013, when it became part of a new National Monument, a significant portion of Iceberg Point was a sprawling young Douglas fir forest that federal land managers ironically attributed to the cessation of Coast Salish burning over 170 years ago rather than removal of sheep 25 years earlier.

We must look to other sources of disturbance to understand succession and recycling of nutrients in San Juan and Gulf Islands ecosystems, in particular floods and wind storms, which can be annual events on a local scale but cannot be manipulated by people like fire. Water and wind are capable of felling trees, dislodging shrubs and moving large volumes of sand and soil. When European settlers arrived, there was significantly more floating woody debris, beach logs and log jams than today (Collins et al. 2002), a reminder that a great deal of forest was blown down and swept away by rivers and tides, rather than burnt.

Conservation

Much of the scientific and conservation interest in the San Juan and Gulf Islands has focused on dry coastal herbaceous meadows or 'grasslands', often in association with scattered trees such as Garry oaks. The terms 'Garry oak ecosystem' and 'grasslands' are misleading because they erroneously imply a dominant native plant species. Garry oaks are rarely abundant in these habitats. Grasses are often abundant in 'grasslands' but they are predominantly Eurasian species; and in less disturbed habitats where native grasses are relatively abundant (e.g. *Danthonia californica, Festuca rubra* and *Festuca roehmeri*), they are usually subordinate to native forbs. A more precise term may be 'xeric coastal meadows'; that is, coastal habitats, relatively dry due to substrate drainage and rain shadow effects, typically salt-influenced. Trees may be present as isolated individuals or small copses, and in the islands are more likely to be Douglas fir or Pacific madrone than oaks. This type of habitat is most likely to include rare native plant species as well as more abundant and widespread native plants generally.

Pre-contact ecosystems that were managed actively by Coast Salish peoples, such as the islands' coastal wildflower meadows, cannot be preserved by leaving them 'natural' (Bjorkman and Vellend 2010). They require ongoing intervention to slow woody succession. However, the management practices of pre-contact peoples cannot simply be resurrected, because the plant community is no longer the same; it has become dominated by Eurasian species, many of which are pyrophilic and thus are promoted by burning, rather than suppressed. Furthermore, nearly a century of regrowth has left island woodlands unusually rich in ladder fuels. Coast Salish people set flashy fires every few years. Burning a forest that has remained undisturbed for a century is an entirely different condition that leads to very hot fires that burn for hours or days, with different results.

A generation ago, conservationists believed that Garry oak grasslands could be restored by persistent burning (Agee and Dunwiddie 1984) but more recent research suggests that this approach may be misguided (Bjorkman and Vellend 2010). A case in point may be the Young Hill unit of the San Juan Island National Park, portions of which have been burned several times in the last 15 years using the practice of stacking woody debris around Douglas firs and madrones but not the oaks. Most canopy is now Garry oak, but the understory is mainly non-native grasses, and native forbs have declined sharply since the authors began documenting this landscape in 2004.

The term 'Garry Oak ecosystems' is misleadingly applied today to all historical parklands in the Salish Sea, whether or not oaks were abundant or even present in the nineteenth century. Indigenous islanders do not appear to have processed acorns for food and did not value the trees, which typically occupy the mid-story in Douglas fir–cedar forests. This is an important distinction between the archipelago, with its shallow soils and extremely dry 'rain shadow' conditions, and the Victoria Capitol Region, where oak-dominant savannah conditions once prevailed. Garry oaks were historically dominant in the moist deep-soiled valleys that stretch south from Puget Sound to the Willamette Valley of Oregon as well, but even there, it is likely that extensive oak parklands were a byproduct of camas cultivation, insofar as Indigenous cultivators suppressed grasses and conifer seedlings to keep their gardens productive. Focusing on having more Garry oak trees is questionable policy (see e.g. Schultz et al. 2011).

Another key policy issue is focusing conservation resources on small, isolated islands, where there has been relatively little recent disturbance. Islands so small as to have escaped boaters and kayakers, fishermen and campers since the nineteenth century are probably also too small to support much plant diversity, especially under conditions of sea level rise and increased storminess. More attention needs to be paid to native species that have been able to survive, however marginally, in disturbed landscapes on the larger islands, where there is space to expand their populations (Bennett 2014). Recent research also suggests that mosaic landscapes on the larger islands, with gardens and hedgerows as well as relatively undisturbed 'natural' meadows, offer greater pollinator diversity, and greater likelihood that pollination services will persist in a changing climate (Parachnowitsch and Elle 2005; Neame et al. 2013; Wray 2013).

Responses to a Changing Climate

The forecast for the Salish Sea as a whole is more frequent extreme precipitation events in winter, and a longer summer drought season beginning as early as May. Shoreline erosion and tidal inundation will increase, more wetlands will dry out earlier and streams will cease flowing before the rains resume. The most urgent threats are to salt marshes impounded by roads and homes, and freshwater fauna such as amphibians, turtles and trout. There is no surplus of ground water in the rain shadow islands, but the protection of existing bogs and other vegetated wetlands, as well as the construction of bog-like reservoirs (as opposed to open-water reservoirs that suffer greater evaporative losses, algal blooms and anoxia) may make it possible to 'bank' some of the winters' rain runoff and release it gradually in summer. Restoring beaver populations should also be considered. Drier summers mean greater demand for irrigation water, which could place aquatic habitats at risk; but well established old orchards that require no irrigation can be restored through pruning and grafting. Indeed, we have found that orchards originally planted between the 1850s and 1910s function like deciduous woodlands and are home to a diversity of native birds, bats, bees and squirrels.

Much of the islands' native flora have already adapted to drought by greening up in winter when there is abundant water (albeit little sunshine), blooming when the rain ceases and going dormant for the remainder of the year. Over the last 25 years or so, peak blooming time has gradually come earlier in spring, which was to be expected. However, the authors have observed that shorter winters and longer, drier summers are associated with a contraction of the blooming season as well. Instead of being spread out over 8–10 weeks, native wildflowers are often finished for the year barely a month after the first flush of blooms in spring. More species now have overlapping blooming periods. Plants are accelerating seed development to get ahead of the impending drought.

Contraction of the native plant flowering season could be a significant challenge for wild bees, as they have fewer days of foraging to provision their nests and must increasingly compete for pollen rather than simply emerge a little earlier or later in the season than potential competitors. We anticipate greater dependence of wild bees on late-flowering Eurasian weeds such as thistles, dandelions and hairy cat's ear that are already well established in coastal meadows, as well as gardens of irrigated Eurasian flowers and vegetables. Thus it may be that non-native landscaping is an essential part of maintaining native pollinator diversity. On the other hand, we find that importing honeybees and mason bees is growing in popularity in the islands, despite the adverse competitive impacts on wild bees (Henry and Rodet 2018). Future loss of sustainable natural pollinator services and their replacement by domestic bees may not be immediately apparent because many native plants in the San Juan and Gulf Islands hedge their bets by spreading vegetatively; for example, by means of rhizomes, bulb splitting or scaling, or tiny corms dispersed by voles and mice. But in the long term, poor pollination would result in a weak adaptive response to further climate change.

Woody succession is proceeding throughout the parts of the San Juan and Gulf Islands that were logged or cleared for farming in the 1850s to 1940s, including most of the xeric coastal and high-elevation meadows on public lands. Severe wind storms, downpours and landslides continue to reopen small patches of woody growth but the soil seed bank, and windblown seeds, are increasingly dominated by weeds. This is the undoing of well intentioned prescribed burns as well: fire alone simply invites colonisation by the most aggressive non-native weedy grasses and forbs. At the same time, the warmer, drier summers forecast for the twenty-first century will create conditions for accidental ignitions and severe wildfires feeding on long-accumulated ladder fuels. As the regional climate changes, fire will facilitate the ascendance of Eurasian weeds, rather than restore conditions in which native herbaceous species can re-establish themselves and thrive. In the authors' opinion, the focus of future fire policy should be reducing accidental ignitions and building living firebreaks (thickets of evergreen or late-fruiting hard-to-ignite shrubs) to contain fires if they begin, rather than starting fires with the aim of restoring 'natural' landscapes or recovering rare native species.

Warmer, drier summers will also improve the success of exotic species introduced intentionally or unwittingly from warmer, drier latitudes. Already we see signs that 'warm water' sport fish introduced by state wildlife agencies and anglers are displacing the islands' native trout and amphibians (Barsh 2010). Commercial fisheries report increasing numbers of Pacific anchovies followed by large cetaceans such as humpback and grey whales, while the ranges of Pacific salmon species appear to be drifting northwards. With greater natural migration of flora and fauna from Oregon and California into the Salish Sea, it will be increasingly important—and increasingly difficult—to draw the distinction in policy and practice between sustainable northward range extensions and 'invasive species'. Newly arriving native species may be very aggressive in competition with established native species. However, actions aimed at saving 'old natives' may delay the assembly of new, sustainable food webs.

The most we can predict with confidence is that late twenty-first century island ecosystems will be very different from those of the late twentieth century, and that conservation of fresh water and management of fire will determine the long-term consequences of climate change in the 'rain shadow'.

Culture, Responsibility and Action

Islands tend to be refugia and reservoirs of genetic diversity and should therefore be a focus of regional conservation efforts in the Salish Sea. The regional importance of the San Juan and Gulf Islands has been recognised officially by the creation of Gulf Islands National Park and the San Juan Islands National Monument, as well as by community-driven declarations such as the San Juan Islands Marine Stewardship Area and Official Community Plans in the Gulf group. Apart from asking whether any of these designations or undertakings will effectively protect biodiversity in the islands—an issue that we do not address here—there is a question whether current conservation efforts in the San Juan and Gulf Islands reflect Indigenous as well as Euro-American cultural perspectives.

At one level, there has been some movement within academic and professional circles to build a transcultural conversation about ecosystem functioning and management practices. One such initiative is the Clam Garden Network, which brings together Indigenous and non-Indigenous knowledge keepers and resource managers from both sides of Haro Strait to discuss shoreline ecosystems collaboratively. More recently, an open-ended group of specialists in botany, food and herbal medicine has been exchanging information as the Indigenous Plants Forum. Hosted by Rosie Cayou-James and the co-authors of this chapter, it is also both transcultural and transboundary and is sponsoring collaborative research and bilingual publication. These conversations involve Coast Salish knowledge keepers and students in their individual capacity, and not as official representatives of First Nations. Interestingly, some non-Indigenous government agencies have issues consulting with individual Indigenous experts independently of the lawyers and elected leaders of officially recognised First Nations or Tribes.

This raises a crucial cultural question: with whom does ecological knowledge and responsibility reside? Current government actions recognise that in Euro-American society, while elected governments assume responsibility, expertise may belong to individuals such as professional scientists and engineers. It seems to be assumed that within Indigenous societies, both responsibility and expertise are inherently collective. This is certainly not the case for Coast Salish culture and society, in which responsibility as well as expertise are intrinsically individual and interlinked (Barsh 2005). Coast Salish traditionally identified a valuable resource, such as a fishing ground or garden site, with an individual custodian whose authority arose from widespread recognition of his/her expertise and successful stewardship.

Ironically, then, when transcultural conversations are organised, it quickly becomes apparent that they are dialogues among experts with no power or responsibility; whereas when governments confer on issues such as the protection of fragile habitats, the participants are too often people that exercise power, but take no personal responsibility for their decisions—and lack deep, locally grounded expertise. As such, the conflicting paradigms of Coast Salish and Euro-American conservation remain a barrier to effective, collaborative action in the San Juan and Gulf Islands.

Acknowledgements

We are grateful for the encouragement and guidance of Coast Salish teachers and colleagues over the four decades we have devoted to understanding the human dimension of island ecosystem histories. Among the most influential have been Helen Sijohn, Laura Edwards, Isidore Tom, Lena Daniels and Victor Underwood Sr, all born before 1900; as well as Mary McDowell Hansen, Earl Claxton, Victor Underwood Jr, Al Cooper, Jack Kidder and our co-workers Rosie Cayou-James and George Adams, with whom we are writing contemporary ecological stewardship materials in Northern Straits Salish and English.

References

Agee, J.K. and P.W. Dunwiddie, 1984. 'Recent Forest Development on Yellow Island, San Juan County, WA.' *Canadian Journal of Botany* 62: 2074–80. doi.org/10.1139/b84-282

Augustine, S. and P. Dearden, 2014. 'Changing Paradigms in Marine and Coastal Conservation: A Case Study of Clam Gardens in the Southern Gulf Islands, Canada.' *The Canadian Geographer/Le Géographe Canadien* 58: 305–14. doi.org/10.1111/cag.12084

Baker, W.L., 2009. *Fire Ecology in Rocky Mountain Landscapes*. Washington, DC: Island Press.

Barsh, R.L., 2005. 'Coast Salish Property Law: An Alternative Paradigm for Environmental Relationships.' *Hastings West-Northwest Journal of Environmental Law* 12: 1–29.

——, 2010. 'Structural Hydrology and Limiting Summer Conditions of San Juan County Fish-bearing Streams.' Lopez Island: Kwiáht.

——, 2016. 'Coast Salish Woolly Dogs.' HistoryLink.org, 22 June. Essay 11243. Viewed 26 June 2023 at: historylink.org/File/11243

——, 2021. 'Coast Salish Reef-net Fishery, Part 1 and Part 2.' HistoryLink.org, 23 May. Essays 21237 and 21238. Part 1 viewed 26 June 2023 at: historylink.org/File/21237. Part 2 viewed 26 June 2023 at: historylink.org/File/21238

Barsh, R.L. and M. Murphy, 2016. 'Coast Salish Camas Cultivation.' HistoryLink.org, 26 April. Essay 11220. Viewed 26 June 2023 at: historylink.org/File/11220

——, 2024. 'Oaks and Acorns in Native American Cultural Landscapes and Diets.' HistoryLink.org, January 28. Essay 22909. Viewed 11 March 2024 at: historylink.org/File/22909

Barsh, R.L., M. Murphy and S. Blair, 2006. 'Iceberg Point Landscape Restoration Study; Progress Report to the Bureau of Land Management.' Lopez Island: Kwiáht.

Bennett, J.R., 2014. 'Comparison of Native and Exotic Distribution and Richness Models Across Scales Reveals Essential Conservation Lessons.' *Ecography* 37: 120–29. doi.org/10.1111/j.1600-0587.2013.00393.x

Bennett, J.R. and P. Arcese, 2013. 'Human Influence and Classical Biogeographic Predictors of Rare Species Occurrence.' *Conservation Biology* 27: 1–5. doi.org/10.1111/cobi.12015

Bennett, J.R., M. Vellend, P.L. Lilley, W.K. Cornwell and P. Arcese, 2013. 'Abundance, Rarity and Invasion Debt Among Exotic Species in a Patchy Ecosystem.' *Biological Invasions* 15: 707–16. doi.org/10.1007/s10530-012-0320-z

Bjorkman, A.D. and M. Vellend, 2010. 'Defining Historical Baselines for Conservation: Ecological Changes Since European Settlement on Vancouver Island, Canada.' *Conservation Biology* 24: 1559–68. doi.org/10.1111/j.1523-1739.2010.01550.x

Brown, J. and J. Kapler, 2000. *Wildland Fire in Ecosystems: Effects of Fire on Flora.* Missoula, MT: Department of Agriculture, Forest Service, Rocky Mountain Research Station (Technical Report RMRS-GTR-42-vol. 2). doi.org/10.2737/RMRS-GTR-42-V2

Carlson, R.L., 1960. 'Chronology and Culture Change in the San Juan Islands, Washington.' *American Antiquity* 25: 562–86. doi.org/10.2307/276639

Cocheba, D.J., R.A. Loomis and E. Weeks, 1973. 'The Land Market and Economic Development: A Case Study of San Juan County, Washington.' *Washington Agricultural Experiment Station Bulletin* 773.

Collins, B.D., D.R. Montgomery and A.D. Haas, 2002. 'Historical Changes in the Distribution and Functions of Large Wood in Puget Lowland Rivers.' *Canadian Journal of Fisheries and Aquatic Sciences* 59: 66–76. doi.org/10.1139/f01-199

Cox, K.D., T.G. Gerwing, T. Macdonald, M. Hessing-Lewis, B. Millard-Martin, R.J. Command, F. Juanes and S.E. Dudas, 2019. 'Infaunal Community Responses to Ancient Clam Gardens.' *ICES Journal of Marine Science* 76: 2362–73. doi.org/10.1093/icesjms/fsz153

Deur, D., 2005. 'Tending the Garden, Making the Soil: Northwest Coast Estuarine Gardens as Engineered Environments.' In D. Deur and N.J. Turner (eds), *Keeping it Living: Traditions of Plant Use and Cultivation on the Northwest Coast of North America.* Seattle: University of Washington Press.

Eis, S. and D. Craigdallie, 1980. 'Gulf Islands of British Columbia: A Landscape Analysis.' Victoria: Environment Canada, Canadian Forestry Service, Pacific Forest Research Centre (Information Report BC-X-216).

Grier, C., 2014. 'Landscape Construction, Ownership and Social Change in the Southern Gulf Islands of British Columbia.' *Canadian Journal of Archaeology* 38: 211–49.

Groesbeck, A.S., K. Rowell, D. Lepofsky and A.K. Salomon, 2014. 'Ancient Clam Gardens Increased Shellfish Production: Adaptive Strategies from the Past Can Inform Food Security Today.' *PLoS ONE* 9(3): e91235. doi.org/10.1371/journal.pone.0091235

Hatch, M.B.A., 2005. Identification of Archeological Sockeye Salmon Remains as a Proxy for Reef-net Fishing. University of Washington (BA (honours) thesis).

Henry, M. and G. Rodet, 2018. 'Controlling the Impact of the Managed Honeybee on Wild Bees in Protected Areas.' *Nature Reports* 8: 9308. doi.org/10.1038/s41598-018-27591-y

Isaac-Renton, M., J.R. Bennett, R.J. Best and P. Arcese, 2010. 'Effects of Introduced Canada Geese (*Branta canadensis*) on Native Plant Communities of the Southern Gulf Islands, British Columbia.' *Ecoscience* 17: 394–99. doi.org/10.2980/17-4-3332

Kennerly, C.B.R., 1860. 'Islands of the Haro Archepelago, Jan. and Feb. 1860.' Manuscript notebook. Smithsonian Institution Archives, Record Unit 7202, Caleb Burwell Rowan Kennerly Papers, 1855–60.

Kondo, M.C., R. Rivera and S. Rullman, 2012. 'Protecting the Idyll but not the Environment: Second Homes, Amenity Migration and Rural Exclusion in Washington State.' *Landscape and Urban Planning* 106: 174–82. doi.org/10.1016/j.landurbplan.2012.03.003

Lepofsky, D., E.K. Heyerdahl, K. Lertzman, D. Schaepe and B. Mierendorf, 2003. 'Historical Meadow Dynamics in Southwest British Columbia: A Multidisciplinary Analysis.' *Conservation Ecology* 7(3): 5. Viewed 26 June 2023 at: consecol.org/vol7/iss3/art5/. doi.org/10.5751/ES-00559-070305

Lepofsky, D., K. Lertzman, D. Hallett and R. Mathewes, 2005. 'Climate Change and Culture Change on the Southern Coast of British Columbia 2400–1200 Cal. BP: An Hypothesis.' *American Antiquity* 70: 267–93. doi.org/10.2307/40035704

Lepofsky, D., G. Toniello, J. Earnshaw, C. Roberts, L. Wilson, K. Rowell and K. Holmes, 2021. 'Ancient Anthropogenic Clam Gardens of the Northwest Coast Expand Clam Habitat.' *Ecosystems* 24: 248–60. doi.org/10.1007/s10021-020-00515-6

Lucas, J.D. and T. Lacourse, 2013. 'Holocene Vegetation History and Fire Regimes of *Pseudotsuga menziesii* Forests in the Gulf Islands National Park Reserve, Southwestern British Columbia, Canada.' *Quaternary Research* 79: 366–76. doi.org/10.1016/j.yqres.2013.03.001

Martin, K.L.M., 2015. *Beach-Spawning Fishes: Reproduction in an Endangered Ecosystem*. Boca Raton: CRC Press. doi.org/10.1201/b17410

McCune, J.L., M.G. Pellatt and M. Vellend, 2013. 'Multidisciplinary Synthesis of Long-term Human–Ecosystem Interactions: A Perspective from the Garry Oak Ecosystem of British Columbia.' *Biological Conservation* 166: 293–300. doi.org/10.1016/j.biocon.2013.08.004

McMorrow, J., S. Lindley and J. Aylen, 2009. 'Moorland Wildfire Risk, Visitors and Climate Change: Patterns, Prevention and Policy.' In A. Bonn, T. Allott, K. Hubacek and J. Stewart (eds.), *Drivers of Environmental Change in Uplands*. Abingdon: Routledge.

Mitchell, D., 1984. 'Predatory Warfare, Social Status, and the North Pacific Slave Trade.' *Ethnology* 23: 39–48. doi.org/10.2307/3773392

Murphy, M., 2016. 'Final Report to the National Fish and Wildlife Foundation, Community Weed Strategy for San Juan Islands National Monument.' Lopez Island: Kwiáht.

Neame, L.A., T. Griswold and E. Elle, 2013. 'Pollinator Nesting Guilds Respond Differently to Urban Habitat Fragmentation in an Oak–Savannah Ecosystem.' *Insect Conservation and Diversity* 6: 57–66. doi.org/10.1111/j.1752-4598.2012.00187.x

Parachnowitsch, A.L. and E. Elle, 2005. 'Insect Visitation to Wildflowers in the Endangered Garry Oak, *Quercus garryana*, Ecosystem of British Columbia.' *Canadian Field-Naturalist* 119: 245–53. doi.org/10.22621/cfn.v119i2.113

Pellatt, M.G. and Z. Gedalof, 2014. 'Environmental Change in Garry Oak (*Quercus garryana*) Ecosystems: The Evolution of an Eco-cultural Landscape.' *Biodiversity and Conservation* 23: 2053–67. doi.org/10.1007/s10531-014-0703-9

Richardson, A. and B. Galloway, 2011. *Nooksack Place Names: Geography, Culture, and Language*. Vancouver: University of British Columbia Press.

Sadler, K.D. and G.E. Bradfield, 2010. 'Ecological Facets of Plant Species Rarity in Rock Outcrop Ecosystems of the Gulf Islands, British Columbia.' *Botany* 88: 429–34. doi.org/10.1139/B10-011

San Juan County, 2016. 'Leave No Trace.' Viewed 26 June 2023 at: www.sanjuanco.com/1124/Leave-No-Trace

Schultz, C.B., E. Henry, A. Carleton, T. Hicks, R. Thomas, A. Potter, M. Collins et al., 2011. 'Conservation of Prairie-Oak Butterflies in Oregon, Washington, and British Columbia.' *Northwest Science* 85: 361–88. doi.org/10.3955/046.085.0221

Sprenger, C.B. and P.W. Dunwiddie, 2011. 'Fire History of a Douglas-Fir–Oregon White Oak Woodland, Waldron Island, Washington.' *Northwest Science* 85: 108–19. doi.org/10.3955/046.085.0203

Spurbeck, D.W. and D.S. Keenum, 2003. 'Fire History Analysis from Fire Scars Collected at Iceberg Point and Point Colville on Lopez Island, Washington State.' Spokane: Department of the Interior Bureau of Land Management.

Stein, J.K., 2000. *Exploring Coast Salish Prehistory; The Archaeology of San Juan Island.* Seattle: University of Washington Press.

Suttles, W.P., 1972. *Economic Life of the Coast Salish of Haro and Rosario Straits.* New York: Garland.

Turner, N.J. and H.V. Kuhnlein, 1982. 'Two Important "Root" Foods of the Northwest Coast Indians: Springbank Clover (*Trifolium wormskioldii*) and Pacific Silverweed (*Potentilla anserina* ssp. *pacifica*).' *Economic Botany* 36: 411–32. doi.org/10.1007/BF02862700

——, 1983. 'Camas (*Camassia* spp.) and Riceroot (*Fritillaria* spp.): Two Liliaceous "Root" Foods of the Northwest Coast Indians.' *Ecology of Food and Nutrition* 13: 199–219. doi.org/10.1080/03670244.1983.9990754

Vellend, M., A.D. Bjorkman and A. McConchie, 2008. 'Environmentally Biased Fragmentation of Oak Savanna Habitat on Southeastern Vancouver Island, Canada.' *Biological Conservation* 141: 2576–84. doi.org/10.1016/j.biocon.2008.07.019

Warren, W.J., 1860. 'Journal of Wm. J. Warren, Sec'y N.W. Boundary Commission of an Expedition in Company with C.B.R. Kennerly, Surgeon and Naturalist, to the Haro Archipelago Jan., Feb, 1860.' Manuscript. US National Archives (College Park, MD), Record Group 76 (Journals of Exploring Surveys), Entry 198 (Northwest Boundary Survey).

Wray, J.C., 2013. The Effects of Natural and Anthropogenic Habitats on Pollinator Communities in Oak-Savannah Fragments on Vancouver Island, British Columbia. Simon Fraser University (PhD thesis).

19

The Absence of Island Thinking on Vancouver Island

Richard Kool

This chapter is a critique of what I feel is a failure of governance processes on Vancouver Island to facilitate a sustainability transition. While there are many grassroots organisations and initiatives focusing on local sustainability issues, I will argue that the lack of Island-level governance, and the relatively constrained vision of municipal and regional governments, keeps the reality that we live on a large island in the Pacific Ocean absent from governmental planning processes and documents.

Islands and Boundaries

As a biologist, I've long thought of islands in terms of places of favourable habitat for a particular species, surrounded by a sea of unfavourable or dangerous habitat. For example, the upper reaches of some of Vancouver Island's mountains are effectively islands for Vancouver Island marmots, a species that is unwilling or unable to live in the forests below their subalpine homes. Even small urban parks can act as critical islands, providing favourable habitat and harbouring an amazing degree of biodiversity that is totally absent from the unfavourable habitat just short distances away.

Growing up in the 1950s and 60s, I followed the space race and the adventures of the astronauts heading into very unfavourable habitat in their little portable island space capsules. When I first read Boulding's now-classic essay, 'On the Economics of the Coming Spaceship Earth' (1966), I recognised the relevance of the spaceship metaphor for islands: Spaceships are islands, Spaceship Earth is an island. Learning how to live on an island, on a bounded piece of geography surrounded by areas that you might travel through but can't live in or on, was going to be important.

The islands I know best, Vancouver Island and its surrounding smaller islands, have nowhere near the hard boundary of a spacecraft, but they do have edges. We could use the language of Waddock and Kuenkel (2019), who talk about:

> permeable containment … that systems need to have 'sufficient' definitional boundaries or 'enclosures' to create some sort of meaningful identity … , in combination with a degree of openness to new inputs and outputs that allow for energetic exchange. (2019: 6)

Islands have semipermeable boundaries; stuff (both physical and psychological) can and does flow into and out of an island. Resources can be extracted and shipped offshore; one can get into a boat, travel out beyond the boundary and bring home resources.

My favourite rocky intertidal interface opens itself up to island dwellers twice a day, and 'when the tide is out, the table is set' according to a traditional phrase found in coastal First Nations lexicon. But when the tide comes in, when the wind is up, when the currents are flowing, when the waves are white-flecked, steep and breaking, those of us that live surrounded by water know that we live on an island with a blue boundary.

Vancouver Island and Sustainability

I feel that most of us on Vancouver Island, and most of the people making decisions about and for Vancouver Island, live as if we are not on an island at all, and I'll assert that Vancouver Island is not being managed in a prudent and responsible manner (Costanza et al. 2000). I hope to demonstrate that our political leadership presents no real awareness of the boundedness of the Island, of the limits to the island's capacity to support a growing population, or even the present one. There seems to be no general sense among the 'settler' (non-Indigenous) community that the Island is like a spaceship, but surrounded by water; and that there are limits to its ability to provide what its inhabitants need, let alone what we collectively want.

Only a truly deluded resident would claim that they don't have to pay attention to the limits island boundaries place on their activities. Populations of any species including humans cannot grow endlessly, nor can productive farmland be endlessly converted into an unproductive state, non-renewable resources be exhausted and renewable resources used at an unsustainable rate. If your island doesn't have what you need or want, you have to bring it in by bridge, ferry, airplane, small boat, pipeline, cargo ship or cable. Island communities use fragile physical tethers to the larger continental landmasses to overcome their biophysical limits.

Vancouver Island is not an island state. If it were, it would be the world's tenth largest island state by area and eighteenth largest by population. Yet there is no one in charge of Vancouver Island as a unit, and those that oversee the Island at a regional or local scale seem to not understand that they live on an island.

Vancouver Island promotes a self-identity of being pretty 'green', with lots of local purchasing, electric cars (e.g. Larsen 2017), organic farmers markets, local seafoods and wines, and protected areas, and home to Canada's first and second elected Green Party Member of Parliament (Elizabeth May and Paul Manley) and first Green Party Member of British Columbia's provincial Legislative Assembly (Dr Andrew Weaver). Indeed, while the Green Party took ≈6 per cent of the national vote in the 2019 Canadian election and ≈13 per cent in British Columbia, it collected ≈26 per cent of the votes on Vancouver Island.

People like living on Vancouver Island; a 2015 survey covering all of British Columbia noted 'Vancouver Islanders are the most likely to say their communities are good or excellent' (Real Estate Foundation of BC 2016: 6), and 'residents of Vancouver Island … are far more likely than average to view present quality of life in their region in positive terms' (McAllister and Noble 2018: 12).

But Are We So 'Green'?

Many of the issues around human unsustainability can be brought into sharp focus on an island. One conceptual tool that can help us is the ecological footprint (EF) measure developed by Bill Rees (a resident of Gabriola Island) and Mathis Wackernagel when they were both at the University of British Columbia in the early 1990s (Wackernagel and Rees 1996; Rees and

Wackernagel 2013). The EF is a widely used sustainability indicator which calculates the area of biologically productive land (mountains or deserts, for the most part, will not count as biologically productive lands) the average member of a community/city/nation appropriates[1] to provide for all that individual consumes per year, plus the land needed to remove what they throw away (including invisible waste such as carbon dioxide [CO_2]).

Best estimates for our Island communities EF (Global Footprint Network 2018) would be based on average Canadian figures that show that each of us uses about seven hectares (ha) of land per year (for scale, a regular sports field is about one ha in size). Richer Canadians have a larger EF as they consume more stuff and make more waste (primarily CO_2 generated through travel) than poorer Canadians. But even poor Canadians have an EF much larger than most of the world's people.

Another estimate of our EF is based on a pilot analysis done with the Municipality of Saanich on southern Vancouver Island, where Hallsworth and Moore (2018) document a per capita EF of 3.3 ha. This is considerably smaller than the national figure and is indicative of the difference between Vancouver Island and the rest of Canada.

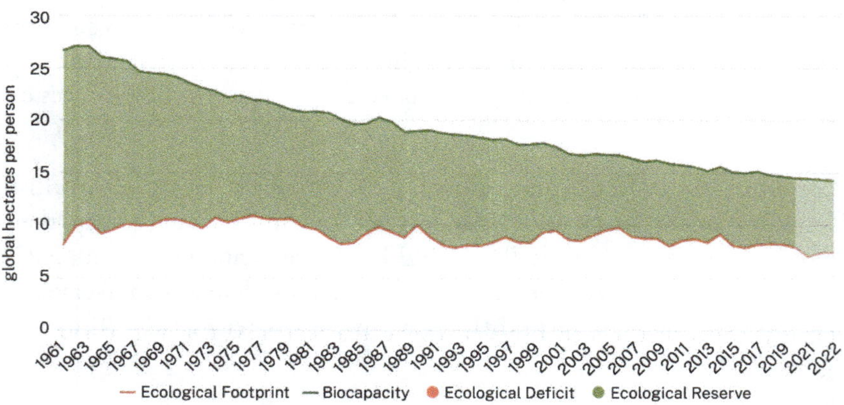

Figure 19.1 Ecological footprint vs biocapacity, Canada
Source: Global Footprint Network (2023).

1 The land used for one's food, clothing, paper etc., land that can't be used by anyone else for anything else because an individual has already 'appropriated' it for their use, no matter where in the world that land exists.

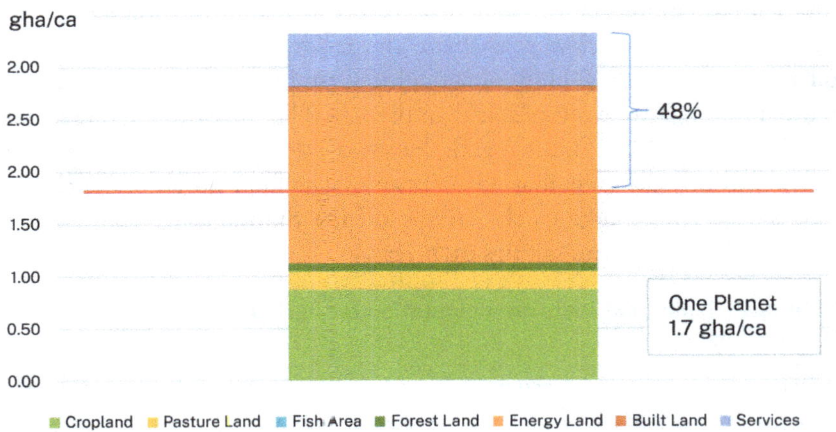

Figure 19.2 Sustainability gap, Saanich 2015

Source: District of Saanich (2018)

Vancouver Island's total land area is about 32,000 km^2, 15 per cent of which is in subalpine and alpine habitats. If each of the ≈850,000 people (estimates from BC Statistics (2020)) on Vancouver Island were to have a standard Canadian EF of 7 ha, our collective EF would be ≈60,000 km^2, nearly twice the total land available on the island. Even using the smaller EF value of 3.3 ha calculated by Hallsworth and Moore (2018), the collective EF of the present population would approximate the entire area of bio-productive land on Vancouver Island (Table 19.1).

Table 19.1. Ecological footprint calculation, total area, Vancouver Island

Total land area (km^2)	32,000
Habitable land area (km^2)	27,200
Total population, 2020	850,000
Total EF (km^2), based on per capita EF of 7.0 ha	59,500
Total EF (km^2) based on per capita EF of 3.3 ha	28,050

Note: 100 ha = 1 km^2.

Source: Author's tabulation, derived from sources in text.

We still have ≈1,000 km^2 (Bullock 2014) of the best agricultural land on the island held within the provincial Agricultural Land Reserve (ALR), primarily on the southeast island between Victoria and the Comox Valley, although between 1954 and 1974, 263 km^2 of the island's prime agricultural land was removed from production (Baxter 1974: 2) and another 100 km^2 or so has been removed from the island's original 1976 ALR allotment.

The math is simple; based on the average Canadian and Saanich EF of ≈1.5 ha/year just for *food production*, the residents on Vancouver Island collectively need ≈13,000 km^2 of agricultural land for our food production, an order of magnitude more land than the 1,000 km^2 existing in Vancouver Island's present ALR (Table 19.2). However, the reality at present, and clearly into the future, is that virtually all of our food provisioning comes from the stream of large trucks driving off the thousands of Seaspan and British Columbia Ferries sailings each year.

Table 19.2. Ecological footprint calculation for agricultural land, Vancouver Island

Productive agricultural land reserve (km^2)	1,000
Per capita ecological footprint for food production, VI (ha)	1.5
Total ecological footprint for food production, VI (km^2)	12,750

Note: 100 ha = 1 km^2.

Source: Author's tabulation, derived from sources in text.

British Columbia Ferries is only one of the many physical tethers that connect us to the mainland. While many of these tethers are not often part of the daily awareness of the average resident, they are quite real. We don't see Fortis Energy's single underwater natural gas pipeline running from the mainland to Vancouver Island. This pipeline terminates at a liquified natural gas storage facility northwest of Ladysmith (Fortis BC 2018) that contains enough natural gas to supply the island for about a month (Hunter 2013) were something to happen to the pipeline. British Columbia Hydro has six hydroelectric generating stations that produce less than 40 per cent of the Island's electricity: the rest travels more than 1,000 km from the Peace River region to Vancouver Island via four underwater cables. While the north and west coast of Vancouver Island have high potential for wind energy production (Sea Breeze Power Corp 2021), and the southeast corner of the region is seeing an increased installation of solar energy, there are no governance structures to facilitate the transition from long-distance hydroelectric power generation to more local energy generation.

Vancouver Island produces virtually no fossil fuels (although historically we were a major producer of coal), yet transportation on Vancouver Island uses on average more than 6 million litres of liquid fuel each day, all of which comes from the mainland through more than 300 shipping deliveries a year (Tanner et al. 2017: 6).

These tethers—gas pipeline, electricity cables, tankers carrying gasoline and ferries carrying food and virtually everything else—are what allow us to live as we do; and yet they are for the most part not salient in the lives of Vancouver Islanders. At least they don't seem salient to the people who are making the plans that will determine the future of the communities on the Island.

Vancouver Islanders live, it seems, in a state of self- and communal delusion. We think we're green and sustainable, while we're not even close to living within the limits the blue boundary puts around us.

Governance and Vancouver Island

Vancouver Island has a wonderful physical reality, but no unitary political or governance reality. Governance on the island is represented by one health region (Vancouver Island Health Authority), 37 municipal governments, seven regional districts and 53 First Nations communities (some under the control of the federal government's *Indian Act,* while others have signed treaties and operate as a third level of government). There also is the Association of Vancouver Island and Coastal Communities (Association of Vancouver Island and Coastal Communities n.d.), which 'deals with issues and concerns that affect large urban areas to small rural communities'.

There is no Vancouver Island government, no level of political authority that can see the entire picture of Vancouver Island. Since the Colony of Vancouver Island joined up with the mainland colony to create British Columbia in 1866, as far as I can tell, there has never been a Minister for Vancouver Island responsible for bringing Vancouver Island issues to provincial or national governments. Based on our size and geographical location, we could be a province or a country (Iceland has about one-third of our population).

We are under-represented nationally, with seven federal Members of Parliament for the ~850,000 Vancouver Islanders, compared to Prince Edward Island's 150,000 citizens having four MPs. Other than the Vancouver Island Health Authority (whose authority leaks over to a bit of the mainland north of Vancouver), there are no organisations, nor are there any politicians, that see the whole of Vancouver Island as their operative domain.

Since 1995, each of our 37 municipal governments and seven regional districts have been encouraged by the province to develop growth management or sustainability plans. Are these plans indicative of an awareness on the part of their drafters that they live on an island? I think not.

Living on an Island from a Municipal and Regional Government Perspective

Several Vancouver Island communities and regional districts have an Official Community Plan, Community Energy Plan, Regional Growth Strategy, Sustainability Plan, or similar documentation. Each plan or strategy is designed for a particular community or region, so the authors are aware of where they are in a geographical sense, but the plans almost uniformly ignore their Island context. Looking at the available plans prior to 2018, one could easily think they are talking about any place in Canada. While the plans all speak to and about their community, virtually none acknowledge that those communities are on an island in the Pacific Ocean.

I searched for indications that the authors of these reports saw their community explicitly as living on an island, but the term 'island', used independently of a particular island name or the name of an island institution, was noticeably lacking. Over the plans from 37 municipalities and seven regional districts, there were only two examples of what I was looking for:

> Capital Regional District (Capital Regional District) Draft Regional Sustainability Strategy: 'Living on an island increases travel costs and limited land availability and the attractiveness of the region for retirees keeps housing costs high for younger workers and families' (Capital Regional District 2014)
>
> Saanich Official Community Plan: 'Economic Vibrancy: … At the same time, a number of challenges in the local economy need to be addressed to ensure continued economic viability. These include the geographic constraints of an island location' (District of Saanich 2008: 62)

Food Security

Other than these two citations, the only other place where I could see any regular effort to talk about the issues of living on an island related to the many places where the concept of 'food security' was presented in both municipal and regional plans. For example, the CRD's RGS, in the chapter concerning 'Food Systems', states that present '...challenges include loss of farmland, lack of farm profitability and financial sustainability, increasing average age of farmers, increasing food prices, limited (72 hour) supply of fresh food in an emergency...' (Capital Regional District 2018, 37). The source of this information is not cited, nor is it clear what is meant by 'an emergency'. And the response RGS offers to the challenges of provisioning food is to propose: 'Increase the amount of land in crop production for food by 5,000 ha to enhance local food security'; apparently, the CRD expects to add 50 km^2 of agricultural land by 2038, somehow imagining the conversion of an area half the size of Saanich, the largest municipal government in the Regional District, into productive farmland (Capital Regional District 2018, 44).

Those few municipal documents that deal with living on an island mainly address concerns about food and food security, and they all seem to be quoting from the same undergraduate paper written by a University of Victoria student (Smith 2010) who cites a report (Macnair 2004) that cites a now-inaccessible article (Haddow 2001) which states 'while fifty years ago farmers on Vancouver Island produced an estimated 85 per cent of the Island's food supply, today Island producers provide about 10 per cent of the food consumed'.

> Tofino Official Community Plan: 'In order to achieve its greenhouse gas emissions targets the District shall: "Minimize reliance on imported food by supporting local food production and consumption."' (District of Tofino 2013: 86)
>
> Comox Official Community Plan: 'When imported feed grains are included, it is estimated approximately 10% of the value of agricultural consumption on Vancouver Island is locally produced.' (Town of Comox 2011: 56)
>
> Campbell River Integrated Community Sustainability Plan: 'Food Self-Sufficiency ... Climate & Energy Growing food locally reduces the need to transport food from distant locations. ... Food Self-Sufficiency: Growing food locally reduces our vulnerability to global

food system disruptions and natural disasters which may prevent food from reaching Vancouver Island.' (City of Campbell River 2011: 59)

Langford Official Community Plan: 'Food self-reliance is an issue gaining some interest in communities on Vancouver Island; 5% to 10% of the food consumed on the island is grown on the island. 95% of the food is therefore imported leading to negative impacts on GHG emissions, traffic, nutritional quality of food and local economy … A recent community energy baseline completed for a local gulf island [sic] found that almost 40% of GHG emissions for that community could be attributed to the consumption of imported food'. (City of Langford 2008: 93)

Saanich Official Community Plan: 'Fifty years ago, farmers on Vancouver Island produced an estimated 85% of the island's food supply. Now, island producers provide less than only [sic] about 10% of the food consumed. Maintaining and enhancing local food production can increase the amount of food, particularly fresh food, available to local residents, decrease or eliminate the need for preservatives, reduce the amount of energy used to transport food, ensure a reliable food source in emergency situations, support the local economy, provide income and employment, and maintain rural and environmentally sensitive areas.' (District of Saanich 2008: 5–4)

Energy Security

The plans I reviewed failed to relate the Island's energy dependency issues to our limited local supply of electricity. For example, in its 2018 Regional Growth Strategy, the Capital Regional District stated:

> Plan for the long term strategic resource needs in the Capital Region—including food (paying specific attention to local food production), energy, water, and aggregate materials consistent … Plans will consider long term demand, security of supply and potential impacts of factors such as long term climate change, fossil fuel depletion and water reclamation where feasible, and make policy and program recommendations to ensure that future needs are successfully anticipated and met. (Capital Regional District 2018: 24)

The chapter on economic development says: 'Finding ways to ensure the long term, affordable supply of strategic economic resources such as water, aggregate and energy' (ibid.: 35). The term 'electricity' doesn't show up in the Capital Regional District Growth Strategy, and neither does 'gasoline' or 'natural gas'. The Regional Growth Strategy treats the energy inputs to the region as a given that do not need planning or consideration.

The Comox Valley Regional District's sustainability plan (2010: 6) says that by 2050, 'the Comox Valley will reduce energy use per capita by 50 per cent and/or will not increase overall energy use from current levels' and will accomplish this through 'decreasing the amount of energy we use and on ensuring that the way we use energy is more efficient'. The provisioning of that energy in the form of gasoline, electricity or natural gas is not mentioned.

The Regional District of Nanaimo (2011) was told by stakeholders to make significant changes to the previous Regional Growth Strategy, 'Including targets and measures for reducing greenhouse gas emissions, reducing energy use, improving air quality, and other regional objectives' (ibid.: 3). One of the plan's goals is to 'Prepare for Climate Change and Reduce Energy Consumption—Reduce greenhouse gas emissions and energy consumption and promote adaptive measures to prepare for climate change impacts' (ibid.: 16). As in the other plans, the provisioning of gasoline, electricity or natural gas is not mentioned.

Solid Waste Disposal

Disposal of waste is another concern:

> Although some regional districts have landfill capacity in the short to mid-term while others—namely Cowichan Valley and Powell River—do not, the reality is that all regional districts have a disposal challenge in the long-term (20 to 40 years from now). Opportunities to site a new landfill are limited, and planning to export waste to the U.S. as a long-term strategy is not without risk. (Tetra Tech EBA 2015: iv)

The Cowichan Valley Regional District is in the unenviable position of having minimal landfill space and must export solid waste by bundling and transporting it via ferries to landfills in Washington State (Tetra Tech EBA 2015: 13). Not being able to deal with our own garbage will increasingly be an issue for Vancouver Island municipal and regional governments, as the authors of the report note that 'waste export may not be reliable in the long term due to border concerns, exchange rates' (ibid.: 30).

Population Growth

None of the regional district plans addressed the impact of continued population growth on a bounded island. Growth of the Island's human population was presented as something that is as uncontrollable as rain. Passive language abounds, such as 'the population is forecast to' or 'population is expected to', which even sounds like the weather report. While those regional districts that mention population growth do talk about the need for compact urban areas and want to consider ways to reduce suburban sprawl, none mention the island population's EF, or how food, energy and waste will be handled. Indeed, British Columbia Government statistics (BC Statistics 2018) see the Island's population going from an estimated 792,000 in 2018 to 924,000 by 2035, as if no one can or should do anything about it. The EF of that higher population, based on Canada's present consumption levels, would be ~65,000 km^2, more than twice the total area of Vancouver Island. While there are numerous citizen and civil society groups concerned about many of these issues, there seems no be no level of government willing and able to step in and confront the difficult issues that a pro-population growth mindset has on the sustainability of Vancouver Island.

To summarise, it seems to me that the Island's existing political boundaries and fragmented governance processes do not genuinely consider Island sustainability concerns. Our blue boundary—the Salish Sea, Strait of Georgia and Johnstone Strait on the east and Pacific Ocean on the west—could be seen as a useful constraint, guiding us towards sustainability at an Island scale. But without Island-wide decision making, or unified Island governance, potential steps towards sustainability are greatly constrained.

I was stunned that these various strategies failed to consider the limitations of the Island: already, our population's EF for food is an order of magnitude greater than the agricultural land available; our energy-producing capacity falls well short of present demand; and our capacity for solid waste disposal is minimal. It is as if the planners and politicians cannot take the concept of 'limits to growth' seriously (Meadows et al. 1992; Turner 2012). The authors of these documents seem reluctant to acknowledge that Vancouver Island is bounded on all sides by water and must consider what limits to growth are.

The structural unsustainability of our communities will not be unravelled over years or perhaps even decades, but our communities, together, can begin to re-weave a form of human endeavour that is far more in keeping with biophysical reality (Robert et al. 1995; Nattras and Altomare 1999). As the Jewish sage Rabbi Tarfon wrote more than 2,000 years ago, 'It is not your responsibility to finish the work, but neither are you free to desist from it' (*Pirqei Avoth* 2: 16).

Finally, there is an important larger lesson in trying to understand sustainability on an island. The earth, like Vancouver Island, is a system with semipermeable boundaries only allowing space dust and energy in the form of sunlight in, and primarily heat out: the planet is a closed system where everything has to come from and go back into the planetary system of stocks and flows. Boulding (1966: 8) wrote:

> The essential measure of the success of the [spaceship] economy is not production and consumption at all, but the nature, extent, quality, and complexity of the total capital stock, including in this the state of the human bodies and minds included in the system. In the spaceman economy, what we are primarily concerned with is stock maintenance.

Just as the residents of Vancouver Island must eventually confront the limitations of our natural resources, the world-dominating societies too must make the conceptual transition to seeing the earth as an island, a spaceship, where we have to take much more care of what we have.

Leadership is needed in all of our communities; it will take courage to recognise that we live on an island; that we far exceed its carrying capacity and are provisioned by fragile tethers; and if we are going to deal in any way with our unsustainability, we are going to have to, at long last, become Vancouver Islanders.

References

Association of Vancouver Island and Coastal Communities, n.d. 'Representing Local Governments on Vancouver Island, qathet, the Sunshine Coast, Central Coast and North Coast.' Viewed 10 May 2019 at: avicc.ca/

Baxter, D., 1974. *The British Columbia Land Commission Act—A Review.* Vancouver: Faculty of Commerce, University of British Columbia.

BC Statistics, 2018. 'Population Projections.' Government of BC. Viewed 31 May 2023 at: gov.bc.ca/gov/content/data/statistics/people-population-community/population/population-projections

——, 2020. 'BC Sub-Provincial Population Projections.' Viewed 3 July 2023 at: gov.bc.ca/assets/gov/data/statistics/people-population-community/population/people_population_projections_highlights.pdf

Boulding, K.E., 1966. 'The Economics of the Coming Spaceship Earth.' In H. Jarrett (ed.), *Environmental Quality in a Growing Economy*. Baltimore: Resources for the Future.

Bullock, R., 2014. 'Provincial Agricultural Land Commission Annual Report 2013/14.' Burnaby: Provincial Agricultural Land Commission.

Capital Regional District, 2014. 'Regional Sustainability Strategy Draft.' Victoria: Capital Regional District.

——, 2018. 'CRD Regional Growth Strategy.' Victoria: Capital Regional District.

City of Campbell River, 2011. 'Campbell River's Integrated Community Sustainability Plan.' Campbell River: City of Campbell River.

City of Langford. 2008. 'City of Langford Official Community Plan.' Langford: City of Langford.

Comox Valley Regional District, 2010. 'Comox Valley Sustainability Strategy.' Courtenay: Comox Valley Regional District.

Costanza, R., H. Daly, C. Folke, P. Hawken, C.S. Holling, A.J. McMichael, D. Pimentel and D. Rapport, 2000. 'Managing Our Environmental Portfolio.' *Bioscience* 50(2):149–55. doi.org/10.1641/0006-3568(2000)050[0149:MOEP]2.3.CO;2

District of Saanich, 2008. 'Sustainable Saanich: Official Community Plan.' Saanich: District of Saanich.

——, 2018. 'Ecocity Footprint Tool Pilot.' Saanich: District of Saanich.

District of Tofino, 2013. 'Vision Tofino Update: Official Community Plan.' Tofino: District of Tofino.

Fortis BC., 2018. 'Liquified Natural Gas.' Viewed 31 May 2023 at: fortisbc.com/NaturalGas/Business/LiquefiedNaturalGas/Pages/default.aspx

Global Footprint Network, 2023. 'Country Trends: Canada.' Viewed 23 June 2023 at: data.footprintnetwork.org/#/countryTrends?cn=33&type=BCpc,EFCpc

Haddow, W., 2001. 'Consumers are King.' FarmSpeak, Summer.

Hallsworth, C. and J. Moore, 2018. *Ecocity Footprint Tool: Ecological and Carbon Footprint Analysis for Achieving One Planet Living*. Saanich: District of Saanich.

Hunter, J., 2013. 'Liquefied Natural Gas Could Be a Boon to B.C.' *Globe and Mail*, 22 April. Viewed 31 May 2023 at: theglobeandmail.com/news/british-columbia/liquefied-natural-gas-could-be-a-boon-to-bc/article11489645/

Larsen, K., 2017. 'Salt Spring Island Lays Claim to Unofficial Title of "Electric Car Capital of Canada".' *CBC News*, 19 June. Viewed 31 May 2023 at: cbc.ca/news/canada/british-columbia/salt-spring-island-lays-claim-to-unofficial-title-of-electric-car-capital-of-canada-1.4150568

Macnair, E., 2004. *A baseline assessment of food security in British Columbia's Capital Region*. Victoria: Capital Region Food & Agricultural Initiatives Roundtable.

McAllister, A. and P. Noble, 2018. *Sustainable Land Use: A Public Opinion Survey of British Columbians*. Vancouver: Real Estate Foundation of BC.

Meadows, D.H., D.L. Meadows and J. Randers, 1992. *Beyond the Limits: Confronting Global Collapse, Envisioning a Sustainable Future*. Toronto, Ontario: McClelland and Stewart, Inc.

Nattras, B. and M. Altomare, 1999. *The Natural Step for Business: Wealth, Ecology and the Evolutionary Corporation*. Gabriola Island: New Society Publishers.

Real Estate Foundation of BC, 2016. *Public Views on Sustainability and the Built Environment*. Vancouver: Real Estate Foundation of BC.

Rees, W.E. and M. Wackernagel 2013. 'The Shoe Fits, but the Footprint is Larger Than Earth.' *PLoS Biol* 11(11): e1001701. doi.org/10.1371/journal.pbio.1001701

Regional District of Nanaimo, 2011. *Regional District of Nanaimo Growth Strategy*. Nanaimo: Regional District of Nanaimo.

Robert, K.-H., H. Daly, P. Hawken and J. Holmberg, 1995. 'A Compass for Sustainable Development.' *International Journal of Sustainable Development and World Ecology* 4: 79–92. doi.org/10.1080/13504509709469945

Sea Breeze Power Corp, 2021. 'Cape Scott-Phase II.' Viewed 30 June 2023 at: seabreezewind.com/?page_id=733

Smith, J., 2010. 'The Agricultural Carrying Capacity of Vancouver Island.' Victoria: University of Victoria.

Tanner, A., H. Dowlatabadi, S.E. Chang, R. da Costa, X. Shen and A. Brown, 2017. *Resilient Coast: Liquid Fuel Delivery to British Columbia Coastal Communities.* Vancouver: University of British Columbia, School of Community and Applied Regional Planning.

Tetra Tech EBA, 2015. *Association of Vancouver Island and Coastal Communities: The State of Waste Management.* Vancouver: Association of Vancouver Island and Coastal Communities.

Town of Comox. 2011. 'Consolidated Town of Comox Official Community Plan.' Comox: Town of Comox.

Turner, G.M., 2012. 'On the Cusp of Global Collapse? Updated Comparison of the Limits to Growth with Historical Data.' *GAiA—Ecological Perspectives for Science and Society* 21(2): 116–24. doi.org/10.14512/gaia.21.2.10

Upland Agricultural Consulting, 2016. 'District of Saanich Agriculture and Food Security Plan: Background Report.' Viewed 6 July 2023 at: saanich.ca/assets/Community/Documents/Planning/BackgroundReportAFSP0525.pdf

Wackernagel, M. and W.E. Rees, 1996. *Our Ecological Footprint: Reducing Human Impact on the Earth.* Gabriola Island: New Society Publishers.

Waddock, S. and P. Kuenkel, 2019. 'What Gives Life to Large System Change?' *Organization & Environment* 33(3): 1–17. doi.org/10.1177/1086026619842482

Wagner, J., 1993. 'Ignorance in Educational Research, or, How Can You *Not* Know That?' *Educational Researcher* 22(5): 15–23. doi.org/10.3102/0013189X022005015

20

Indigenous Co-management of Salish Sea Protected Areas

Nelly Bouevitch, Soudeh Jamshidian and John R. Welch

> Co-management is actively participating. We want to actually participate. We want to be fully involved in management. We need to be involved in how federal regulations will interact with our constitutionally protected food, social and ceremonial fishing rights. We should have the base data available to us as First Nations managers, including statistics and surveys, in order to be involved with the management of the park. These issues need to be conveyed back to the communities. Any kind of closure through zoning or otherwise, is a serious concern to them.
>
> Warren Johnny, representative for Stz'uminus First Nation (Hul'qumi'num-GINPR Committee 2005: 25)

Since time immemorial, Indigenous Peoples of the Salish Sea have taken care of their territories *so their territories can take care of them.* Other authors in this book document continuous Coast Salish occupation of the Salish Sea as evidenced in archaeological records dating back 14,000 years (see Chapter 7), in co-evolved cultural and ecological landscapes and habitats (Chapter 18), and in the rich history and knowledge held in communities today (Chapter 11). The expropriation of Indigenous territories was part of the European powers' efforts to colonise land and people for economic and political gains. Beginning in the mid-1800s in far western Canada, the state (first the British 'Crown' then Canadian authorities) imposed artificial boundaries over Indigenous lands and waters, eliminated local customs and rights, redefined lands and resources as private property and maximised wealth extraction for the benefit of distant power centres. The division of land, divorced from Indigenous social, cultural and economic contexts, sought to remove Indigenous peoples, rights and titles from most of their territories (Scott 1998; Pinkerton and Silver 2011).

Despite these impositions, First Nations of the Salish Sea have sustained interdependencies on fishing, gathering, hunting and cultural uses of specific sites and vast land and seascapes. Above all, the region's First Nations have defended their territories and resisted non-Indigenous encroachments on their livelihoods and cultural integrities (Union of BC Indian Chiefs 2005). Deep, well documented Indigenous histories, uninterrupted presence on the land and strings of legal victories confirming never-ceded titles to Indigenous Territories, continue to inspire Indigenous peoples and governments to assert interests in and authority over the Salish Sea. The convergence of Indigenous histories with assertions of cultural, economic and legal interests has set the stage for initiatives aimed at sharing rights and responsibilities for portions of the Salish Sea. This chapter examines Indigenous-led conservation and co-management experiments as both important dynamics in recent decades and as windows into possible futures of collaborative governance of lands and waters co-owned by Indigenous and non-Indigenous peoples.

Co-management in Protected Areas

The Crown's historic disregard for Indigenous owners and assertion of authority translated to almost unbridled resource extraction and exclusion of Indigenous concerns from conservation movements. Partitioning of parks into pockets of 'pristine' wilderness, marketed to wealthy, refuge-seeking tourists, separated 'nature' from lands dominated by human occupation and use (Cronon 1996; Martin 2006; Clapperton 2010; Campbell 2011; Neufeld 2011; Sandlos 2011). These conservation approaches limited Indigenous Peoples' access and use of the land and contradicted their worldview, which asserts their inseparability from and place in the land (Murray and King 2012).

Shifts in parks and protected areas management since the 1990s have opened pathways for co-management. Co-management is defined as the negotiated sharing of rights and responsibilities for territory, specific resources or both (Pinkerton 2003). Co-management varies in type, degree and level of success. Prospects for more complete and successful co-management are greater where (1) the respective parties' rights and responsibilities are formally recognised, (2) the parties use adaptive management or another approach that involves co-learning, transparency, trust and experimentation, and (3) there are clear and justly apportioned material *and* intangible benefits from management actions.

Spurred by shifts in the international conservation movement and by Indigenous people's assertions of their interests in their territories, parks agencies across Canada started to recognise Indigenous-led governance, knowledge systems and cultural values. In 2001, the International Union for Conservation of Nature (IUCN) diversified protected area governance by recognising four types: state, shared, private and governance by Indigenous peoples and local communities (Borrini-Feyerabend et al. 2013). In 2008, IUCN updated its guidelines to acknowledge cultural values and reclassify protected areas as 'a clearly defined geographical space, recognized, dedicated and managed, through legal or other effective means, to achieve the long-term conservation of nature with associated ecosystem services and cultural values' (Dudley 2008: 8).

In 2010, the tenth meeting of the Convention on Biological Diversity (CBD) Conference of the Parties (COP10), adopted 20 biodiversity targets for the 2011–2020 period—the Aichi targets (CBD 2010). These targets included requirements for parties to enhance conservation efforts through participatory planning, knowledge management and capacity building (CBD 2010: 1). The Aichi targets also led to expanded conservation goals, with countries expected to conserve 20 per cent of their land by 2020 and 30 per cent by 2030.

These targets accelerated Canada's efforts. Canada responded by creating its own conservation goals, known as Canada Target for biodiversity conservation. Target 1 aimed to conserve at least 17 per cent of terrestrial areas and inland water, and 10 per cent of marine and coastal areas of Canada, through networks of protected areas and other effective area-based measures by 2020. This target, in conjunction with Canada's endorsement of the United Nations Declaration of the Rights of Indigenous People (UNDRIP), led to formation of the Indigenous Circle of Experts (ICE) in 2017. ICE demonstrated that Indigenous knowledge systems, legal traditions and customary and cultural practices deserved recognition on par with Western frameworks for conservation. ICE also provided recommendations for reconciliation through creation of Indigenous Protected and Conserved Areas (IPCAs) (ICE 2018).

These global commitments and Canada's national targets laid foundations for collaboration among federal and provincial agencies with Indigenous peoples and governments. British Columbia (BC) is the first province to legislate the implementation of UNDRIP at the provincial level (BC Legislative Assembly 2019). Legal authorities now recognise First Nations' rights and title to parks and protected areas: the constitutional protection of aboriginal rights and title (*Constitution Act* 1982, s. 35), the Supreme Court decision in *Tsilhqot'in Nation v. British Columbia*, other legal victories and the National Park Reserve designation. National Park Reserves recognise the contested status of parks territories, obliging the federal government to manage national park lands without extinguishing aboriginal rights or title to the reserves (*Canada National Parks Act*, S.C. 2000, c. 32). Canada's legal and political evolution has slowly come to recognise IPCAs and to enable steps toward co-management, including in the Salish Sea.

Co-management Frameworks and Standards

Most co-management arrangements address common pool resources (CPRs), meaning goods that are difficult to exclude users from and which may be degraded by individual users (Dietz et al. 2002: 1–11). Because every tract of land or other CPR involves a distinctive suite of ecological, jurisdictional and sociocultural factors, every co-management arrangement is unique. Frameworks for describing and classifying these institutional arrangements help to assess the level of co-management 'completeness'. 'Completeness' of co-management arrangements refers to the level of participation and the equity power sharing. In this context, *institutional arrangements* refer to rules—the rights, duties and powers to organise activities, make decisions and produce outcomes (Ostrom 1992).

Co-management may include various types and degrees of power-sharing and partnership arrangements between the state government's centralised management and community management (Pomeroy and Berkes 1997; Armitage 2008). Arrangements are more complete where the allocation of rights creates incentives for collaboration, where co-managers recognise the legitimacy of one another's rights and where the resource is sustainably managed (Schlager and Ostrom 1992; Pinkerton 2003). Pinkerton's co-management framework distinguishes and enables the evaluation of four suites of conditions affecting co-management: community, natural resources, government agency and institutional arrangements (Pinkerton et al. 2014; Rocha and Pinkerton 2015).

'Ethical space' is another important concept for co-management in Indigenous contexts. Ethical space enables partnership models that encourage a cooperative spirit between Indigenous peoples and Western institutions and that promise to transform archaic ways of interaction and the legalistic discourse that dominates land and resource management (Ermine 2007). Complete co-management infused with updated ethics ensures Indigenous world-views and knowledge systems are treated on par with Western knowledge systems. The goal of open-minded and open-hearted sharing of diverse values, interests and preferences among co-managers is made explicit.

The translation of the lofty concept of ethical space into practice means that any proposed co-management initiative or IPCA must be developed and 'assessed through the respective systems of Indigenous and non-Indigenous Peoples' values and preferences. (ICE 2018: 17). Complete co-management is achieved where owners/managers 'cross-validate their respective decisions and considerations' (ibid.). Such dialogues can be further guided by engagement standards provided in the UNDRIP (especially free prior and informed consent), the Canadian Constitution and applicable Treaties. Where these carefully constructed and negotiated standards are followed, complete co-management is more likely.

Case-Study Approaches to Co-management

Applying co-management standards and frameworks to case studies provides inclusive descriptions of resource management systems, identifies characteristics impeding and contributing to effective management, enables comparisons with other cases and highlights leverage points to assure allocations of benefits commensurate with owner and shareholder rights (Pomeroy et al. 2011; Rocha and Pinkerton 2015).

Additional historical context is useful to understanding Salish Sea case studies. The expansion of constitutional protections for aboriginal rights and title has obliged the state to share access, management and use rights with Indigenous peoples. At the same time, escalating non-Indigenous occupation and visitation has resulted in crowding and resource degradation, which often subtracts from uses and benefits for First Nations rights holders. Surges in residential and recreational uses of the Salish Sea linked to the COVID-19 pandemic have brought unprecedented pressures on the fragile ecosystems and cultural landscapes of the Salish Sea archipelago (see Chapter 18, this volume).

Private timber companies own the majority of land in the Salish Sea Region, and fiscal and economic stakes are too high for the state to shift from its approach of non-recognition (Thom 2019). Parks and protected areas are apt contexts for advancing co-management because state governments control resource access and consumptive uses of these lands and waters. This is especially true in the Salish Sea Region, where federal and BC governments have yet to agree to include privately held lands or resources in negotiations of treaty rights and titles, and where there is general consensus among Indigenous and non-Indigenous owners and stakeholders that protected areas need and deserve conservation of diverse cultural, ecological, educational and scientific values. National parks with 'reserve' designations are among the only areas where the state recognises Indigenous land claims and is willing to participate in at least limited co-management.

National park reserves fall under simultaneous, non-exclusive ownership by both Indigenous governments and the Crown (*Canada National Parks Act* 2000). While Parks Canada establishes zoning restrictions within parks, which can limit access and use, the ability to exclude users is contingent on the capacity of parks staff and the perceived legitimacy of parks among locals (Pinkerton and Silver 2011). Thus, power-balance and property rights are needed to prevent CPR dilemmas, especially degradation from unrestricted access, often referred to as the 'tragedy of the commons' (Dietz et al. 2002: 1–11).

Looking at lands and waters in the Salish Sea as CPRs clarifies the rationales and stakes for First Nations and federal government agencies to collaborate. Application of the standards discussed above in the specific contexts described below reveals that good CPR management is contingent on equity in decision making by local users and the state, in allocating benefits and in fostering resiliency (Pinkerton 1989; Agrawal 2002). Conservation and management in the Gulf Islands National Park Reserve (GINPR) and the Tla-o-qui-aht Tribal Parks offer examples of how co-management can develop in the Salish Sea. Where implemented deliberately and meaningfully, co-management has the potential to proliferate and expand into more and different jurisdictional and ecological contexts.

Case Study 1: Gulf Islands National Park Reserve

In the Gulf Islands National Park Reserve (GINPR) a partnership between Parks Canada and six Coast Salish First Nations—the Stz'uminus First Nation, Cowichan Tribes, Halalt First Nation, Lake Cowichan First Nation, Lyackson First Nation and Penelakut Tribe—is opening pathways to co-management by revealing the time depth of Indigenous resource management and mobilising Indigenous world-views, values and ecological knowledge to inform conservation (Bouevitch 2016). The partners are working together on 'Listening to the Sea, Looking to the Future', a seven-year initiative (2014–present) working to restore and manage two clam garden sites according to the traditional practices of Hul'qumi'num and W̱SÁNEĆ peoples. Elders and knowledge holders guide this work while Parks Canada scientists monitor changes to the intertidal ecosystem (GINPR 2021).

Clam gardens are boulder alignments constructed by First Nations near the zero-tide line to enhance clam harvest in the Pacific Northwest (Haggan et al. 2006). The landward side of these terraces create ideal conditions for staples in coastal First Nations diets, including cockles and butter, littleneck and horse clams (Harper et al. 1995; Williams 2006; Groesbeck et al. 2014). Active maintenance of the beaches contributed to the abundance of harvested food (Caldwell et al. 2012; Groesbeck et al. 2014).

First Nations assert that human presence on the land contributes to both social and ecological health, and academic research has corroborated this world-view (Cuerrier et al. 2015). Clam gardens supported First Nations' cultural, social and ceremonial practices while contributing to governance and territory tenure systems (Haggan et al. 2006). Clam gardens served as places for elders to pass on cultural and educational knowledge (HTG 2005; Haggan et al. 2006). Historical records suggest that clam garden beaches contributed to economic resilience by providing consistent sources of food in years when salmon and other key species were scarce (Williams 2006: 66).

GINPR proposed to restore a clam garden as an eco-cultural restoration project using the underlying principle that social, cultural and ecological integrity and well-being are interdependent (GINPR 2013). The restoration proposal originated in the Clam Garden Network, a coalition of First Nations, academics, researchers and resource managers from coastal BC,

Washington State and Alaska (Clam Garden Network 2015). Plans called for mobilising the results of studies that documented dramatic reductions in the roles of shellfish and other traditional foods in Hul'qumi'num diets since the 1800s, reductions that negatively affected community health, food security and economic welfare (Fediuk and Thom 2003; HTG 2005). Acting on these concerns, GINPR proposed to restore a clam garden within its jurisdiction (GINPR 2013).

The clam garden restoration project exists in a contemporary context; GINPR does not aim to re-establish the clam gardens as they existed in the past (Bouevitch 2016). Eco-cultural restoration challenges the colonial ideals of wilderness that are embedded in protected area history and management (Cronon 1996). Eco-cultural restoration advances the proposition that cultural knowledge, when applied at relevant contexts and scales, can sustain and promote biological diversity, cultural diversity and positive feedback between the two. To this end, the clam garden restoration project is applying local conservation values and practices to create ecological and social benefits (Augustine and Dearden 2014; Cuerrier et al. 2015: 443).

GINPR comprises 60 km^2 of land, intertidal areas and marine areas, including 15 islands and many islets (Figure 20.1). In 2003, after a full decade of land claims negotiations with the HTG, the federal government created GINPR to promote ecological protection and sustainability in an area increasingly impacted by growth in the surrounding urban areas. 'At the stroke of a pen', the establishment of GINPR effectively removed these lands from the treaty negotiation table (Egan 2008: 21). GINPR co-management is important to HTG because most of their territory is in private ownership (Morales 2006), a result of the Crown's transfer of millions of acres of Vancouver Island and other accessible terrain into private ownership to finance the Canadian railroad in the late 1800s (Union of BC Indian Chiefs 2005: 13–20). The establishment of GINPR, therefore, exacerbated communities' concerns that the federal government was further inhibiting their access to their traditional territories and limiting their ability to maintain their preferred ways of life and associated cultural traditions (Hul'qumi'num-GINPR Committee 2005).

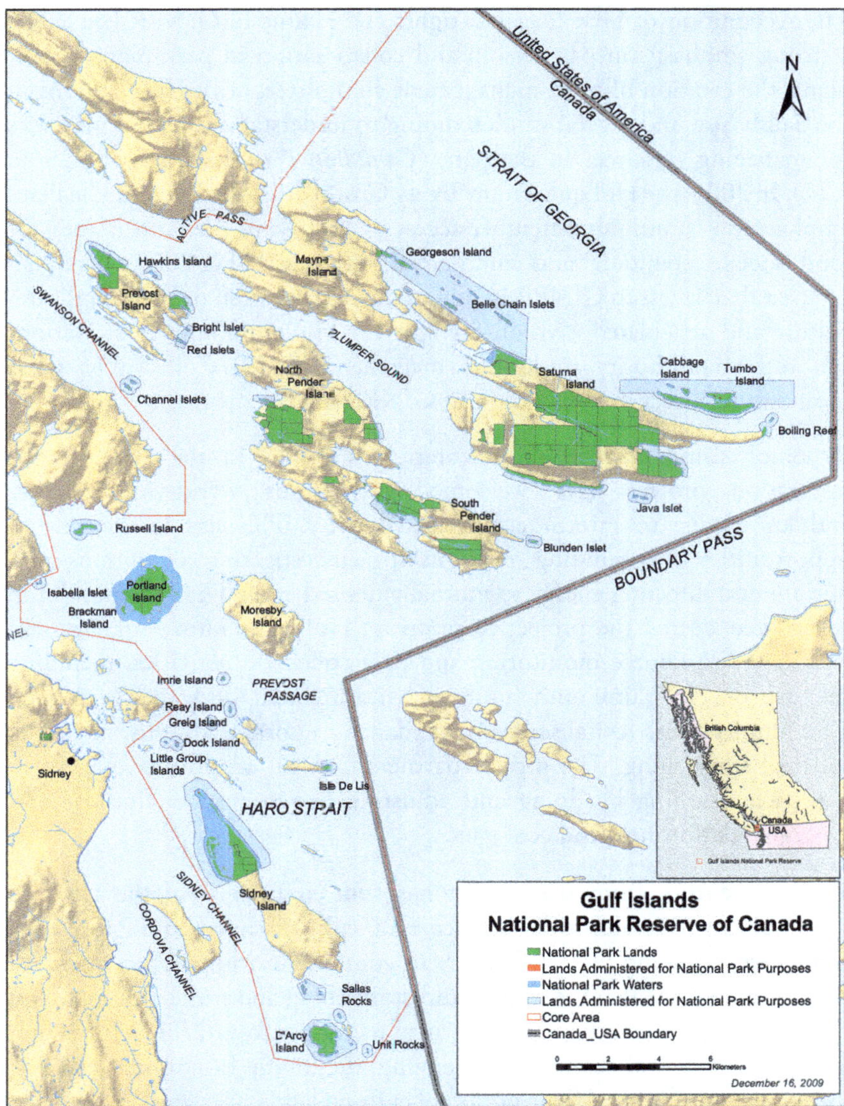

Figure 20.1 Gulf Islands National Park Reserve

Note: The reserve includes 15 islands, numerous small islets, reef and intertidal areas, and a small portion of Vancouver Island.

Source: GINPR (2022).

The recognition of First Nations' rights and claims to GINPR has led to a strong emphasis on partnership and collaboration in park management plans, the creation of joint management committees with First Nations in the Salish Sea, and several studies aiming to understand the local priorities for managing resources in the park (*Canadian National Parks Act* 2000, s. 12). In 2005, the Hul'qumi'num Treaty Group developed a Land Use Plan, emphasising community member access to and use of the Gulf Islands for food, social, ceremonial and commercial purposes (HTG 2005). The plan asserted that access to GINPR is vital to the perpetuation of Hul'qumi'num culture and articulated a vision for how the Hul'qumi'num First Nations can reclaim authority in local management (ibid.). Collectively, these mechanisms direct the focus of joint GINPR management.

By emphasising eco-cultural restoration, partners in the clam garden restoration project have welcomed Indigenous world-views, values and knowledge to inform ecological integrity. This has led to diverse opportunities for building relationships, investigating conditions and options and directing energy to mutually desired ends. The Hul'qumi'num committee steered the project to prioritise Hul'q'umi'num' language and teachings in resource monitoring and data collection activities, including the use of Hul'q'umi'num' units of measurement and species names. Community feasts revitalised culinary, dance and oral traditions, including seasonal storytelling. The project partners took an adaptive management approach, learning by doing and adjusting in response to findings and evolving relationships and ecologies.

The momentum from the project has spurred interest in the work in neighbouring communities. Science and culture camps brought youth and Elders together with researchers to combine learning about intertidal ecology with traditional teachings and storytelling about the clam gardens. Overall, the project is contributing new narratives to GINPR natural and cultural heritage. These are challenging the myth of 'hunter-gatherer' societies, communicating the notion of cultural landscapes and emphasising the importance of shellfish to Coast Salish culture (GINPR 2012; Lyons et al 2021).

Ultimately, Hul'qumi'num communities are interested in restoring beaches to a state where sustainable, long-term harvest is possible (HTG 2005: 10). This interest reveals the co-management arrangement at GINPR to be both promising and incomplete. Urban and industrial development in surrounding regions has resulted in ecologically depressed conditions for shellfish in GINPR (Fediuk and Thom 2003; Ayers et al. 2012). The Department of Fisheries and Oceans (DFO) and the Canadian Food Inspection Agency (CFIA) often deem harvest unsafe, leading to closures throughout the park reserve (GINPR 2012). At the same time, communities have expressed concerns that federal monitoring is ineffective or absent in areas critically important to cultural and subsistence harvests (HTG 2005: 31). Because the clam garden restoration project focuses on shellfish harvest, GINPR employees sampled water quality and shared the results with DFO and CFIA. This led to openings for shellfish harvest in areas that would have otherwise remained under restriction (Bouevitch 2016).

Harvest openings and other developments at the interface of Hul'qumi'num and Western knowledge and management systems are demonstrating how 'Listening to the Sea, Looking to the Future' can offer alternatives to centralised, state-based systems that are divorced from the coupled human–natural ecologies that have created the Salish Sea region. GINPR operates at a much more localised level than DFO and CFIA. Albeit small, these openings have shown that conservation efforts driven by local values can create benefits beyond the initial scope of conservation.

At the same time, co-management in GINPR remains incomplete because of an overarching power imbalance between the federal government and First Nations. First Nations-led parks committees can steer priorities on a project-by-project basis, but the federal government has yet to share power for broader policy and priority settings for parks management. Formal policy affirmations of Indigenous rights in parks management would support Hul'qumi'num community priorities, for example, to restore long-term shellfish harvest. Table 20.1 sets the stage for the second case study and any comparisons by other researchers, by laying out general conditions that contribute to and inhibit progress toward complete co-management.

Table 20.1 Factors supporting and limiting co-management in GINPR

Factors	Supporting co-management	Constraining co-management
Historical context	Early and continuous consultations among GINPR personnel and representatives of affected First Nations Clam garden project grounded in community needs and interests	GINPR established without due recognition of First Nations interests First Nations are sceptical about local benefits and state agency motives
Nature of the community	HTG vision for co-management across their territory Community knowledge of clam digging and clam garden restoration Project activities facilitate HTG use of clam beaches Increasing project participation by Hul'qumi'num and W̱SÁNEĆ members	HTG population is spread out, with few communities located adjacent to beaches HTG members cannot easily access beaches because islands are scattered and boat access is difficult
Nature of the resources	High cultural salience Shellfish are stationary Shellfish habitat can be restored	Increasing frequency and duration of fishery closures due to contamination, climate change Areas surrounding protected areas are highly populated, ecologically compromised
Nature of the state agencies	Parks Canada allows local management Parks staff dedicated to improved relationships with First Nations Parks staff are able to respond to local management priorities Parks participation in Clam Garden Network is opening new management paradigms, learning opportunities Parks staff work well with BC Parks and with other agencies	Parks mandate does not extend beyond GINPR boundaries, but urbanisation and other region-scale changes affect GINPR Final decision making is centralised, outside local control
Nature of the co-management arrangement	Constitution guarantees First Nations' rights in unceded Territories HTG effectiveness in influence collective choice rights Formalised, legal, multi-year collaboration via terms of reference with Traditional Knowledge working groups Clam garden restoration responsive to local priorities and dynamics Adaptive management supports co-learning, trust	Parks retains collective choice rights and authority over management priorities conservation actions GINPR lacks authority to delegate power to Clam Garden Network or other co-management entities

Source: Authors' tabulation.

Case Study 2: Indigenous Protected Areas in the Nuu-chah-nulth Territory

As of 2022, GINPR remains one of the few examples where co-management is developing in protected areas in the Salish Sea. However, there are examples from Indigenous Protected and Conserved Areas (IPCAs) in the Salish Sea region where informal co-management is more 'complete'. IPCAs are lands and waters where Indigenous governments have primary roles in protecting and conserving ecosystems through Indigenous laws, values and governance and knowledge systems. IPCAs vary in terms of their governance and management objectives, while sharing three essential elements: (1) Indigenous leadership; (2) long-term conservation commitments; (3) Indigenous rights and responsibilities-driven management processes (ICE Report 2018).

Tla-o-qui-aht Tribal Parks (Figure 20.2) are among the first IPCAs established in Canada. Located on the west coast of Vancouver Island, the Tla-o-qui-aht Tribal Parks is home to the 15 First Nations groups, including Tla-o-qui-aht Nation, comprising the Nuu-chah-nulth people (Vancouver Island Economic Alliance 2014). Tla-o-qui-aht Tribal Parks were formed amid the struggle against clearcut logging of old growth primary rainforests in unceded Tla-o-qui-aht territory (Townsend 2022). The history of Tla-o-qui-aht Tribal Parks reveals articulations between complete co-management arrangements and Indigenous-led conservation.

In 1970, Pacific Rim National Park Reserve was established through an agreement between the provincial and federal governments, but without the consent or input of affected Nuu-chah-nulth Nations. In 1983, MacMillan Bloedel created a plan to clearcut the majority of the old growth temperate rainforest on Meares Island, based on a tree farm licence granted by BC's government nearly 30 years earlier (Wren 1985). In November 1984, loggers attempted to land at C'is-a-qis. Elected Chief Moses Martin told them that they're welcome ashore, but they must leave their chainsaws on the boat because 'this is not a tree farm—it is our garden; it is a tribal park'. After this exchange, the loggers stopped attempting to enter Meares Island and left (Joe Martin, Nuu-chah-nulth elder and master carver, personal communication, 2021).

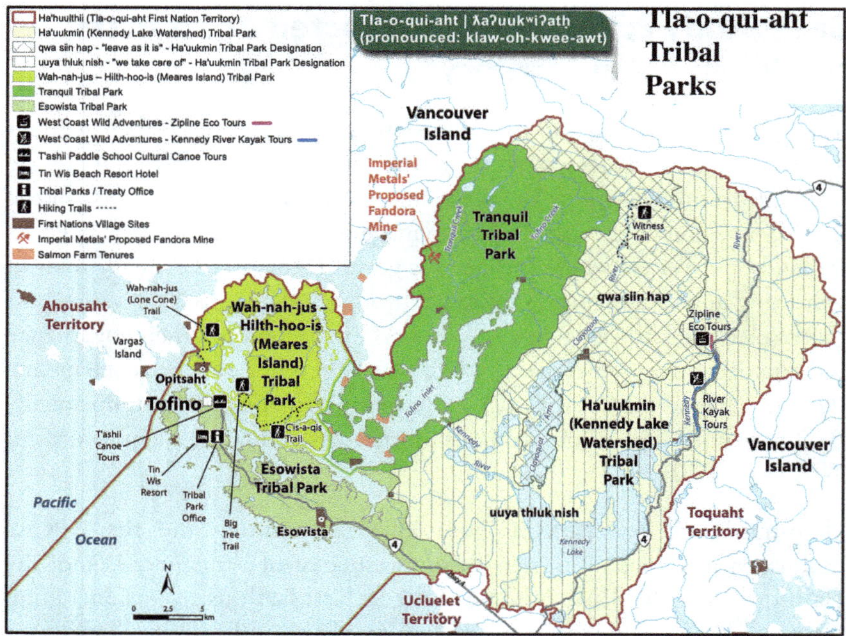

Figure 20.2 Tla-o-qui-aht tribal parks
Source: Senichenko (2013).

In 1984, Tla-o-qui-aht and Ahousaht leadership issued the Meares Island (Wanachus-Hilthuu'is) Tribal Park Declaration (Murray and King 2012; Joe Martin, Nuu-chah-nulth elder and master carver, personal communication, 2021). The declaration was linked to a direct-action strategy: a peaceful blockade by Tla-o-qui-aht members and their allies to prevent loggers from bringing their equipment ashore. While the blockade was enforced, Tla-o-qui-aht leaders launched legal proceedings, ultimately successful, that resulted in a court injunction preventing logging of Meares Island, pending resolution of Indigenous rights and title (Orozco-Quintero et al. 2020).

In 1985, Nuu-chah-nulth Tribal Council launched the Meares Island court case. The British Columbia Court of Appeal granted an injunction, halting all plans to log Meares Island until the land issue could be resolved (Orozco-Quintero et al. 2020). However, in 1993, the BC Government issued permits to MacMillan Bloedel to clear-cut more than two thirds of Clayoquot Sound's rainforests, sparing only Meares Island. Protests drew about 11,000 people, leading to nearly 900 arrests as protesters blocked the logging roads (Dawes 2020). The protests succeeded, and the BC provincial government announced the creation of the Scientific Panel for Sustainable Forest Practices in Clayoquot Sound.

Nuu-chah-nulth advisers were included to develop protocols for ecosystem-based forestry that combine Indigenous and scientific knowledge systems (Eli Enns, Nuu-chah-nulth activist, personal communication, 2021). The old growth forests of Meares Island remain standing (Murray and King 2012).

As the success of Indigenous resources conservation and management was revealed, in 2003, Tla-o-qui-aht First Nation, Indian and Northern Affairs Canada and Parks Canada signed a memorandum of understanding to remove 86.4 ha of land from Pacific Rim National Park Reserve and add this land to Esowista Indian Reserve. Since 1984, the Tla-o-qui-aht have established three additional Tribal Parks: Ha'uukmin (Kennedy Lake Watershed), Tranquil Tribal Park and Esowista Tribal Park (Eli Enns, Nuu-chah-nulth activist, personal communication, 2021). Collectively, the four Tla-o-qui-aht Tribal Parks encompass all of Tla-o-qui-aht First Nation territory. The Tribal Parks are a prime example of Indigenous leadership in conservation in Canada (IISAAK OLAM Foundation 2021). They include 500-year plans for stewardship, ecological restoration and community economic development.

The newly established Tribal Parks are a modern-day application of traditional governance values, processes and structures such as *hishuk ish tsa'walk*, meaning everything is one and everything is interconnected. Tla-o-qui-aht Tribal Parks have pioneered an Ecosystem Stewardship Contribution program entitled 'Tribal Parks Allies', which partners with local businesses to fund Tribal Parks Guardians and ecological restoration initiatives (Indigenous Guardians Toolkit n.d.). Similar to other protected areas, the Tribal Parks case study features a range of 'extractive' and 'economic' activities (Murray and King 2012). The Tribal Parks also include community economic development solutions such as ecotourism, minimal impact run-of-river hydroelectric projects and Indigenous housing solutions (Eli Enns, Nuu-chah-nulth activist, personal communication, 2021).

The Tla-o-qui-aht First Nations overseeing the governance of the Tla-o-qui-aht Tribal Park have reimagined parks management within their knowledge system and values (Murray and King 2012). This co-management arrangement could develop because of the authority, power and leadership the Nuu-chah-nulth people acted to protect their territory. The Nuu-chah-nulth have repeatedly made efforts to renegotiate their relationship with the Canadian government (Goetze 2005). Nuu-chah-nulth people exercised their de facto authority which led to gaining de jure rights.

These de jure rights were gained through a long process, including decisive assertions of rights, civil disobedience, court battles and negotiations, and engaging in consultation processes and projects to build relationships (Orozco-Quintero et al. 2020). The support of non-Indigenous allies and the Nuu-chah-nulth people's leadership in the court and throughout negotiations obliged the provincial–federal governments to understand and accept their authority and move towards a complete co-management scenario. This power encouraged the provincial–federal governments to add a part of the Pacific Rim National Park Reserve to the Tribal Park. The Tla-o-qui-aht Nation has since signed several agreements with the Federal and Provincial governments for revenue sharing, forest and range management and sustainable land stewardship to protect cultural heritage (Government of British Columbia n.d.).

Concluding Remarks

The historical developments discussed throughout this book underscore at least four broad themes. First, natural and cultural histories are inseparable. Despite colonial perceptions and insistences to the contrary, since time immemorial the Salish Sea region has hosted and co-evolved in accord with extensive human occupation and use. Evidence presented throughout sections II and III, and elsewhere, indicate that until about the 1870s, human engagement with Salish Sea and landscapes added resilience and stability.

Second, Indigenous knowledge systems (see Chapters 6 and 13, this volume) extend beyond the biophysical dimensions of ecosystems and warrant consideration in the governance of relationships and institutions. The adoption of Indigenous values, interests and preferences in the management of GINPR and the Tla-o-qui-aht Tribal Parks set the pace and tone for region-scale compliance with international, national and provincial mandates to develop and implement innovative co-management arrangements grounded in non-exclusive territorial co-ownership. The GINPR clam garden restoration project and the Tla-o-qui-aht Tribal Parks can help envision how co-management might develop and evolve in the

Salish Sea. Eco-cultural restoration in GINPR, especially clam garden rehabilitation, opens new possibilities for co-management. Recognition of the great time depth of Indigenous resource management and of the validity and applicability of Indigenous world-views, values and ecological knowledge in GINPR, the Tla-o-qui-aht Tribal Parks and other IPCAs demonstrates the benefits of applying the two-eyed seeing approach (Wright et al. 2019) in policies, practices and economic and community development.

Third, the privatisation and commodification of much of the Salish Sea region has imposed a capitalist–industrialist institutional veneer. Cascading effects from this process compromise efforts to build sustainable and resilient societies (Chapters 15 and 19, this volume). The time is ripe to learn as much as possible about collaborative management and to translate this learning into practice. While neither GINPR nor the Tla-o-qui-aht Tribal Parks demonstrate complete co-management, both illustrate that the management of CPRs driven by diverse interests is possible, beneficial and scalable. The shared governance arrangements emerging in both case studies are living examples of what the Supreme Court of Canada has called for, namely shared, non-exclusive governance grounded in dual value systems and directed toward mutually desired futures. In this sense, the case studies presented in the chapter, and the Salish Sea region itself, embody and advance mandates to address multiple social and ecological crises facing Canada and the rest of the world, by surpassing the construct of superimposed governance systems that are removed from the local cultural, economic and historical context.

Fourth, the 150-year experiment with colonialism has failed. Increased residential and visiting human populations, coupled with uncertainties linked to climate change (see Epilogue), indicate that land and resource planning and management for the Salish Sea region need to be rescaled by devolving decision-making authority to the communities who rely on and possess direct knowledge about their homes. Governance based on shared interests, not commodities, opens paths to resilience, sustainability and reconciliation. Proliferation and enhancement of co-management at GINPR and elsewhere depends largely on the willingness of non-Indigenous owners and managers to recognise First Nations' legal prerogatives and duties, and apply lessons from protected areas to co-management of fisheries, forests and other critical resources. The Salish Sea needs and deserves management systems that match and sustain the diversity and resiliency of this magnificent biocultural region and its Indigenous owners.

References

Agrawal, A., 2002. 'Common Resources and Institutional Sustainability.' In E.E. Ostrom, T.E. Dietz, N.E. Dolšak, P.C. Stern, S.E. Stonich and E.U. Weber (eds), *The Drama of the Commons*. Washington, DC: National Academy Press.

Armitage, D., 2008. 'Governance and the Commons in a Multi-Level World.' *International Journal of the Commons* 2(1): 7–32. doi.org/10.18352/ijc.28

Augustine, S. and P. Dearden, 2014. 'Changing Paradigms in Marine and Coastal Conservation: A Case Study of Clam Gardens in the Southern Gulf Islands, Canada.' *The Canadian Geographer* 583: 305–14. doi.org/10.1111/cag.12084

Ayers, C., P. Dearden, and R. Rollins, 2012. 'An Exploration of Hul'qumi'num Coast Salish Peoples' Attitudes Towards the Establishment of No-Take Zones Within Marine Protected Areas in the Salish Sea, Canada.' *The Canadian Geographer* 562: 260–74. doi.org/10.1111/j.1541-0064.2012.00433.x

BC Legislative Assembly, 2019. 'UNDRIP Legislation Enacted.' Viewed 31 May 2023 at: leg.bc.ca/dyl/Pages/2019-UNDRIP-Legislation-Enacted.aspx

Borrini-Feyerabend, G., M. Pimbert, M.T. Farvar, A. Kothari and Y. Renard, 2013. *Sharing Power. Learning-By-Doing in Co-Management of Natural Resources throughout the World*. Cenesta, Tehran: IIED and IUCN/CEESP. doi.org/10.4324/9781849772525

Bouevitch, N., 2016. Eco-Cultural Restoration as a Step Towards Co-Management: Lessons from the Gulf Islands National Park Reserve. Simon Fraser University (thesis project).

Caldwell, M.E., D. Lepofsky, G. Combes, M. Washington, J.R. Welch and J.R. Harper, 2012. 'A Bird's Eye View of Northern Coast Salish Intertidal Resource Management Features, Southern British Columbia, Canada.' *The Journal of Island and Coastal Archaeology* 72: 219–33. doi.org/10.1080/15564894.2011.586089

Campbell, C.E., 2011. 'Governing a Kingdom.' In C.E. Campbell (ed.), *A Century of Parks Canada, 1911–2011*. Calgary: University of Calgary Press. doi.org/10.2307/j.ctv6cfrjf

CBD (Convention on Biological Diversity), 2010. 'Aichi Biodiversity Targets.' Viewed 31 May 2023 at: cbd.int/sp/targets/

Clam Garden Network, 2015. 'About Us.' The Clam Garden Network. Viewed 31 May 2023 at: clamgarden.com/about-us

Clapperton, J., 2010. '"Who Opposes Parks, After All?" Sliammon First Nation, BC Parks, and Settler Conservation.' Presentation at *Place and Replace: A Joint Meeting of the Western Canadian Studies and the St John's Collage Prairie Conference*, University of Manitoba, 16–18 September.

Cronon, W., 1996. 'The Trouble with Wilderness: Or, Getting Back to the Wrong Nature.' *Environmental History* 1(1): 7–28. doi.org/10.2307/3985059

Cuerrier, A., N. Turner, T.C. Gomes, A. Garibaldi and A. Downing, 2015. 'Cultural Keystone Places: Conservation and Restoration in Cultural Landscapes.' *Journal of Ethnobiology* 353: 427–48. doi.org/10.2993/0278-0771-35.3.427

Dawes, C., 2020. 'How First Nations-Led Protests in Canada Sparked a Conservation Movement.' *National Geographic*, 9 October. Viewed 31 May 2023 at: nationalgeographic.com/travel/article/how-vancouver-island-protest-launched-first-nations-conservation-movement

Dietz, T., N. Dolsak, E. Ostrom, P.C. Stern, 2002. 'The Drama of the Commons.' In E. Ostrom (ed.), *National Research Council U.S. Committee on the Human Dimensions of Global Change*. Washington, DC: National Academy Press.

Dudley, N. (ed.), 2008. *Guidelines for Applying Protected Area Management Categories*. Gland: IUCN. doi.org/10.2305/IUCN.CH.2008.PAPS.2.en

Egan, B., 2012. 'Sharing the Colonial Burden: Treaty-Making and Reconciliation in Hul'qumi'num Territory.' *The Canadian Geographer*, 564: 398–418. doi.org/10.1111/j.1541-0064.2012.00414.x

Ermine, W., 2007. 'The Ethical Space of Engagement,' *Indigenous Law Journal* 611: 193–4.

Fediuk, K. and B. Thom, 2003. 'Contemporary and Desired Use of Traditional Resources in a Coast Salish Community: Implications for Food Security and Aboriginal Rights in British Columbia.' Paper presented to the *26th Annual Meeting of the Society for Ethnobiology*. Seattle, Washington, May.

GINPR (Gulf Islands National Park Reserve), 2012. 'First Nations Cooperative Planning and Management Committees.' Parks Canada. Viewed 31 May 2023 at: parks.canada.ca/pn-np/bc/gulf/plan/e

——, 2013. 'Clam Garden Research Project: Detailed Project Description.' Parks Canada (Internal Document).

——, 2021. 'Sea Garden Restoration.' Viewed 31 May 2023 at: pc.gc.ca/en/pn-np/bc/gulf/nature/restauration-restoration/jardins-de-la-mer-sea-gardens

——, 2022. 'Gulf Islands National Park Reserve of Canada.' Viewed 18 July 2023 at: parks.canada.ca/pn-np/bc/gulf/plan/f

Goetze, T.C., 2005. 'Empowered Co-Management: Towards Power-Sharing and Indigenous Rights in Clayoquot Sound, BC.' *Anthropologica* 47(2): 247–65.

Government of British Columbia n.d. 'Tla-o-qui-aht First Nations: Agreements.' Viewed 31 May 2023 at: gov.bc.ca/gov/content/environment/natural-resource-stewardship/consulting-with-first-nations/first-nations-negotiations/first-nations-a-z-listing/tla-o-qui-aht-first-nations

Groesbeck, A.S., K. Rowell, D. Lepofsky and A.K. Salomon, 2014. 'Ancient Clam Gardens Increased Shellfish Production: Adaptive Strategies From the Past Can Inform Food Security Today.' *PloS One* 93: e91235. doi.org/10.1371/journal.pone.0091235

Haggan, N., N.J. Turner, J. Carpenter, J.T. Jones, Q. Mackie and C. Menzies, 2006. '12,000+ Years of Change: Linking Traditional and Modern Ecosystem Science in the Pacific Northwest.' Vancouver: UBC (Fisheries Centre Working Paper #2006-02).

Harper, J.R., J. Haggert and M.C. Morris, 1995. 'Broughton Archipelago Clam Terrace Survey: Coastal and Ocean Resources.' Sidney, BC (Final Report).

HTG (Hul'qumi'num Treaty Group), 2005. *Shxunutun's Tu Suleluxtst In the Footsteps of our Ancestors: Interim Strategic Land Plan for the Hul'qumi'num Core Traditional Territory.* Viewed 31 May 2023 at: hulquminum.bc.ca/pubs/HTG_LUP_FINAL.pdf?lbisphpreq=1

Hul'qumi'num–GINPR (Gulf Islands National Park Reserve) Committee, 2005. 'HTG Park Committee Report and Recommendations on Interim Management Guidelines.' Viewed 31 May 2023 at: www.hulquminum.bc.ca/pubs/IMG_Communications_Final_Report.pdf

ICE (Indigenous Circle of Experts), 2018. 'We Rise Together: Achieving Pathway to Canada Target 1 Through the Creation of Indigenous Protected and Conserved Areas in the Spirit and Practice of Reconciliation.' Viewed 31 May 2023 at: publications.gc.ca/site/eng/9.852966/publication.html

Indigenous Guardians Toolkit. n.d. 'Tla-o-qui-aht Tribal Parks Allies—Reconciliation in Action.' Viewed 31 May 2023 at: indigenousguardianstoolkit.ca/story/tla-o-qui-aht-tribal-parks-allies-reconciliation-action

Lyons, N., T. Hoffmann, D. Miller, A. Martindale, K. Ames and M. Blake, 2021. 'Were the Ancient Coast Salish Farmers? A Story of Origins.' *American Antiquity* 86(3): 504–525. doi.org/10.1017/aaq.2020.115

Martin, T., 2006. 'Co-Management of a National Park: The Wapusk National Park's Experience'. *Canadian Studies* 61: 139–47.

Morales, R., 2006. 'New Treaty, Same Old Problems.' *Cultural Survival Quarterly*, 301: 22.

Murray, G. and L. King, 2012, 'First Nations Values in Protected Area Governance: Tla-o-qui-aht Tribal Parks and Pacific Rim National Park Reserve.' *Human Ecology* 40(3): 385–95. doi.org/10.1007/s10745-012-9495-2

Neufeld, D., 2011. 'Kluane National Park Reserve, 1923–1974: Modernity and Pluralism.' In C.E. Campbell (ed.), *A Century of Parks Canada, 1911–2011*. Calgary, Canada: University of Calgary Press.

Orozco-Quintero, A., L. King, and R. Canessa, 2020. 'Interplay and Cooperation in Environmental Conservation: Building Capacity and Responsive Institutions Within and Beyond the Pacific Rim National Park Reserve, Canada.' *Sage Open* 10(2): 1–8. doi.org/10.1177/2158244020932683

Ostrom, E., 1992. *Crafting Institutions for Self-Governing Irrigation Systems*. San Francisco, California: ICS Press.

Pinkerton, E., 1989. 'Introduction: Attaining Better Fisheries Management through Co-Management—Prospects, Problems, and Propositions.' In E. Pinkerton (ed.), *Co-operative Management of Local Fisheries: New Directions for Improved Management and Community Development*. Vancouver: UBC Press.

——, 2003. 'Towards Specificity in Complexity: Understanding Co-Management From a Social Science Perspective.' In D.C. Wilson, J.R. Nielsen and P. Degnbol (eds), *The Fisheries Co-Management Experience, Accomplishments, Challenges and Prospects*. doi.org/10.1007/978-94-017-3323-6_5

Pinkerton, E., E. Angel, N. Ladell, P. Williams, M. Nicolson, J. Thorkelson and H. Clifton, 2014. 'Local and Regional Strategies for Rebuilding Fisheries Management Institutions in Coastal British Columbia: What Components of Co-Management Are Most Critical?' *Ecology and Society* 19(2): 72. doi.org/10.5751/ES-06489-190272

Pinkerton, E and J. Silver, 2011. 'Cadastralizing or Coordinating the Clam Commons: Can Competing Community and Government Visions of Wild and Farmed Fisheries be Reconciled?' *Marine Policy* 35(1): 63–72. doi.org/10.1016/j.marpol.2010.08 002

Pomeroy, R.S. and F. Berkes, 1997. 'Two to Tango: The Role of Government in Fisheries Co-Management.' *Marine Policy* 21(5): 465–80. doi.org/10.1016/S0308-597X(97)00017-1

Pomeroy, R.S., J.E. Cinner and R.J. Nielsen, 2011. 'Conditions for Successful Co-Management: Lessons Learned in Asia, Africa, the Pacific and the Wider Caribbean.' In R.S. Pomeroy and N.L. Andrew (eds), *Small-Scale Fisheries Management: Frameworks and Approaches for the Developing World*. Oxfordshire: CABI Publishing. doi.org/10.1079/9781845936075.0115

Rocha, L.M. and E. Pinkerton, 2015. 'Co-Management of Clams in Brazil: A Framework to Advance Comparison.' *Ecology and Society* 20(1): 7. doi.org/10.5751/ES-07095-200107

Sandlos, J., 2011. 'Nature's Playgrounds: The Parks Branch and Tourism Promotion in the National Parks, 1911–1929'. In C.E. Campbell (ed.), *A Century of Parks Canada, 1911–2011*. Calgary, Canada: University of Calgary Press.

Schlager, E. and E. Ostrom, 1992. 'Property-Rights Regimes and Natural Resources: A Conceptual Analysis.' *Land Economics* 68(3): 249–62. doi.org/10.2307/3146375

Scott, J.C., 1998. *Seeing like a State.* New Haven: Yale University Press.

Senichenko, G., 2013. 'Tla-o-qui-aht Tribal Park Map.' Wilderness Committee 32(6). Viewed 18 July 2023 at: wildernesscommittee.org/sites/default/files/publications/2013_tla-o-qui-aht_Paper-Web-2.pdf

Thom, B., 2019. 'Leveraging International Power: Private Property and the Human Rights of Indigenous Peoples in Canada.' In I. Bellier and J. Hays (eds), *Indigenous People and the Law: Scales of Governance and Indigenous Peoples' Rights*. London: Routledge. doi.org/10.4324/9781315671888-8

Townsend, J., 2022. Indigenous and Decolonial Futurities: Indigenous Protected and Conserved Areas as Potential Pathways of Reconciliation. University of Guelph (PhD thesis).

Union of BC Indian Chiefs, 2005. *Stolen Lands, Broken Promises. Researching the Indian Land Question in British Columbia*. Vancouver: Union of British Columbia Indian Chiefs. Viewed 31 May 2023 at: ubcic.bc.ca/stolenlands_brokenpromises

Vancouver Island Economic Alliance, 2014. 'Vancouver Island First Nations.' Viewed 31 May 2023 at: viea.ca/business-living-on-vancouver-island/first-nations

Williams, J., 2006. *Clam Gardens: Aboriginal Mariculture on Canada's West Coast*. Point Roberts and Vancouver: New Star Books.

Wren C.S., 1985. 'Canadian Battle Rages Over Lovely Timbered Isle.' *New York Times*, 17 May.

Wright, A.L., C. Gabel, M. Ballantyne, S.M. Jack and O. Wahoush, 2019. 'Using Two-Eyed Seeing in Research With Indigenous People: An Integrative Review.' *International Journal of Qualitative Methods* 18: 1–19. doi.org/10.1177/1609406919869695

Epilogue: Reflecting on Change

Richard Hebda

Over and over, we feel the relentless beat of change: rapid social change, economic change, technological change and, looming above all, climate change. As the maxim goes, change is a constant, but today it is the *rate of change* that challenges us. I reflect on these changes as an ecologist, educator and long-time resident of the region, frequently called upon by Salish Sea Islanders to look to and plan for the future.

I cannot speak for Indigenous people, though I have learned much from them, and I am not a cultural geographer or a sociologist. The changes Islanders and all of us face, however, are broad and transformative and I hope I can provide a meaningful perspective on how we as Islanders can think about and prepare for the future.

Scales of Change

Salish Sea islands have experienced change on many time scales. As described in the chapters of this volume, the islands owe their very existence today to climatic changes in the not-so-distant past. Geological processes and changes acting over millions of years created and uplifted bedrock, setting the platform upon which the islands formed. Climatic changes generated land-transforming glaciers, lowered and raised sea levels, and sculpted the islands and waterways that we see today. These climatic and physical changes led to the complex of species and ecosystems of the region, some of them unlike those anywhere else on earth. These changes also took many millennia.

On a similar time scale the first people came, learned from and adapted to the natural environment and created evolving, but well-suited ways of living. As new opportunities arose, and new resources presented themselves, people adopted plant and animal species to their needs. They learned to live with these many species, enhanced their value to the culture, but as far as can be determined avoided widescale disruptive transformation to the islands and surrounding waters. The result was a dynamic equilibrium, a give and take. Human communities prospered yet respected the 'values' of the land that supported them.

Settlers brought outside ways but also biological invaders on land and in the sea that inserted themselves into island ecosystems and began to transform them. They cleared the land, fished the sea and began generating wide-ranging human-derived ecosystems, while overwhelming Indigenous cultures. Equilibrium was disrupted in the islands and the sea around them. Yet the islands maintained a unique insular character. Until recently, island human communities still depended on and maintained a give-and-take relationship with the land and sea, albeit a different one than in previous centuries and millennia.

Over the past few decades, during the time I have lived in the region, population growth, technological development and a seeming relentless drive to turn everything into a commodity have led to increasingly rapid social and ecological pressures on the islands, on adjacent large population centres and on the planet.

We are now at a time of disequilibrium at a planetary level. Four global systems, the hydrosphere, geosphere, atmosphere and biosphere shaped and supported the planet and Salish Sea region for millions of years, functioning together as part of a varying equilibrium. This fluctuating balance has for the first time in planetary history been disrupted by a new system, the *sociosphere* (the sum of human activity and thought).

Commodity vs Community

In historic times, the Salish Sea Islands, like many islands around the world, were subjected to the buffeting winds of society but remained at the edges, to a large extent reliant on their own communities and ways. Islands and communities were connected by water and by economic and social systems that were strongly shaped by the places where they lived. Now and increasingly so in the last four decades, the scope and rate of change has become unprecedented at the island, regional and planetary scale.

The islands had for millennia been part of local and regional communities even in historic settler times. In the past decades, however, natural resources have become commodities. Tangible commodities were taken first: fish and trees. It was formerly possible to go down to the dock and buy seafood at a reasonable price. Now local folks cannot compete with the demands of the export market for the limited catches that remain. Even the once ubiquitous and culturally important herring stocks are of great concern.

Pink salmon is still readily available but other salmon species are in serious decline. Many factors are at play, but gross overharvesting is a major one. During the same period, many Canadian Gulf Islands were stripped of their timber a second time. A timber contractor offered a much-needed local financial injection, then relentlessly removed whatever was available, moving from island to island, leaving behind ecological and economic consequences.

Today even the land, water and 'island experience' have become major commodities. Local and regional economic activity is of course vital to island communities. Now, however, everything is for sale, driven by the relentless outside pressure of neocapitalism to make a profit in the guise of making a living for local people. This commodity way of life is sold as 'community' and the impacts are challenging.

Intense development is widespread on many islands. Land and housing prices are beyond reach of the children of local residents, and service employees cannot afford or even find places to live. New structures for summer homes and for investors keep going up. Islanders have to wait hours for ferry sailings on Gabriola Island because the spaces are taken up by construction vehicles. Going for a quiet break on a peaceful Gulf Island sometimes involves as much stress as the stress you are trying to escape.

Water supply is becoming a progressively contentious issue because of inadequately managed growth and demand. Waves of tourists and seasonal visitors sometimes empty local store shelves because they fail to bring enough for their needs. And of course, for islanders the prices go up! Even the sea among and around the islands becomes congested by commodity super-ships that supply the continent with consumables.

Ironically these impacts are occurring just as we finally recognise the bioregional distinctiveness and identity of the islands. A National Park Reserve has been established to conserve some of the distinctive ecological and cultural values of the region. The concept and name of the Salish Sea, for example, is only a few decades old. The idea that the international border is ecologically and culturally divisive is gaining traction.

Coast Salish Indigenous People have long recognised the islands and adjacent coast as part of one region. Settler residents in Canada are increasingly shedding outsider roots and developing a unique, hybrid identity. Recognition of the rights and importance of the Indigenous people and traditions are increasingly acknowledged at the start of public gatherings. The premise, content and chapters of this volume recognise these new and positive developments.

Adapting to Change

The pressure of a newly emergent sociosphere will result in transformation of all the islands in a similar and predictable way. The ecological and social impacts are becoming evident. Thirty years ago, I wrote about the possible impacts of climate change, particularly on the culturally vital western redcedar tree (*Thuja plicata*). Since then, more quickly than I had imagined, the decline and death of cedars has become painfully evident on the islands of the Salish Sea.

Alarmingly so, even drought-adapted Douglas fir (*Pseudotsuga menziesii*) trees are dying on stony dry hilltops on southern Vancouver Island. This consequence of summer drought was not on my horizon three decades ago but is now a stark reminder of the rapid pace of change. Conifer needles are shed quickly to conserve moisture, resulting in a more open forest canopy. Ecological transformation is under way and accelerating.

These climate change effects are exacerbated by ceaseless arrivals and establishment of invasive species. The daphne or spurge laurel (*Daphne laureola*) is choking out the native understorey. The result of these processes, without adaptive action, will be homogenised novel and depleted ecosystems stressed by climate change and lacking ecological resilience. These ecosystems will lose keystone and iconic species and undergo structural changes beyond those experienced in past millennia.

These challenges are not insurmountable. Islanders may not be able to reduce global carbon dioxide emissions, but they can adapt to and reduce their impacts on island landscapes and human communities. Important responses already exist such as the conservation and restoration of Garry oak (*Quercus garryana*) ecosystems, conservation initiatives on Galiano Island, and restoration projects such as Helliwell Provincial Park on Hornby Island.

A Community Endeavour

Adaptation to climate change is a community *and* an ecological endeavour. I have made several suggestions at Salish Sea Island community presentations on the topic of adapting to climate change:

1. Begin with reliable climate change projections for your island and consider them in a broad regional and global context. Identify clear ecological and community principles and objectives. Think about what your island's values have been and what they might be in the future. Establish baseline conditions by identifying and mapping key characteristics such as ecological communities, keystone and rare species.
2. As you contemplate specific projects, consider multiple and flexible outcomes and targets which involve multiple sites and varied strategies. Specific impacts of climate and social change on geographically small and semi-isolated sites such as islands will vary. Multiple sites and approaches help spread the overall risk of loss and transformation in a similar manner to a diversified investment portfolio.
3. Practice the full scope of approaches including preservation, conservation and restoration. A wide scope enables actions that accommodate community social values. Restoration is vital since so many of our natural areas are highly degraded and at risk from catastrophic climate change impacts. It also helps prepare island land and water for future climates and associated extreme weather events.
4. Continue initiatives involving key flagship ecosystems and species. Identify and inventory potential rapid change sites such as drought-sensitive shallow soils on bedrock. In these and other dry sites, pre-adapt them by fostering or introducing keystone species suited to future climatic and ecological conditions. Planting Garry oaks around rock balds and on shallow soils is an example of this.

5. Consider radical species preservation and conservation to provide the raw material for future plant communities on your island and for your archipelago. This could involve growing keystone and rare species on institutional grounds, in boulevards, along roadsides and on green roofs. All places in your community have a vital role in adapting nature to climate change.
6. Consider also assisted migration as a strategy. It involves introducing species beyond their current geographic range to new sites in anticipation of future climatic conditions. Some islanders believe adaptation must involve only local island species or even genetic types. The rate of climate change and degree of impacts are so great that 'pulling up the drawbridge' is an inadequate strategy. Assisted migration should involve only regionally native species or genetic types generally growing in more southerly climates. From a biological diversity and ecosystems perspective, we are all going to have to think beyond our communities and our islands.

Two recent examples illustrate innovative approaches on adaptation to climate change at the community level. The handbook *Islands in the Salish Sea: A Community Atlas* (Harrington and Stevenson 2005) demonstrates how islanders can describe their home places and provide a basis for facing and tracking change. *Transition Saltspring* (2023) is another forward-looking, community-driven example, with components including responding to climate change, restoring ecosystems and reimagining community.

Beyond Isolation

By their geographic nature, islands have been considered as separated or isolated (sharing the same Latin root word, 'insula' = island) places. Isolation is no longer the case today as global factors such as climate change, biodiversity loss and the pressures of the sociosphere challenge us all.

Can Salish Sea Islands and Islanders adapt to these planetary onslaughts? Can initiatives and experiences within each of our communities provide lessons for society and the planet at large? Can adaptation within our archipelago provide paths and tools for the expected changes ahead?

From an ecosystem perspective, we are all going to have to think beyond our communities and our islands. Our experiences will play a vital role in sharing knowledge on our islands and beyond on the initiatives we must undertake in the face of climate change and catastrophic biodiversity loss.

The Salish Archipelago may provide us with an example not only for other islands but for all on *Island Earth*:

Honour and learn from the past and anticipate and adapt for the future.

Acknowledgements

Thank you to:

Moshe Rapaport, for an opportunity to write this contribution and to reflect upon and develop some island ideas and for thoughtful feedback; the late Violet Williams and Elsie Claxton, Indigenous Salish elders for teaching me the deep meaning of the Salish lands and waters; Adam Olsen, Member of the Legislative Assembly of British Columbia for perspectives of a younger generation of Indigenous Salish Sea Islanders; Patrick Nunn for constructive review of the text; my sister Lucy Hebda for sharing Gabriola Island experiences and challenges; Countless Salish Sea Islanders for sharing their understandings and finally, my wife Elaine Hebda, a born Islander, for providing perspectives and sharing island adventures over the past 40 years with a once young mainlander.

References

Harrington, S. and J. Stevenson (eds), 2005. *Islands in the Salish Sea: A Community Atlas*. Victoria: TouchWood Editions.

Transition Salt Spring, 2023. 'Responding to Climate Change. Restoring Ecosystems. Reimagining Community. TOGETHER.' Viewed 5 July 2023 at: transitionsaltspring.com/

www.ingramcontent.com/pod-product-compliance
Lightning Source LLC
Chambersburg PA
CBHW070646150426
42811CB00051B/757
* 9 7 8 1 7 6 0 4 6 6 3 7 4 *